Developments in British Politics 2

Developments in British Politics 2

Henry Drucker *General Editor*
Patrick Dunleavy, Andrew Gamble, Gillian Peele *Editors*

MACMILLAN
EDUCATION

Publishers' note
This book is designed as a direct replacement for Developments in British Politics *(see next page)*

First published 1986
Reprinted 1987
Reprinted with corrections 1987

This book replaces *Developments in British Politics*:* First published 1983, Reprinted 1983, Revised edition 1984, Reprinted with corrections 1985

Published by
MACMILLAN EDUCATION LTD
Houndmills, Basingstoke, Hampshire RG21 2XS
and London
Companies and representatives
throughout the world

Typeset by Wessex Typesetters
(Division of The Eastern Press Ltd)
Frome, Somerset

Printed in Hong Kong

ISBN 0-333-41368-7 (hardcover)
ISBN 0-333-41369-5 (paperback)

*Single or multiple copies of the 1984 revised edition of *Developments in British Politics* may be ordered while stocks last from Henry Drucker, Department of Politics, University of Edinburgh, Edinburgh EG9 1JT, at £8.50 per copy (incl. p. & p.).

Contents

SECTION II: GOVERNMENT

Part Three: The Political Science of British Politics

Preface

Developments in British Politics 2 is the successor volume to *Developments in British Politics*. The first volume in this series was published in 1983; and an updated edition to take account of a number of minor changes and the election of June 1983, was published in 1984. The current work differs markedly from its predecessor; it is not just a second edition. None the less, the same intention and structure, as well as editorial team, evolved for the first volume, have produced it. When the editors came to plan this work we thought it was particularly important for us, having emphasised the contemporaneousness of our subject, not to reproduce earlier work. We had to avoid chapters that were just marginal revisions – the same argument with some up-to-date facts for decoration – of papers written previously. We have therefore divided the material differently and have added a chapter to each of the first two parts. Against our original intention to use the same team of authors for the Part One chapters, we have brought in some new authors and changed the remit of the authors who remained. No chapter is just a rewritten version of a first-volume paper.

The result, even more than its predecessor, is the result of careful and detailed discussions among the editors and with Steven Kennedy of Macmillan. Our authors deserve thanks for their cheerful willingness to make alterations in their text and for meeting a host of points from four very demanding editors. An enterprise of this kind would founder if even one chapter were late; but none was. We are grateful to Paul Arthur, John Barnes, John Bennett (who compiled the index), Nick Bosanquet, Ivor Crewe, John Fairley, Matthew Fletcher, Martin Holmes, Richard Layard, Brendan O'Leary, and Adrian Sinfield, all of whom helped beyond reasonable expectation. Last, but not least, we should like to thank all those undergraduates who have aided us, both by using the original *Developments in British Politics* and by showing in their essays where we had gone wrong.

Comments from readers of the first book in this series helped us plan its successor. We would be very pleased to hear from readers about the current book.

Henry Drucker

Patrick Dunleavy
Andrew Gamble
Gillian Peele

List of Contributors

Kevin Boyle is Professor of Law at University College, Galway. He is co-author of *Ten Years on in Northern Ireland* and *Ireland: A Positive Proposal*.

Henry Drucker is Director of Development, Oxford University, and author of *Doctrine and Ethos in the Labour Party*.

Patrick Dunleavy is Reader in Government at the London School of Economics, and author of *Urban Political Analysis*.

Michael J. Elliott writes for *The Economist*. Before that, he was Lecturer in Law at the London School of Economics, and he has worked in the Central Policy Review Staff. He is author of *The Role of Law in Central/Local Relations*.

Geoffrey K. Fry is Senior Lecturer in Politics at the University of Leeds, and author of *The Changing Civil Service*.

Andrew Gamble is Professor of Politics at Sheffield University, and author of *Britain in Decline*.

Tom Hadden is part-time Professor of Law at Queen's University Belfast, and co-author of *Ten Years on in Northern Ireland* and *Ireland: A Positive Proposal*.

Martin Harrop is Lecturer in Politics at the University of Newcastle-upon-Tyne. He is co-author of *Comparative Government: An Introduction*.

Christopher T. Husbands teaches Sociology at the London School of Economics. He is the author of *Racial Exclusionism and the City: The Urban Support of the National Front*, and is conducting research on racist political movements in western Europe. He is also co-author of *British Democracy at the Crossroads*.

Michael Moran is Senior Lecturer in Government at the Victoria University of Manchester. He is author of *The Union of Post Office Workers: A Study in Political Sociology*, *The Politics of*

Industrial Relations, *The Politics of Banking*, and *Politics and Society in Britain*.

Paul Mosley has been an Economic Adviser in the Government Economic Service, and is now Professor and Head of the Department of Administrative Studies in the University of Manchester. He is author of *The Making of Economic Policy* and various articles in economics and politics journals.

Peter Nailor is head of the Department of History and International Affairs at the Royal Naval College, Greenwich. His principal research interests are in British defence policy and European security.

Kenneth Newton is Professor of Political Science at the University of Dundee, and author of various books on local and urban politics.

Richard Parry is Lecturer in Social Policy, University of Edinburgh. He is author of articles and chapters on public employment and Scottish government.

Gillian Peele is Fellow and Tutor in Politics at Lady Margaret Hall, Oxford. Her publications include *The Government of the United Kingdom: Political Authority in a Changing Society*, and *Revival and Reaction: The Right in Contemporary America*.

Raymond Plant is Professor of Politics at Southampton University, and author of *Political Philosophy and Social Welfare*.

R. A. W. Rhodes is Lecturer in Government at Essex University, and author of *Control and Power in Central/Local Relations*.

Acknowledgements

The author and publishers wish to thank the following who have kindly given permission for the use of copyright material:

Penguin Books Ltd for maps and tables from *Ireland: A Positive Proposal* by Kevin Boyle and Tom Hadden (1985).

Wheatsheaf Books Ltd for extracts from *Making of Economic Policy* by Paul Moseley.

Every effort has been made to trace all the copyright-holders, but if any have been inadvertently overlooked the publishers will be pleased to make the necessary arrangement at the first opportunity.

Acknowledgements

The author and publishers wish to thank the following who have kindly given permission for the use of copyright material:

Penguin Books Ltd for maps and tables from *Ireland: A Positive Proposal* by Kevin Boyle and Tom Hadden (1985).

Wheatsheaf Books Ltd for extracts from *Making of Economic Policy* by Paul Mosley.

Every effort has been made to trace all the copyright-holders, but if any have been inadvertently overlooked the publishers will be pleased to make the necessary arrangement at the first opportunity.

Introduction

Developments in British Politics 2 seeks to explain rather than to describe the political world which has been emerging since the beginning of the Thatcher government in 1979. It is designed primarily as a text for undergraduate students of politics, but should prove useful to students of other subjects such as economics, public policy, sociology and public law, to school teachers and pupils and, we hope, to a more general readership. In editing it we have been conscious that the undergraduates for whom it is primarily intended will have no personal memories of any election before 1979 and only hazy recollection of any Prime Minister other than Margaret Thatcher. For these readers events during the Heath or Callaghan years lack the vivacity and the potential to illuminate that those of us who have lived through them take for granted.

In choosing authors we sought people who were experts in the part of the discipline about which they were writing. We have not chosen them to reflect any single ideological or academic school; on the contrary, we have gone out of our way to select authors with a diversity of standpoints and readers will notice the differences between them. Therefore, the themes of the book have arisen out of their work and have not been imposed on them by us. In Part One the main theme which emerges concerns the course of the Thatcher experiment. To most of us that experiment seems now to have run out of steam. Many of its attempted reforms have been broken by events. However, this is not to say that others would have had more success; neither is it to suggest that there is an emerging consensus for a return to the old world; still less is it to imply that we think we know what is to happen next.

Developments in British Politics 2 has three parts. Part One, entitled 'The Evolving Political System', covers the central ground of most first-year undergraduate politics courses. It consists of nine major chapters and is subdivided into three sections: The Mobilisation Of Bias, Government, and Public Policy. Part Two consists of five shorter chapters on Current Issues. Part Three, entitled 'The Political Science of British Politics', is divided into two chapters.

The New Right ideology may have lost confidence in its critique but, as Raymond Plant shows in Chapter 1, that ideology has shaken the previous Keynesian consensus of all the parties. Few people now advocate 'big state' solutions to Britain's problems. The argument is about how to fill the gap left between the ailing state and the weak market. Ideological confusion among political leaders is matched by voter volatility. In Chapter 2 Martin Harrop dissects the three main explanations of volatility among the voters. Electoral volatility blocks the return to the once reliable alternation in government between the two great parties, but whether we are on the way to a multi-party system or a dominant party system is, as Henry Drucker and Andrew Gamble show in the third chapter, not clear. Much will depend on how the opposition parties prepare for the next General Election.

The three chapters in the Government section begin with Geoffrey Fry's analysis 'Inside Whitehall'. Contrary to many, Fry is not convinced that the Thatcher government's ideas about the organisation of central government have been particularly radical; the dramatic impact of Thatcherism in Whitehall is a product, in part, of the previous orthodoxy. By contrast, Patrick Dunleavy and R. A. W. Rhodes argue in 'Government Beyond Whitehall' that the government has been thoroughgoing in its attempts to control local authorities and the other parts of 'sub-central' government but has failed to appreciate the limits of its own authority. They argue that its repeated attempts at control have often resulted in more interference in but little achievement of the government's policy goals. Gillian Peele shows in 'The State and Civil Liberties' that despite the Conservatives' commitment to reducing the economic functions of the state, their policies have created a new balance between the state and the individual which has raised alarm about civil liberties.

The electoral success of a government, Harold Wilson once

said, depends on the success of its economic policy. Paul Mosley argues in Chapter 7 that the intellectual foundations of the Thatcher government have been so eroded that the government and the Treasury now have no coherent strategy for managing the economy. Nor has the Tory search for a new approach to social policy proved more successful. In Chapter 8 Richard Parry argues that paradoxically the Conservatives are taking a political risk in attacking the welfare state, but have not secured the substantial spending cuts which alone might give them the economic reward they seek. For most of the post-war era, foreign and defence policies have not been as politically salient as economic and social policy. Their prominence in recent years has been unusual. Peter Nailor points out in Chapter 9 that the Ministry of Defence and the Foreign Office had learned to like their relative obscurity and have had difficulty in adjusting to their more exposed position.

Some of the most important changes in British politics have occurred outside the conventional areas of an introductory year's study. In Part Two we focus on five of the most important current issues: Northern Ireland; the role of law in politics; industrial relations; race and gender issues; and the changing role of the mass media. If, against the record of recent 'solutions' to the problems of Northern Ireland, the 1985 Hillsborough initiative on Anglo-Irish relations can be made to stick, it will be a great achievement for both the British and Irish governments. Kevin Boyle and Tom Hadden analyse the problem and assess some possible remedies in Chapter 10. Through three case-studies – of industrial relations, administrative law, and the legal profession – Michael J. Elliott shows in Chapter 11 that law and politics are deeply intertwined. With increasing numbers of citizens and organisations such as local authorities and trades unions turning to the courts for redress against the actions of government, it is possible that a Bill of Rights might be adopted. In Chapter 12 Michael Moran argues that the government's attack on the trade unions has had considerably less success than is widely believed. The normal organisation of British interests along lines of class division make racial and gender based issues difficult to raise. Christopher T. Husbands examines the effects of blacks and women on the conventions of

British politics and the reactions to them by whites and males, in Chapter 13. He shows that both blacks and women have had some success in getting their concerns taken seriously but neither has been able to place them near the top of the national political agenda. Political issues are presented to the public by the press and broadcasting organisations. Ken Newton points to the problems in the way these powerful institutions are organised.

Part Three, by Partrick Dunleavy, is a distinctive feature of the *Developments in British Politics* series. Most textbooks meander unselfconsciously between the various political science traditions or, in some cases, give the students an uncritical snapshot from one perspective. Here, Dunleavy maps the debates among academic authors about political institutions and processes. In Chapter 15, eleven topics, all of which have been raised in earlier chapters, are reviewed. Finally, in Chapter 16, the three main theories of the state which sustain the views raised in the previous chapter are discussed. Thus, controversies about each of the substantive chapters is fitted into the main arguments about that subject and about British politics as a whole. The route to further study of politics is thus signposted.

References in this book are given in the Harvard system. The author and date of each reference is cited in the text in parentheses. A complete list of all the works used and referred to in all the chapters is found at the end of the book. A guide to further reading for each chapter, and a full name and subject index, conclude the book.

PART ONE

The Evolving Political System

Section I
The Mobilisation of Bias

1

Ideology

RAYMOND PLANT

State, Markets and Decentralisation

Over the past ten years we have seen a growing disillusionment with the role and nature of government and the state in British politics. Right across the political spectrum political parties are looking for ways in which to replace government as the major agent of allocation of values, goods, services, benefits, burdens and costs. In the Conservative Party under Mrs Thatcher's leadership there has been a determined attempt 'to roll back' the frontiers of the state and to substitute market forces whenever possible for state action. Where it has not been possible to replace state provision, as in health and education, considerable encouragement has been given by government to enable market forces to play a role – perhaps with the hope that with more widespread growth of, for example, private hospitals and health insurance, more and more people will choose to have their needs met in the market rather than within the health service. Similarly the Alliance of Liberals and Social Democrats has emphasised the market as guaranteeing freedom of choice and promoting efficiency, while at the same time this perspective has been combined with a vigorous critique of the state and what is seen as the over-extensive role of central government. However, the Alliance does not place as much reliance as the Conservatives on private markets – there is also emphasis on decentralised political institutions and a secure role for community politics. Thus, through their proposals for constitutional changes, the emphasis on markets and more decentralised forms of politics, we can see a distrust of the role of the state as it has developed over the past few decades. The Labour Party too, perhaps inherently the most statist party, has been exploring ways in which decentralisation and a market orientation can be made compatible with socialist goals.

These developments are remarkable in the sense that across parties there seems to be a tacit admission of the failure of the British state to govern efficiently, justly and authoritatively, and parties of all ideological stripes have within them movements and groups which are looking for ways of achieving the policy goals of their parties that do not presuppose the sort of role for the state that has been characteristic of British politics since the war – although its roots go back much further (Greenleaf, 1983).

What, then, are the grounds for disillusionment with the state? There is no widespread agreement about this and answers depend largely upon political standpoints.

The New Right

On the Right, the last ten years have seen the growth of neo-liberalism, or the New Right (Plant and Hoover, 1986), which has embodied a critique of the growth of British government together with prescriptions for its ills. This free-market form of conservatism has deep roots in the Conservative Party (Greenleaf, 1983) but has been eclipsed since the 1930s. It owes a good deal to the writings of Milton Freedman and F. A. Von Hayek, to the research efforts of the Institute of Economic Affairs, the Adam Smith Institute and the Centre for Policy Studies, and to the speeches and writings of Sir Keith Joseph (1975; 1976) and Mrs Thatcher's *Let Our Children Grow Tall* (1977). We need to grapple with various aspects of the neo-liberal critique of British government not all of which are represented in any one book. It is rather a matter of pulling different threads together.

Rejection of Central Government Planning

The first is a rejection of the possibility of economic planning for both technical and moral reasons. The technical reasons have their roots in debates between Hayek and Mises with Marxists in the 1930s. Crudely speaking, the Hayek/Mises view is that economic transactions are far too complex to plan to secure any particular outcome in terms of productivity, growth and pattern of production and conception. As such, economic planning must be sufficient because it would have to depend upon constructed ideas of prices in the absence of an overall free market in goods and services, and could not use in its planning the widely dispersed and non-propositional forms of knowledge and information which individuals use in making their own economic choices. It is not just that planning is complicated, because, if it were, planning problems could be solved at least in principle by computers. The problems with planning are qualitative. Governments cannot (in principle) co-ordinate the types of information which individuals use in making their own choices into a plan. Such knowledge is too dispersed and based upon knowledge of people, local conditions and circumstances, of the fleeting moments, not known to others. Planning is inefficient, and free markets will do a better job of promoting growth, relieving poverty, at least in absolute terms, and co-ordinating individual decisions than any amount of planning.

Second, planning is to be rejected on moral grounds. The classic place for this argument is Hayek's *Road to Serfdom* (1944), in which he argues that the degree of information to be gathered and the degree of direction of labour, investment and resources which would be necessary for planning would inevitably lead to an authoritarian organisation and pose a grave threat to individual liberty. Far from being able to plan for the future of the economy and thereby take responsibility for growth and the level of employment, government can only provide the framework of law and financial stability within which individuals can make their own economic decisions. If growth is to come and unemployment to go down then these will only happen when individuals have the confidence to invest and risk their capital. They will do this only if the rates of taxation are regarded as fair, if the currency is stable and the economic investment as free as possible of arbitrary interference by government. Thus government, on this neo-liberal analysis, has to abandon its assumed, but mistaken, role in planning, and with it the concomitant responsibility for particular economic outcomes such as unemployment and growth levels, and welfare.

Critique of Corporatism

Allied to this is a critique of economic policy arrangements, particularly corporatist and tripartite arrangements. Corporatist arrangements between government and the major interest groups in the field of economic management emphasise a mistaken collective responsibility for planning the economy, and in the view of the radical Right the state should play no role in seeking a consensus about economic policy between major interest groups. Decisions about prices and investment cannot be made by such groups but would have to rest upon market mechanisms. There is also a moral objection in that in such arrangements between the government and individual interest groups, which reached their high point in the last two years of Mr Heath's premiership and in the 1974–8 Labour government, Parliament is effectively by-passed as decisions are taken by the government in consultation with interest groups in an undemocratic way (Thatcher, 1977). Certainly the Thatcher government has been persistent in keeping the CBI and the TUC out of economic policy-making, and it seems clear that the NEDO, while it has not been abolished, has been turned into more of a 'talking shop' than an institution which is likely to affect policy. Indeed the government has on occasion forbidden discussion of its macro-economic policy as being outside its sphere of responsibility. Hence the thrust of policy has been to turn away from planning, to emphasise the free market, to stress what government cannot do, to indicate the harm that can be done by over-optimistic assumptions about the capacity of governments to manage the economy and, as a consequence of this, to disdain responsibility for the management of the economy in terms of achieving specific targets for employment, growth and welfare.

Critique of Welfare State

The growth of the welfare responsibilities of government has also been a major concern of the New Right. The constrained economic environment within which governments have had to operate since 1974 have, so it is agreed, thrown into relief the problems of providing the resources for the maintenance, let alone the extension, of welfare rights within a mixed economy. The seemingly deep-seated difficulties of securing growth in the economy has led many to see the welfare state as a brake on economic development and to confront what has come to be called the 'fiscal crisis of the state'. The fiscal crisis lies in the fact that in a constrained economy with a welfare state, the government has two imperatives. The first is welfare expenditure, seemingly necessary to secure legitimacy and meet the expectations raised in this field; the second is to secure the profitability of private industry necessary for development, investment and paying shareholders a decent return on their investment. High rates of growth might allow these imperatives to be pursued at once, securing both private profit and a fiscal dividend for welfare expenditure. Low or zero growth makes this very difficult. The New Right has a double strategy here: to cut back as far as possible on state expenditure in this field – a difficult task with high levels of unemployment and an ageing population; the second stand is to encourage the development of market-based forms of provision, private hospitals and insurance in the health field, encourage private education via the assisted places scheme, and occupational pensions by ensuring their portability. The strategy seems to be to create an environment more favourable to market allocation in this sphere, in the hope that, over time, people will look to the state less and less for welfare, taking more responsibility for their own lives.

It is very important to remember that this approach does not necessarily imply a lack of concern with the plight of the worst off. Indeed, if combined with a vigorous defence of capitalism and free markets it is argued that the poor will get richer in an absolute sense than they would under a more constrained system. Their relative position might decline but the New Right is inclined to see relative poverty as another name for inequality and thus not in their view a proper concern of government.

Coupled with these arguments is the view that if not tightly constrained the welfare state has an inbuilt tendency to grow. The critic rejects the view that there is a definite range of needs for the welfare state to meet at a definite level and which could mean that welfare spending would, when those needs are met, reach a plateau. Needs are open-ended, and if they are seen as the basis of welfare rights there is no intrinsic limits to the extent of such rights and the claim on the tax and borrowing resources of the state. The transformation of extended needs into rights turns the welfare system into what Mrs Thatcher has called the entitlement society which inevitably breeds resentment against the state when such rights cannot be met, as they cannot. This resentment at the failure to meet excessive welfare expectations is a powerful political phenomenon which has stopped the government from cutting back welfare spending as much as it would have liked. The Central Policy Review Staff's paper sent to the Cabinet with the blessing of the then Chancellor, Sir Geoffrey Howe, proposed the wholesale privatisation of welfare services. Its leak caused an uproar, and in the 1983 election Mrs Thatcher was forced to reject its ideas and claim that the NHS was safe in Conservative hands.

The problem about the open-endedness of needs is linked to another feature of the welfare state which the New Right regards as central and baneful: namely the connection between the welfare state and pressure groups. The elastic idea of 'needs' encourages the formation of pressure groups to secure funding to meet the needs which have been identified. This pressure-group activity, including pressure from the producers of such services such as doctors, nurses, social workers, teachers and ancillaries of all sorts in health and education, is unregulated, and political parties are subject to enormous electoral pressures to increase welfare. This pressure is fuelled by both producers and consumers of such services. According to radical Conservatives there are only two ways around this. The first is to try to reduce more and more the role of the state in the sphere of welfare distribution, and make it more and more a matter for the market, either by direct substitution or by encouraging the environment for the market to attract people away from state services. The other is to show, as often as possible, that most thinking about welfare is the victim of what Brittan (1983) has called the 'Wenceslas myth' – that welfare provision is a matter of generosity or stinginess on the part of government when in fact it comes from taxation and borrowing which affect the climate in which private business operates. It has been suggested, for example by the Institute of Directors, that the tax part of pay slips should be disaggregated so that taxpayers become more aware of how much they are paying for different parts of government expenditure. If they know this in detail, so the argument goes, the taxpayers would vote for parties offering reduced welfare programmes. As it stands, there is no way in which the cost of the welfare state can be brought home to the average taxpayer, and in these circumstances his voting may not be wholly responsible.

Preference for 'Market' Solutions

In the case of the economy and the welfare state, therefore, the radical Right takes a market-oriented line and has tried to weaken the view that the state has responsibilities for planning, growth, employment levels and for specific levels of welfare for its citizens. This, combined with the privatisation of British Airways and British Gas to make them much more responsive to market forces, has been part of the more general strategy of questioning the role of the state in spheres in which its function seemed entrenched, in an attempt to secure greater choice, efficiency and responsibility in making the voter more aware of the relative costs and benefits of programmes.

No Political Decentralisation

However, this attempt at what might be termed economic decentralisation – that is, reducing the power of the state and encouraging individual initiative and responsibility via the market – has not been matched in terms of political decentralisation. In fact the reverse seems to have been the case. Of all the parties the Conservatives seem the most wedded to the present constitutional settlement in Britain, and in particular have shown no real interest in devolution either to the nations of Scotland and Wales or to the regions, and critics will point out the extent to which recent legislation has placed a set of severe financial and legal constraints on local government. In this sense it might be best to say that while a major thrust of the government is neo-liberal it is far from being libertarian except in the economic sphere. In the non-economic spheres of political, social and personal relationships there has been some emphasis on trying to strengthen traditional values – for example, the Cabinet study of family life – and, as noted above, a failure to be interested in forms of decentralisation other than economic.

One can speculate about the reasons for this. In the first place, the obvious argument about political decentralisation is that if decentralised political units such as regional/national Assemblies, or local governments, were able to have a spending programme for what were seen by the people themselves as urgent matters, then this would have an enormous impact upon the economic climate locally. Examples are frequently quoted of how increased demands for local services which drive up rates actually put local firms out of business. So, given that the government does have a responsibility for securing a stable financial and political climate against the background of which individuals can make economic decisions, then without severe constraints on local politics and the levels of local funding required to sustain political demands, these could have a severe effect on the economy. There is a further possible reason that has in fact been advanced by critics of the Thatcher government, and it is that the freeing of market forces in the economy is bound to create a need for a strong state in a narrower sphere of activity. The attempt to reassert the rigours of a less constrained form of capitalism is bound, on this view, to create social tensions between individuals, classes, localities and regions which it will take a good deal of centralised authority to keep contained. In the view of such critics the imperative to roll back the frontiers of the state in the economic sphere may well involve having to strengthen it in other spheres, particularly law and order, rather than weakening it by a process of political decentralisation to parallel economic decentralisation.

Toryism and the Authoritarian Right

There are, of course, other strands within current conservatism which ought to be mentioned. For example, Julian Critchley, the MP for Aldershot, holds a strongly anti-ideological view of conservatism that would eschew all attempts to formulate a coherent philosophical vision of conservatism whether of the 'wet' or 'dry' variety. Indeed, he argued recently in the television series 'The Writing on the Wall', that conservatism was at its strongest when it demonstrated little interest in politics and chose to govern by the seat of its pants. By 'politics' here he clearly means an ideological form of politics. This attitude is perhaps represented at a more serious level in the party by John Biffen, who is an avowed sceptic about theories in politics and is the Cabinet's leading 'consolidator'. There are, in addition, Heathite modernisers such as Peter Walker who, unlike Sir Keith Joseph and Mrs Thatcher, have never repented their allegiance to the Macmillan/Heath style of conservatism, and who believe that the government has a direct role to play in the economy, in job-creation and in encouraging modernisation and growth, rather than the neo-liberal monetarist belief that all government can do is to secure the background conditions for them to take place as the result of the initiatives of markets, employers and investors. Institutionally this group is most clearly represented by the Tory Reform Group.

In addition there has been a growth of interest in a more philosophically minded anti-liberal view of conservatism, associated with the Salisbury Group and its publication, *The Salisbury Review*, edited by Professor Scruton of London University. The Group and the *Review* are committed to the development of an articulate and coherent conservative philosophy but they are strongly critical of neo-liberal preferences of the Tory Party. They do not place such a high value on the market as a guarantor of individual liberty, and are much more interested in developing ideas about the state as the symbol of political identity, the nature of Nationalism and the cultural identity which they believe should underpin a sense of nationalism, the role of religion in political life and the place of tradition in maintaining social and political solidarity.

Their work represents a major theoretical attack on the tenets of individualistic liberalism which they believe is alien to the conservative instinct. They are more interested in Hegel and Burke than Mill, Locke and Hayek. While their work is mainly theoretical, articles in the *Review* have had some impact. John Casey, a Cambridge English don, has argued in favour of repatriation of immigrants on the grounds that they cannot be assimilated into the dominant culture and thus pose a threat to overall identity; Ray Honeyford, the former Bradford headmaster, has written critically about the impact of multicultural education. In both of these contexts one can see the influence of dominant themes for the group such as Nation, State, Indentity, Community. These are values which have considerable salience for the conservative tradition.

It may be that part of Mrs Thatcher's skill as a politician lies in the fact that in economic policy she appeals to the individualistic, free-market, neo-liberal interests of the party, whereas in social and constitutional terms she is much more of a traditional Tory anxious to preserve a unified national state and traditional values.

Among the critics of the New Right on the other wing of the Party are Francis Pym, James Prior, Peter Walker, Norman St John Stevas and Ian Gilmour. It is perhaps Gilmour among these who has developed a wide-ranging critique of the ideology of neo-liberalism, which he regards as incompatible with traditional Toryism particularly of the One Nation, Disraelian sort. Perhaps the first, and in some ways the most telling, point is that there is something profoundly unconservative about a Tory government pursuing a particular set of ideological commitments in an attempt to break out of what have come to be regarded as traditional assumptions about the role and responsibility of the state. Critics of this type tend to emphasise the place of tradition in Conservative thought and the idea of politics as a sceptical attempt to modify an inherited tradition of political life to changing circumstances rather than pursuing a set of overall, and theoretically grounded, goals.

In addition, this view of Conservative thought places great emphasis, as did Disraeli, on the idea of Britain being a national community and the importance for Conservative politics of seeking to integrate people into it and pursuing policies which will sustain it. Gilmour, for example, sees the emphasis on the market as anti-communitarian. The welfare state is central in making for the integration and the sustaining of relationships which the free market may threaten or destroy. As he argues in *Inside Right*, loyalty to the state:

> will not be deep unless they gain from the state protection and other benefits. Homilies to cherish competition and warnings against interference with market forces will not engender loyalty . . . Economic liberalism because of its starkness and its failure to create a sense of community is likely to repel people from the rest of liberalism. (Gilmour, 1978, p. 118)

While the neo-liberal conservative tries to move social choice away from parties and governments onto the impersonal market so that by narrowing the sphere of government it can avoid the problem of overload and strengthen government in a narrower sphere, it runs the risk of destroying the sense of national community which is central to traditional conservatism. Conservatives such as Harold Macmillan and R. A. Butler were not afraid to use the power of the state to check and constrain the rigours of the market, and the views of Gilmour and others that modern Conservatives should feel the same.

The Liberal Party

In the other political parties there has been an attempt, which as we have seen is lacking in the neo-liberal strand of the government policy, to weld together economic decentralisation and political decentralisation. There is the same suspicion of the state; its efficiency, its arbitrariness in economic management, its bureaucracy and the threat to freedom which it poses; but certainly in the Alliance the commitment to decentralisation is much more thoroughgoing.

Keynesianism

While the Liberal Party accepts the centrality of the free market, indeed it is on the nineteenth-century tradition of the Party, particularly Cobden, Bright and Gladstone, on which some neo-liberals rely, its policy of economic decentralisation is more thoroughgoing than that of the Conservatives in that it does not believe that the liberation of market forces is sufficient for a free and efficient economy. The Liberals welcomed Keynesian techniques as freeing government from the need for direct physical interference with the economy while allowing for overall economic management. Their own particular emphasis was less on freeing the market from direct planning and more upon changing social relationships in firms and industries by attempting to decentralise the economic power of large-scale enterprises. So, for example, there is a long commitment to industrial democracy in which there would be councils at each workplace with rights of conciliation and information – rights which would be guaranteed by the state – election of boards with common voting rolls for employees and shareholders and increased participation in investment, and shareholding and profit-sharing by employees. These changes which would affect the distribution of power in the workplace and in firms operating in the market were necessary in the Liberal view to complement a free-market economy. Merely to move over to a free-market economy and remove state regulation will do little to increase individual liberty without changes in the distribution of power within the market. The introduction of industrial democracy would, on this view, diversify power and new responsibilities would be put on employees, and their representatives would act more responsibly in the market place, particularly in relation to wage demands.

Proportional Representation

Such a view presupposes the falsity of the class politics which Liberals believe the other two major parties represent. There are not irreconcilable differences between workers and shareholders. There is a possibility of a sense of community, of interest to be found between these groups, but it is continually threatened at the political level as the result of the adversary politics of the other two parties. A changed electoral system would represent a more balanced representation of political forces to be achieved, and against the sort of background, so it was argued, the proposals for industrial democracy would help to engender a greater sense of community and co-operation.

Alongside economic decentralisation in the form of the market and industrial democracy, together with electoral reform, was to be set community politics and an emphasis upon political decentralisation, which, as we have seen, is absent from 'conservative capitalism'. There is a constitutional and a more directly political aspect to these reforms. The constitutional point follows to some extent from proportional representation which would be likely to result in power-sharing in government, and to that extent diversifying power in the executive. In addition there would be a transfer of some powers to a provincial or regional level, together with a strengthened local government, so that Britain would become more like a federal state with much more thoroughgoing devolution than characterised the attempt by the Callaghan government, with Liberal support, to grant devolution to Scotland and Wales. It is central to the Liberal vision here that diversification and decentralisation of power will create social harmony and a sense of community rather than centripetal tendencies. The key stance to this will be proportional representation and the idea that it will reveal the underlying progressive consensus in British politics hitherto split between different wings of the other parties.

Community Politics

However, there is also a strong commitment to community politics within the Liberal Party although it is not universally popular among MPs. The community approach pioneered in Liverpool consists in trying to bring political issues down to earth and enable Liberal councillors to stay close to the needs of their communities. Community politics reflects a distrust of national political leadership, its perceived elitism and distance from the political concerns of individuals at the grass-roots level, together with a distrust of the idea of political reform coming from above through the existing power structure. Apart from anything else, a community approach was electorally helpful in that it seemed as though a community approach concentrating upon local issues and needs could gain the party support in unlikely areas such as inner cities, usually thought to be Labour bastions but thought by Liberal community activists to be ripe for conquest by exploiting years of decline and neglect. Indeed the greatest successes of community politics has been in northern towns and cities, and the trail was blazed by the election in 1969 of a community activist in the unlikely inner-city constituency of Birmingham Ladywood. Similar successes have followed in Alton and Bermondsey. The exploitation of local issues and grievances looked to be potent weapons against the existing party machines, particularly Labour, and particularly in inner-city areas. Again, then, we see the emphasis upon decentralist politics, reform from below rather than above, a distrust of national leadership and party structures.

There are, however, unanswered questions about the community approach to politics. Because there has not been a Liberal government since the rise of community politics, it is not at all clear what the relationship between community politics and a national government would be. On the face of it, its thrust seems to be oppositional and opportunistic, using deeply felt grievances against existing parties and power structures for perfectly understandable electoral purposes. However, how would this work with a Liberal government in power? It has been argued by critics as diverse as Roy Jenkins and Harold Wilson that political leadership has to be national and synthesising, whereas the whole point about community politics is its roots in the locality and in the specific needs and aspirations of the local community. Community-politics theorists had a view about this and it was focused largely on the role of the Member of Parliament and his/her representativeness. It is argued that a recurrent feature of British politics is the distrust and low esteem for politicians of all parties, and this is thought to be a response to their distance from the lives and needs of their constituents. In community politics the MP would have to be much more involved in the day-to-day life of the community, not making fortnightly or monthly journeys for surgeries; it would involve much more commitment than that. This greater involvement would engender more trust, and this would in its turn make the MP communicate with his local electorate more clearly and on a more trusting basis about issues of national and international politics (Mole, 1983).

However, all this is very theoretical and there is no clear way in which a community approach to politics can be clearly integrated into a national approach. It is ideal for opposition, less easy for a potential party of government. Community can be a beguiling idea full of warmth and identity, but frequently it thrives on opposition and a sense of grievance. Often there can be different communities even within a single constituency. One need think no further than Tottenham or Handsworth. A national Party with national goals and policies has to have some way of linking the national and the local into more of a clear synthesis than perhaps the community politicians in the Liberal Party have achieved.

However, we see again in the Liberal Party the distrust of the state, the central place for the market; but what clearly differentiates their view from that of the Conservatives is the idea that the market itself is no defence of freedom without other forms of economic decentralisation, political reform and a reinvigoration of local and community politics.

Some senior Liberals viewed the emergence of the SDP with suspicion given the party's emphasis on the diversification of power and community-based politics, because the leaders of the SDP – Jenkins, Owen, Williams, Rodgers, etc. – were regarded as social democrats who, having become a minority in their own party, wanted a new vehicle for the old preoccupations of Gaitskell/Crosland social democrats, with their emphasis on economic management, greater material equality and redistribution, a hostility, at least in Crosland, to wider political participation, and the commitment to reform from above using the apparatus of the state.

The Social Democratic Party

However, this overlooked the early Social Democrats' emphasis on revising the view which hitherto social democrats had taken of the state. In a review of a memorial volume to Crosland, David Marquand, a sometime Labour MP and founding member of the Social Democratic Party, argued that Crosland had a rather closed mind on the nature of the British state:

> Crosland took the traditional structure of the British state for granted and failed to see that the centralist, elitist logic underlying it was incompatible with his own liberal and egalitarian values. (Marquand, 1981)

Wary of the State

This point, and perhaps it is the main point of difference with the 1960s' social democrats in the Labour Party, was vehemently argued by Dr Owen in *Face the Future*, the founding text of the new social democracy as C. A. R. Crosland's *Future of Socialism*, 1956, was the founding text of the old:

> The argument that Britain has begun to move irrevocably to a corporate state is not new and in recent years the critique has been vigorously mounted from the Hayek Right. The year 1979 saw the return of a Conservative government which for the first time, meant a government which was openly critical of the corporate state. The challenge to Social Democrats is not to allow this critique to be mounted only from the viewpoint of the Right. (Owen, 1981, p. 31)

This is part of the general tenor of writing by social democrats in the early 1980s, when it seemed that their enthusiasm for the kind of state which the post-war consensus had spurned and in which social democracy had played a major role, was under attack. In 1979, Evan Luard, sometime Labour MP and subsequent member of the SDP, wrote *Socialism without the State*, and in 1981 Shirley Williams wrote *Politics is For People*, both of which, along with Owen's book, provide an internal critique of the centralism and statism of what they now saw as the traditional Labour Party view. This critique of centralism expressed itself in a number of ways, and perhaps the most important is a greater willingness to embrace the market as a distributive mechanism than Labour Party social democrats were prepared for. Dr Owen's 'tough and tender' approach first developed at the party conference in Salford in 1983 set the seal on this.

Importance of Equality

The idea is that the free market should be encouraged in the interests of freedom of choice and the greater efficiency which would come from competition together with a greater commitment to the social services and welfare than they believed the Conservatives offered together with redistribution of income and wealth in the interests of greater equality. It is perhaps this last point which most clearly differentiates the SDP approach to the market from that of the Conservatives. While the Conservatives have encouraged markets, they have not sought to diminish inequality so that people enter the market on a more equal basis. On the contrary, in fact, tax changes have increased inequalities of income. However, in the Conservative view this is necessary to provide incentives for entrepreneurs, who will help to maximise the growth of the economy, and this will in turn improve the absolute standard of living of the worst off. The SDP has argued, however, that if markets are to be encouraged in preference to bureaucratic forms of state allocation, there has to be some redistribution (the extent is not clear) in order to equalise access to the market, and a generous welfare state to cope with all of those who will never gain their full potential in the market. These ideas also include an exclusion of industrial democracy and rights in the workplace.

Support for a Bill of Rights

As is the case with the Liberals, these ideas about decentralisation via economic markets have been linked with proposals for constitutional reforms, and in the case of the SDP this involves the devolution of power to regional assemblies together with electoral reform.

One further constitutional reform favoured by both the Liberals and the SDP is a Bill of Rights, probably based on the incorporation of the European Convention on Human Rights into British law. This proposal in itself shows the extent of the disenchantment with British government, so that it is now assumed that the protection of individual rights and liberties requires an entrenched constitutional provision. How this entrenchment could be achieved and the form of judicial review that would go with it is unclear, but nevertheless it is a development of the utmost importance that it should be proposed to put such a constraint upon the power of government.

The general issues here, at least as they concern the SDP, have been well stated by David Owen and in a way which perhaps adds a commentary on what has gone before:

> it is that in the absence of a theory of the state and a comprehensiveness of its role and its philosophical relationship to society politicians have been buffeted by many transient pressures and have prepared piecemeal ad hoc changes which have lacked both coherence and conviction. (Owen, 1981, p. 165)

So far, as we have seen, British politics is characterised by a widespread disillusionment with the role of the state developed since 1945 and by the search for different complementary ways of dealing with the distribution of resources in society. The basic Liberal and Social Democratic ideas have been a rediscovery of the importance of market forces as an aid both to efficiency and the expansion of choice, as well as helping with the depoliticising of issues which have threatened to become unmanageable, together with a search for forms of political decentralisation. How then do these changes affect the Labour Party?

The Labour Party

There has also been a growing emphasis on the importance of local politics within the Labour Party. Obviously to some extent this is dictated by political tactics. Given the size of the majority of the Conservative government following the 1983 election, the only real advances which Labour could make were in those areas and councils under Labour control rather than in Parliament. In addition, the political salience of local government in the light of the government's attempts to restrict local spending in terms of its overall economic policy, has meant considerably more concentration upon the local base of politics than has been the case at any time since the war.

'The New Urban Left'

However, it would be wrong to give the impression that this interest is purely a response to the current political circumstances. The development of what has come to be called the 'new urban left' does rest upon a number of principles and values and involves an attempt to rethink some aspects of the Labour Party's view of socialism. The emphasis of the local politics in the Labour Party is on the attempt to devolve power to smaller decision-making units in the locality and the region, to make bureaucracies, and particularly welfare bureaucracies, more accountable to claimants and more decentralised in their operation, and to encourage greater participation in local efforts to secure socialist advances within the limits available to a local authority. Only if a strong participatory local socialism can be built will socialism be developed overall. Socialism on this view cannot come from reform from above; it has to be built upon commitment from below. There is both a philosophical and a practical point here. The practical point is that if socialist policies can be seen to work, meet needs and secure people's loyalty at the local level, then this will give non-committed voters confidence that Labour provides a genuine and workable alternative. The philosophical point is that socialism does require a change of attitude and more of an orientation to the idea of a common or social good, and cannot be produced by central political reforms however important they may be.

Rethinking Nationalisation

If socialism is to last it has to be built upon a secure change in values. This will only occur if people are encouraged to play an active part in the design and implementation of socialist goals at the local level. The philosophical and the practical issues here are well exemplified in the following argument from Hodgson:

> The participating socialist society of the future has to be prefigured in the capitalism of the present. Without examples to point to, people will never be generally convinced of the validity of socialism. (Hodgson, 1984, p. 153)

However, despite all this – and it is very important to the understanding of the modern Labour Party as we shall see in more detail later – on the face of it the Labour Party still seems to have a strong vested interest in the sort of role for central government which has typified the period since 1945. The Labour Party is committed by its constitution to public ownership, and indeed gas, electricity, coalmining, the railways, etc. were nationalised by the 1945–51 Labour government. Nationalisation in this context meant state ownership. Labour was also committed to planning as a way of managing the economy in a rational way in order to eliminate waste and gain overall socialist objectives. It also has a heavy commitment to the welfare state. In addition, it is the party linked with the trade-union movement, which, it might be thought, has a particular interest in re-establishing corporatist procedures for economic management so that it could regain its place of entrenched influence on government which has been undermined by Mrs Thatcher's vigorous repudiation of tripartite procedures. The Labour Party has also a commitment to greater equality, not just in income and wealth, but also to greater social equality through maintaining things such as comprehensive education, a high standard of care in the Health Service and a high level of state pension. It looks therefore as if the public spending implications of these factors and the managerial role of the state would imply a less critical view of the role of the state in British politics and less of a search for non-centralist, non-state institutions.

New Forms of Planning

However, this is not entirely the case, and many of the same features which we have seen in other parties in relation to their view of the role of government are rejected in the Labour Party. First of all nationalisation is being rethought, so that it could come to mean the wider dispersal of share ownership of major companies, tax incentives to encourage individuals to keep the shares, and legal changes to make share ownership a more powerful weapon in influencing company policy. This, it is argued, is compatible with the basic socialist idea of equalising, and therefore diversifying power. State ownership of industry has, on this view, done nothing to equalise power; it has in fact done the opposite, concentrating it in the state, in the hands of ministers and civil servants.

New ways are also being sought to develop an approach to planning which does not involve such centralised direction, and work has recently been published in this field by socialists such as Geoff Hodgson in *The Democratic Economy* (1984) and Peter Hain in *The Democratic Alternative*. As Robin Murray, the Director of Industry for the GLC, has argued (Murray, 1984, p. 219): 'Since 1979, socialist initiatives in economic policy have shifted from the national to the local level. Faced with large increases in unemployment . . . more and more Labour Councils have been extending their economic role.' On the Left this kind of development is linked to the effort to establish local enterprise boards in labour-controlled authorities, the most extensive being the Greater London Enterprise Board under the auspices of the GLC. Such boards could assist industries by providing premises, incentives and advice, but beyond this some boards have attempted wholesale efforts at generating industrial restructuring – for example, a case which Murray discussed is that of the furniture industry in London over which the GLEB took an important initiative. In addition, it could be argued that markets have an important role to play in the local economy just because they seem to embody a degree of decentralised decision-making:

> Markets are a necessary complement of any planned economy which aims at devolving decision-making power to the regions, the locality, the community, and the marketplace. (Hodgson, 1984, p. 185)

However, in terms of its official policy documents, the party does still seem to be committed to an updated form of the sort of planning tried in the 1960s and 1970s. In the party document *Planning and Full Employment* published in October 1985, a central planning ministry is proposed along with a British Enterprise Bank and a National Investment Bank. Also, of course, Labour economic policy-making is based upon attaining some kind of understanding with the unions, so that money for new investment would not be taken out by high wage claims.

In these senses, therefore, the Labour Party seems, at least at the level of policy-making for the economy, to be committed to some form of revamped Tripartism together with a changed form of public ownership of nationalised industries. This seems to be borne out when one considers the kinds of relationships which might be envisaged to exist between the local enterprise boards and a Labour government's overall economic policy. So, for example, at a local government conference before the 1983 election, Gyford reports that 'delegates were warned that such boards could not be given freedom to act in ways which went against the national economic planning framework which Labour is promising as part of its overall economic strategy' – as Gyford remarks, 'a warning which suggested a rather low degree of attachment to ideas of local autonomy and a somewhat centralist attitude to problems of national–local relations' (Gyford, 1985, p. 114). This point raises some general issues to which I shall return.

Workers' Co-ops

There has been, in addition, some emphasis on the role of workers' co-operatives, too, as indeed there has been in the economic thinking of the Alliance Parties, and this again reflects a growth of interest in less-centralised economic forms. In the context of the Labour Party this should come as no surprise. Although the Labour Party after the war, and perhaps purely as the result of the war, adopted a statist strategy, there has always been a major decentralist co-operative anti-state and even syndicalist stand in British socialism. Both Guild Socialism before the First World War, inspired by G. D. H. Cole, and the co-operative movement generally have been important for the socialist tradition in Britain, and currently there is a good deal of interest in rediscovering these values, and trying to compensate for the inevitable larger role for the state in a socialist society dedicated to greater equality and social justice. The 1974–9 Labour government sponsored some workers' co-operatives such as Merridan motorcycles but developing out of declining industries they were not a success and thereby hangs part of the problem with encouraging growth of co-ops. How are they to be encouraged to develop in industries which are not on the point of failing? Clearly at present there is very little incentive in a successful industry for workers within it to seek a fundamental change in its organisation. To provide such incentives there might be a need to develop a wider range of *state* institutions, such as a co-operative development agency or some kind of central institution for lending capital to co-operatives at favourable rates. Without such institutions it is difficult to see how a greater role for co-operatives could grow spontaneously.

Decentralisation

This issue might be thought to raise in a particular case a number of more general problems about decentralisation in politics. It is now frequently argued on the Left that socialism requires more in terms of the decentralisation of power than Labour governments have achieved since 1945 given their penchant for centralised solutions to political and economic problems. However, this very attempt to decentralise power produces dilemmas of its own, at least for socialists. In the first place a commitment to equality may be thought in some ways to favour decentralisation of power – to make power more equal it has to be dispersed more widely – but if it is dispersed more widely then of course there is more freedom of choice to exercise power in the interests of those who now hold it, and as a result the outcomes of the exercise of decentralised power may in some respects militate against greater equality. An example may make this more clear. The Labour Party has favoured comprehensive education as a mechanism for generating greater equality and a common culture, but it required the exercise of power by central government to achieve this. If there is a genuine decentralisation of power to local authorities, for example, would it be possible for them to preserve grammar schools of their own which might conflict with what the central government thought necessary to secure greater social equality overall? In this sense greater equality in the distribution of power might lead to greater inequality in the overall outcomes for society when such power is exercised.

The problem is endemic in any attempt to decentralise economic power, for example, to co-operatives. While worker-owned co-operatives will embody a higher degree of equality of income, power status, etc., this does not address the question of relations between such co-operatives and the extent of possible inequalities between them. In any system of decentralised enterprises, differences are bound to arise out of differences between internal efficiencies, skills of workers and managers, accessibility to and relations with suppliers and consumers, the age and quality of equipment, consumers' choices and demands, and decisions as to how the earnings of the enterprise are to be allocated between wages, bonuses, services, increasing employment opportunities, depreciation and investment. In short, without some redistribution *between* enterprises, the outcomes will embody the same sorts of inequalities which arise within ordinary markets, with co-operatives, for example, replacing individuals. If socialists criticise the market for being indifferent to the distributive outcomes between individuals, then similar inequalities arising out of decentralised groups within a socialist economy will have to be addressed. Such inequalities would have to be rectified by the state. These arguments would also apply to decentralised political institutions and to social and public services. Given the differences in prosperity and revenue bases between regions and localities, it is inevitable that inequalities in the provision of services will arise, and to attempt to equalise these can only realistically be achieved by the exercise of government power.

It could be argued that the issue of decentralisation poses a question about what sort of equality the Labour Party favours. The decentralist thrust seems to favour a commitment to a sort of procedural equality – attempting to equalise the power of individuals to choose their own way of life; however, this decentralised form of power is likely to yield highly unequal outcomes, which either have to be accepted because they have arisen as a democratic result from a situation of equal power, or have to be criticised because the concept of equality being used is not a procedural one but an outcome or patterned one – namely that equality is not just to do with procedures, but more to do with the nature of results. This point has been well made by Barry Hindess (1983, p. 81):

> Once new democratic mechanisms have developed there can be no guarantee that socialists will approve of the decisions they generate. Conquests of this kind are an inevitable feature of a democratically organised society, and an extension of democratisation may also extend the opportunities for them to occur.

The dominant strand of Labour politics since its inception despite the importance of Guild Socialism, the Co-operative movement and the importance of municipal socialism particularly in London before the war, has been parliamentary-based centralism. At the moment the Labour Party is characterised by a debate which has been submerged for nearly two generations, over whether socialism is to come from reform from above in the pursuit of particular outcomes such as equality and social justice, or whether socialism is more properly thought of as improvement from below.

No Return to Butskellite Consensus

As we have seen, therefore, the role of government is at the very centre of the agenda of debate about British politics. There is a sense that the role of the interventionist Keynesian post-war state has run its course and the social democratic, Butskellite consensus associated with it is no longer relevant. What chance is there of a new consensus around one or other of the sorts of strategies which I have outlined? There do seem to be elements for such a consensus: for example, the scepticism about the role of the state among the main political parties, a scepticism which emerged clearly for the Labour Party in Kinnock's address to the 1985 Labour Conference, in which he talked about the state as 'servant' and as 'enabling'. But there are still deep differences between the parties about the role of the state. There is some consensus, too, about the role of markets, and to some extent market-based ideas have gained intellectual and political ascendency in recent years, to the extent that even the Left, which is least congenial to market ideas, have been busy considering ways in which markets could play an effective role in a socialist economy (Nove, 1983; Hodgson, 1984; Hattersley, 1985; Kellner, 1984). But there is clearly no consensus about the extent to which markets should be constrained by macro-economic planning (as would be the case for Labour), or the extent to which an extension of markets ought also to involve redistribution of resources to give more economic power in the market to the worst-off members of society. The view of both Alliance and Labour is that such redistribution is necessary; the Conservative government believes the opposite – namely that the poor will benefit in absolute terms from a free market without redistribution. There is, in addition, little consensus about political decentralisation, and very little thought seems to be being given to the relationship between local decision-making bodies and central institutions in these areas in which potential conflicts may arise. Until some of these issues are resolved it is very doubtful that there could be a consensus around the idea of shifting basic choices onto the market and decentralised institutions and rendering them depoliticised.

2

Voting and the Electorate

MARTIN HARROP

Volatility

The 1983 election seemed to confirm that electoral volatility had become an established feature of the British political scene. There is no question that the electorate *has* become more volatile – but the change has not been a slow, steady one and its magnitude depends on how exactly 'volatility' is defined.

There are three main forms of electoral instability. The first involves changes in vote shares at by-elections and in mid-term opinion polls. Volatility within a single Parliament is now well established; it has not increased much since 1974. Its electoral (if not its political) significance is in any case exaggerated; mid-term flirtations rarely lead to long-term changes in commitments. The troughs and valleys recorded in opinion polls and by-elections during a government's life have become a less adequate guide to underlying trends.

The second form of volatility is *net volatility*: the change between one election and the next in the parties' shares of the vote. There is no doubt that swings from one election to the next have increased, though again February 1974 was the turning-point. Net volatility is measured by Pedersen's Index, which is obtained by summing the percentage point change in the Conservative and Labour Parties' share of the vote and dividing by two. Since February 1974, the index has only fallen below 6.0 on one occasion, in October 1974. The figure for 1983

(11.8) is exceeded only by February 1974 (Crewe, 1985, p. 9). Net volatility has grown.

The third form of change is *gross* volatility – the total amount of vote-switching by individuals between one election and the next. There is no convincing evidence that it is becoming more common. Crewe (1985) shows that there was only the slenderest increase in gross volatility over the 1960s and 1970s; no reliable evidence is available for 1983 although there was probably some increase in this election. As a rule, about two-thirds of the electorate vote for the same party or abstain consistently across a pair of elections. This proportion does not decline sharply when a longer sequence of elections is considered. This is because people who change their minds once are more likely to do so again. Many 'explanations' of growing volatility assume that gross volatility is the only sort; it is not and neither is it increasing rapidly.

How can net volatility grow without a parallel increase in gross volatility? The answer is that vote-switching is only one cause of electoral change. The other sources are (i) moves from voting to abstention (and vice versa), and (ii) the physical replacement of the electorate (comings-of-age and deaths; immigration and emigration). In general, electoral change proceeds not through changes of heart among existing voters but through the distinctive views of new voters. Even in the 1983 election, when Labour was hit by massive defections, its weakness among young voters and previous non-voters still contributed to its overall defeat (Crewe, 1986). So we should not expect any one-to-one relationship between net and gross volatility.

There are three main explanations for the increase in net volatility, explanations which correspond to the models of voting distinguished in Chapter 15. According to the *party identification* model, volatility is caused by a decay in voters' attachments to the major parties, a decay which itself reflect changes in socialisation and a decline in the class cleavage. According to the *radical* model, volatility reflects the emergence of new lines of cleavage which cut across traditional class divisions, yielding a blurring of issue attitudes and a growing dependence by voters on the media (Dunleavy and Husbands, 1985, p. 25). According to the *issue-voting* model, volatility is a natural consequence of an increasingly instrumental electorate confronting the ideological lurches of the major parties in the last ten years.

This chapter examines each of these explanations, focusing not just on interpretations of volatility but also on the general value of these models in accounting for electoral patterns in the 1980s.

The Party Identification Model

Both the extent and the strength of electors' general attachments to political parties have weakened over the last twenty years. Whereas eight in ten electors thought of themselves as Conservative or Labour in 1964, only seven in ten did so in 1983 (Figure 2.1). Furthermore, the intensity of allegiance among major party identifiers has declined even more sharply. Whereas 46 per cent of Conservative and Labour identifiers said their allegiance was 'very strong' in 1964, only 33 per cent did so in 1983. Even party loyalists are less committed than they used to be.

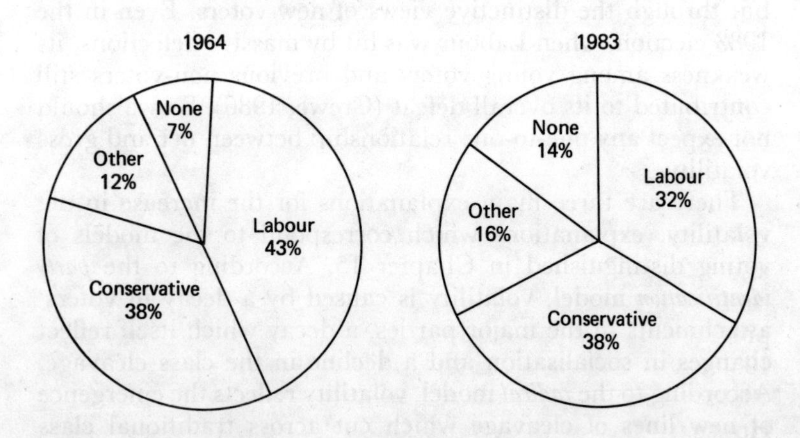

FIGURE 2.1 *Party identification, 1964–83*

Source: Crewe (1984), table 3.

There is no doubt that this partisan dealignment is linked to electoral volatility. Indeed, some critics of the concept of party identification (for example, Himmelweit *et al.*, 1985, pp. 194–6) suggest that survey respondents interpret questions about party allegiance with reference to their own voting histories, in which case electoral volatility would cause the drop in party identification rather than vice versa. Whether this is justified or not (and I return to the point later) it is clear that trends in party identification provide a smoothed redescription of volatility rather than an explanation for it.

However, the concept of party identification is rooted in a general model of voting which stresses the importance of family socialisation as a mechanism linking social cleavages to individual choice. The model suggests that party identification is passed on from parents to child along with other loyalties. Partisanship strengthens with age, sustained by interaction with people of similar views, as electors come to rely on their allegiance to 'solve' the problem of interpreting complex political information. Given family influences on partisanship, it is therefore possible that changes in socialisation patterns could account for partisan dealignment. For example, Crewe (1985, p. 128) considers the role of such factors as the growth of social and residential mobility, the earlier leaving of the parental home, the rise of television as a source of values and information, and the rapid growth of single-parent and second-parent families. But he dismisses all these factors, since the fall in partisanship is common to all age-groups and is not just restricted to younger voters who have been most affected by these changes in socialisation.

Within the party identification model, class dealignment is the most common explanation for weakening party loyalties and electoral volatility. Implicit in some discussions of this is a three-generation interpretation of electoral alignments. The present class alignment, it is argued, was forged during the 1930s and 1940s, with the second generation of the 1950s and 1960s inheriting the strong partisanship of its parents. However, the ripples from the Depression and the war become weaker as time passes, so that the third generation of the 1970s and 1980s responds in a volatile way to new political events and social divisions. It is less responsive to class appeals. This three-generation account also fails to fit the facts of weakening partisanship among all age-groups but the general thesis of a weakening in class voting does merit closer scrutiny.

A Decline in Class Voting?

An initial reading of the evidence certainly supports the proposition of a decline in class voting. Figure 2.2 shows a slow decline since 1951 in the proportion of middle-class voters supporting the Conservatives and a more rapid decline since 1966 in the proportion of working-class voters supporting Labour. In 1983, for the first time, less than a half of all voters (and a mere third of the electorate) cast a ballot for their 'natural' class party.

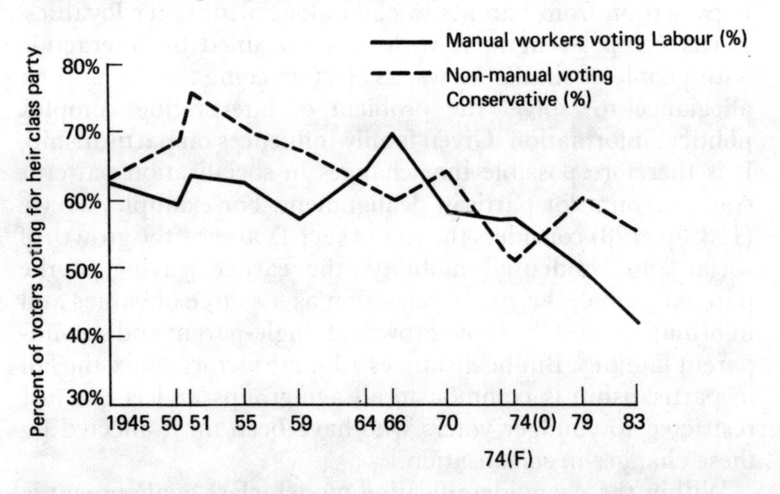

FIGURE 2.2 *Class and voting 1945–83*

Source: Heath *et al.* (1985), table 3.1.

In an important contribution, Heath *et al.* (1985) challenge the thesis of a decline in class voting. Evidence such as Figure 2.2 is, they say, misleading for two reasons. First, it ignores changes in the relative size of social classes. Over the last twenty years Britain has become a white-collar society (Figure 2.3). The shrinkage of the working class would be expected to produce a fall of 7 per cent in Labour's share of the vote, even if the proportion of working-class people voting Labour had remained constant. Second, Heath *et al.* argue that excessive attention has been paid to a party's *absolute* level of support in a given class. What really matters is whether a party's support in that class has changed *relative* to its support from other classes. The real question is whether Labour's decline in the working class has proceeded at a faster rate than its decline in the middle class. And to this the answer given is No! The Conservatives *did* become relatively more popular among the working class in 1979 and 1983 but this is attributed to short-term political factors, not to any long-term decay in the class structure. In summary, Heath *et al.* suggests that commentators have confused the decline of the Labour Party (probably brought about by political factors) with the decline of class voting (brought about by social changes).

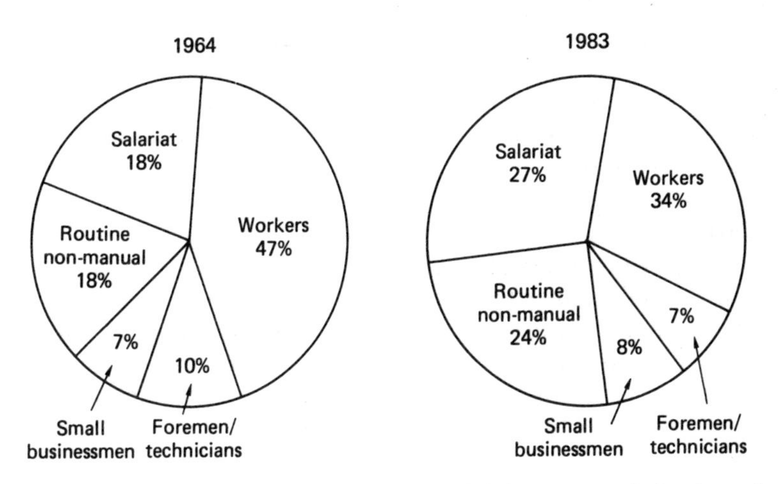

Note: The 'salariat' comprises managers, administrators, professionals and supervisors of non-manuals.

FIGURE 2.3 *The class structure 1964–83*

Source: Heath *et al.* (1985), table 3.3.

Heath *et al.* perform a valuable service in pointing out ambiguities in the concept of class voting. Their conclusion is also consistent with the fact that there has been no rapid weakening in class identities among the electorate (Crewe,

1985, p. 130). None the less there are several reasons why it would be wrong to conclude that there has been no decline in class voting even in the relative sense. First, the authors define social class by the respondent's occupation rather than by the head of the household's. This yields a very large decline in the size of the working class since 1964 because many newly employed women in routine white-collar jobs are excluded from the working class, even though their husbands may have worked in manual jobs for years. Other definitions produce a somewhat smaller fall in the size of the working class. Second, even when class voting is defined in a relative sense, there is evidence of decline. For example, in 1959, Labour's percentage support among the working class was 2.7 times as great as its support among the middle class; by 1983, this ratio had fallen to about 2.3. There is just a hint of the *ad hoc* in the authors' explanation for this. Third, Heath *et al.* do not consider the possibility that the decline of the Labour Party among *both* classes reflects the fragmentation of the class structure. As a party with a strong class basis, it would be logical to expect Labour to show the first symptoms of decay if this base were in fact fragmenting.

In conclusion, then, there is no question that the Labour Party is less attractive to the working class now than, say, twenty-five years ago; but it is equally true that Labour has also lost ground among the middle class though not at the same rate. On an absolute basis, the decline in class voting is dramatic; on a relative basis less so. By the same token, the growth in volatility is itself often exaggerated; the likelihood of a connection between the two remains high.

The Value of the Model

That said, there is no doubt that the party identification model explains electoral stability more successfully than change. So what is the value of this model in an era of volatility? In my opinion party identification still provides the most useful departure-point for understanding voting behaviour. The fact is that almost nine in ten electors still have a general loyalty to a particular party; that this loyalty often develops early in life and shapes the development of attitudes to policies and leaders;

that this loyalty generally strengthens with age; and that party identification accounts for the one in two electors who still vote for the same party time after time. These points explain the difficulties facing the Alliance in effecting a realignment in the electorate when there is just a steady trickle rather than a dramatic flood of newly enfranchised people onto the register.

The concept of party identification also has two useful analytic properties. First, it smooths out short-term fluctuations in voting patterns, providing a better guide to underlying trends. As we have seen, trends in party identification give a clearer picture of dealignment than the more confusing trends in electoral volatility. Second, and most important, party identification provides an indispensable base-line from which to assess specific election results. Consider for instance the 1983 election. In terms of votes, the Alliance was only two percentage points behind Labour – but in terms of party identities the deficit remained enormous (Alliance 19 per cent; Labour 37 per cent). This is the crucial evidence that 1983 was a deviating rather than a realigning election. It enabled analysts to identify Labour's defectors and also to predict that many, though probably not all, of them would soon return to the fold.

This did indeed happen. In the European elections of June 1984, the distribution of the vote fell much closer to levels of party identification in 1983 than to the General Election result. The average of the parties' monthly opinion poll ratings in the two years after the 1983 campaign also came very close to the 1983 figures for party identification. However, the Alliance vote in by-elections and local elections stayed above its share of party identification. A by-election, in particular, enables the Alliance to concentrate its limited resources on a single constituency, to appeal to electors to vote tactically and to attract protest voters who none the less retain an identification with another party to which they return at the next general election. But Alliance over-performance at by-elections (and the growing evidence that Labour does better at local elections than would be expected) can only be demonstrated by reference to a suitable base-line. Thus party identification remains an indispensable tool of electoral analysis.

The Radical Model

Proponents of this approach (Dunleavy and Husbands, 1985, section 1.3) believe that people vote in accordance with the interests of their social location, though political parties and the mass media both influence the interpretation of these interests. These factors operate directly on how people vote and not just indirectly by influencing voters' attitudes. Hence the focus of this model is on the complex patterns of interests in society, rather than on the individual elector; it is a sociological rather than a social-psychological approach.

Advocates of this approach believe that the party identification model exaggerates the importance of social interaction as a mechanism for the transmission of partisanship – whether from parent to child, husband to wife or neighbour to resident. In particular, the emphasis on family socialisation leads advocates of a model to underestimate the importance of new cleavages such as the divide between the public and private sector, and to overestimate the significance of older divisions such as occupational class. Advocates of the party identification approach also interpret differences in voting patterns between areas as reflecting a natural tendency towards electoral homogeneity among people who interact regularly. But often voting patterns differ from place to place because underlying cleavages also have a spatial aspect. Thus, strong Labour voting on council estates reflects the real political interests of the tenants rather than some mysterious process of 'contagion' on the estate (Dunleavy, 1979). This criticism is valid to the extent that few studies have actually answered the question 'who says what to whom with what electoral effects?' But conformity is such a powerful influence on behaviour, and the number of electoral facts which can be interpreted within this framework is so great, that this criticism is really an argument for conducting more research within the model, rather than for rejecting it altogether (Franklin, 1985, ch. 2).

Sectors

The radical model explains volatility with reference to new

social divisions which cut across the traditional class cleavage. The crucial factor, it is argued, has been the growth of the state in the post-war period. This has led to divisions between (i) the public and private sectors – a production sector effect, (ii) those who use public services (principally in housing and transport) and those who use private services (home-owners and car-owners) – a consumption sector effect, and (iii) the state-dependent population (pensioners, the unemployed and claimants of supplementary benefit) and the rest. Many working-class people who own their own homes and cars and who work in the private sector find themselves on the conservative side of these sectoral cleavages. Such voters become more responsive to the mass media, more attracted to third parties and more volatile in their electoral behaviour.

It is a neat scheme and only two reservations are in order. The first is that at least some sectoral divisions are neither new nor increasing in importance. Consider housing tenure, which is far and away the most important of all the sectoral variables. In 1983, the Conservatives led Labour among home-owners not only among all voters but also within each social class (see Table 2.1). But this difference is well established; Heath *et al.* (1985, p. 57) show that it has not increased since the 1960s. What has changed is the size of the various tenure-groups.

TABLE 2.1 *Conservative lead over Labour in votes, by class and housing, 1983*

Class	Housing		
	Owners	Private tenants	Council tenants
Salariat	+44	+49	−3
Intermediate classes	+45	+31	−25
Working class	+2	−19	−42

Note: Each entry indicates the percentage of voters voting Conservative minus the percentage voting Labour. Alliance support was more evenly spread across all categories.

Source: Adapted from Heath *et al.* (1985), table 4.2.

Owner occupation has become the predominant tenure as the private rented sector has gone into decline. But as yet this has not really helped the Conservatives. Heath *et al.* point out that working-class home-owners formed the same proportion of the electorate in 1983 as in 1964. The expansion of home-ownership among manual workers was offset by the contraction of the working class itself; the new home-owners were also the new middle class. These authors believe the Conservatives have benefited far more from occupational trends than from housing trends. The proportion of the electorate whose housing tenure cuts across their class has not increased substantially. Thus housing is a continuing division rather than a new cleavage capable of explaining electoral volatility and partisan dealignment.

A similar point applies to the employment sector, the other important sectoral variable. There is nothing new about public-sector radicalism. In 1965, for example, Parkin (1968) found that CND supporters were concentrated heavily 'in the employment of state and local authorities'. What has happened is that the public services, though not the nationalised industries, have expanded their share of the workforce. The balance between the public and private sectors, rather than changes in the impact of employment sector on voting, probably accounts for Labour's relatively satisfactory performance among the middle classes over the last twenty-five years.

The second reservation about the radical model is that the interests associated with particular sectors – whether home-owners, government employees or car-drivers – probably only explain a small portion of the statistical relationship between sectors and votes. Take housing again. Do home-owners vote Conservative because they are home-owners or because they are affluent, middle-aged and living in predominantly Conservative environments in the Midlands and South? And do real Socialists buy their own homes anyway? Between 1979 and 1983, for example, people who bought their council house did not defect heavily to the Conservatives; they already were a relatively Conservative group. Heath *et al.* conclude that, in the short run, housing tenure reflects rather than causes differences in party preferences. Thus the interpretation of sectoral differences in voting raises many questions. The weakness of the radical model is that it plumps for one interpretation of these differences while ruling out of court the evidence of the voter's own attitudes. These attitudes provide a basis on which a choice between competing interpretations can be made.

Exactly the same point can be made about sector of employment. There is some evidence of an emerging base for the Alliance among middle-class professional graduates in the public sector, a group from which left-wing parties have always gained a disproportionate share of their membership. But do such people move to the left as a *result* of working in the public sector or are they attracted to jobs such as teaching and social work *because* of their radicalism? It is probably a mixture of the two. Certainly much middle-class radicalism is either inherited through the family (in accordance with the party identification model) or based on the experience of higher education, particularly in the arts and social sciences. For such people, the public sector provides a favourable environment for strengthening their political values but it does not create them. Again, careful interviewing can give evidence of whether employment sector is cause or consequence of political values.

The Media

The mass media play a major role in the radical model of voting (see also Chapter 14). Television and the press are believed to define what are to count as acceptable political views, to influence citizens into similar modes of thinking and also to create a stream of messages which can favour one particular party (Dunleavy and Husbands, 1985, pp. 110–17). By contrast, the party identification model views the media as just a reinforcing influence on voting decisions, providing material which is selectively interpreted by voters whose standing commitment to a party insulates them from the effects of the media. In the issue-voting model, the media are seen more as a source of information than values.

The reinforcement interpretation of the media is no longer (if it ever was) an adequate guide to media effects. Here the party identification model is rather weak. For four main reasons, the media must now be considered full members of the family of influences on voting behaviour:

(1) The media are more important when party loyalties are weak – and party loyalties have diminished in Britain since the 1960s.
(2) The media are more important when there are new subjects to cover – and British politics has seen major developments since the 1960s.
(3) The media are more important when coverage is credible – and the arrival of television has strengthened the credibility of the media.
(4) The media are more important when people rarely discuss politics – and television has supplanted conversation as a major channel of political communication.

Television in particular is an important channel of electoral change in campaigns. As Table 2.2 shows, people who changed their vote between 1979 and 1983 were twice as likely to say television had helped them decide as people whose vote was stable. Though voters' reports on what influenced them are not

TABLE 2.2 *Influence of television on voting behaviour, 1983*

	All	Vote			Stability of vote		
		Conservative	Labour	Alliance	Same as 1979	Different from 1979	New voters
	%	%	%	%	%	%	%
Has television coverage helped you decide who to vote for?							
Yes	18	17	19	30	13	25	31
No	79	81	79	67	85	73	63
Don't know	3	2	2	3	1	2	6

Note: All N_s exceed 100, except last column (N = 46). Voting behaviour in 1979 is subject to recall errors.

Source: Gunter *et al.* (1984).

always reliable, there is no doubt that television is the vital arena within which new parties must establish themselves. Newspapers, by contrast, are much less important. Their open political bias means they are less credible than television. Given the Conservative predominance on Fleet Street, I have estimated that the 'press effect' is worth a swing of about 1 per cent from Labour to the Conservatives between two elections (Harrop, 1986). Not to be sniffed at but hardly decisive either. Television is what matters.

None of this means the radical view of the media can be accepted in full. Television matters but there is no evidence that how television 'frames' news (by the selection, tone and style of its coverage) is the key factor. In all likelihood television impact derives mainly from the content of its coverage. Television did not cause Labour's incompetence in the 1983 campaign; at most it reported the amplification by the other parties of Labour's bountiful failings.

The Issue-Voting Model

This model treats the voter as an informed consumer, choosing between parties on the basis of which comes closest to (or is least distant from) the voter's own conception of what governments should do. As Himmelweit *et al.* (1985, p. 70) say, 'Basic to the consumer model of voting is the assumption that people actively search for the party and/or candidate which seems most likely to implement policies which they favour and whose style of government they can respect.' This model sees voters acting individually and instrumentally, not collectively or expressively.

Advocates of this model claim that party identification is not an *independent* influence on voting behaviour. They argue that party identification does not add to our ability to predict electoral choice once the voter's policy preferences are taken into account. They suggest that party loyalties flow from issue attitudes just as much as votes. So party identification is a confusing concept, partly reflecting policy preferences and partly reflecting voting habits, which obscures the fundamental connection between what electors want and who they vote for (Himmelweit *et al.* 1985, pp. 192–4).

The problem with this criticism is that it ignores the powerful role played by party identification in shaping policy preferences (Miller, 1980). Indeed the psychological function of party identification is precisely to enable electors to make sense of complex political issues in which they do not have much interest. The fact that party identification develops among children before policy preferences is further evidence of the priority of basic party loyalties. So dispensing with the concept of party identification would certainly lead to an overestimate of the impact of issues on votes. None the less, the issue-voting model does at least bring politics back in. As with the radical model, it views the voter as responding to the substance of political debate. It is therefore a natural starting-point for the analysis of electoral *change* though less useful in understanding *levels* of party support.

The broad concept of 'issues' can be divided into three parts: first, the party's ideology – for example its attitude to equality; second, the party's specific policies – for example, on nationalisation; and third, the party's competence – its ability to implement its proposals if elected. This third theme of credibility falls on the borderline between issues and image. The impact of each of these dimensions is considered separately; in each case there is disagreement among political scientists of the importance of the dimension.

Ideology

In the 1960s, Butler and Stokes (1974, ch. 15) suggested that most voters did not think or vote in ideological terms. They had at most a nominal understanding of terms such as 'left' and 'right' and did not organise their attitudes to policies in a way that suggested any clear structure. Parties not principles were the lenses through which voters viewed politics.

This perspective has been disputed by several recent writers. For example, Himmelweit *et al.* suggest that political ideology is an increasingly important feature of electors' political evaluations, underpinning their attitudes to specific issues (see also Scarborough, 1984). But these conclusions are drawn from elaborate statistical analyses of what are still modest correlations between elector's views on various issues; they do not provide evidence, independent of the policy preferences themselves, of abstract reasoning or underlying principles. A similar problem attaches to the use of the term 'ideology' by Heath *et al.* (1985, ch. 8). These authors believe that voters form an overall perception of what the parties stand for – for example, Labour for nationalisation, the Conservatives for tax-cuts – and that these perceptions can be regarded as 'ideological' interpretations of politics by voters. But this definition does not capture the abstract, systemic nature of an ideology. Butler and Stokes's contention that most voters do not think ideologically still stands.

The baby should not be thrown out with the bath-water. Statistical analyses of the correlations between voters' policy preferences have yielded interesting results, even if they do not prove that the electorate thinks ideologically. Nearly all these investigations reveal at least two dimensions to popular political thinking. As described by Robertson (1984), these are a *left–right* and an *authoritarian–liberal* dimension. The first and more important left–right dimension covers socio-economic radicalism attitudes to a more equal distribution of wealth, power and opportunity. The second dimension of liberalism covers 'moral' issues such as immigration, capital punishment and law and order. These dimensions are independent of each other, so that a left-winger on socio-economic issues is not necessarily a liberal on moral issues. Indeed, working-class Labour voters tend to be authoritarian on moral issues while middle-class Labour voters are the most liberal group of all.

This raises two important questions for contemporary electoral behaviour. The first is whether the Alliance can replace Labour as the natural home for middle-class liberals. The evidence is that to an extent it already has. Heath *et al.* found some base for the Alliance in 1983 among 'moral' liberals, a base which was not characteristic of Liberal support at earlier elections. Miller and Taylor (1985) also found heavy support for the Alliance among middle-class liberals. They report that the Alliance scored between 35 per cent and 45 per cent among each of these sections of the middle class: those who opposed nuclear power, opposed nuclear weapons or wanted a cut in defence expenditure, opposed the sale of council houses or supported comprehensive schooling and public transport, opposed the death penalty and stiff sentencing generally, denied the importance of law and order or inflation, and stressed the importance of unemployment or the health services, or were classified as post-materialists. None of this means the Alliance is set to replace Labour as Britain's party of the left. The Alliance's social base among professional graduates in the public sector is too narrow to win elections though it can provide a respectable number of members. In 1983, most Alliance support came from protest voters unconcerned with broad questions of social liberalism. The Alliance received more votes in 1983 from working-class people with no O-levels than from salaried professionals with degrees.

The second question raised by Robertson's analysis is whether Labour can retain the support of those working-class people who adopt an authoritarian stance on moral issues. This is particularly crucial since the most socially conservative sections of the working class – older men with minimum education living on council estates (Harrop, 1982) – have historically been the backbone of the Labour vote. In contrast to the early 1970s, issues such as immigration and law and order were not to the fore in 1979 and 1983. But given continuing publicity to problems of order in the inner cities, the Conservatives have a ready-made weapon to deploy in future, closer contests.

This discussion shows that public opinion on broad economic and moral questions does illuminate electoral behaviour. In particular, authoritarian attitudes among Labour voters show that there are limits to the power of party identification to shape voters' attitudes. But whether this proves that most voters think ideologically is another matter; the concept needs to be defined more carefully, and investigated by more sensitive techniques, before this conclusion would be justified.

Policies

Just as there is disagreement about whether voters think ideologically, so too is there a debate over whether people have views about the detailed policies which form the small print of political debate. Butler and Stokes concluded in the 1960s that many people responded to questions about policy issues which divide the parties as if they were tossing a coin. And where attitudes do not exist, they cannot influence votes. Not surprisingly, advocates of the issue-voting model disagree. They argue that the impact of issues on votes has increased since the 1960s.

Is issue-based voting really increasing? Franklin (1985) argues that it is. He suggests that the decline in class voting has opened the way to choice between parties on the basis of issue preferences rather than class loyalties. He finds that the direct effect of issue preferences on votes more than doubled between 1964 and 1983 though party identification remained the strongest influence of all. Indeed as party identification has cast off from its class anchors since the 1960s it has ceased to be a simple transmission-belt between class and vote. It may now have become a more independent influence on votes. Franklin's estimates of the absolute impact of issues on votes may be too large because his analysis does not allow for a link *from* votes *to* issue preferences, but this is unlikely to affect trends. The probability is that issue voting has increased.

Just as the impact of issues on votes has varied over the years, so have issue preferences themselves. As Crewe (1985, pp. 138–40) notes, trends in public opinion since the 1960s on issues that divide the parties

> reveal a quite exceptional movement of opinion away from Labour's traditional positions amongst Labour supporters over the last twenty or 30 years. There has been a spectacular decline in support for the 'collectivist trinity' of public ownership, trade union power and social welfare.
>
> By contrast the convergence between the policy positions of the Conservative party and its supporters has remained strong and stable.

These trends show that a sufficient portion of the electorate must have real views on these issues for their changes of heart (or their replacement in the electorate by people with different views) to affect the overall distribution of replies. This portion is not necessarily a majority of the electorate; in fact, there is still considerable instability in policy attitudes when the same individuals are asked about the same issues at intervals of a year (Lievesley and Waterton, 1985). Party identification is much more stable than issue preferences. And falling support for Labour's policies is doubtless a reflection as well as a cause of its electoral decline. But the fact is that analysis of policy preferences does reveal trends which go some way to explaining electoral volatility. The issue-voting model says little about the origin of these preferences and whether these lie in the social structure or the mass media; but it does at least direct our attention to the issues themselves.

Competence

British elections are more often about competence than policies. They are about the parties' abilities to achieve shared objectives (high employment, low inflation, real economic growth), not the objectives themselves. They are about the parties' capacity to govern – their coherence, their direction, their implementation skills. Because competence is a managerial rather than a philosophical characteristic, perceptions of a party's ability can change over time and thus help to account for electoral change. Ideological policy preferences change more slowly, even though we have seen they are far from static; hence such factors are more likely to explain votes than swings.

By post-war standards, the elections of 1979 and 1983 offered unusually clear choices to the voters. But many political scientists argue that even in 1983 competence (or rather Labour's lack of it) was the key factor. What sunk Labour, suggests Miller (1984), was not its specific policy proposals but its lack of credibility:

> The issues were not inherently against Labour. Instead, disunity and incompetence in the Labour leadership destroyed Labour's credibility on its own natural issues and allowed its opponents to put their issues on the agenda . . . Labour was not ready for government in 1983 and the electorate knew it.

Though pitched at a more sophisticated level, this analysis resembles the defensive interpretation of the result offered by Labour's left, an interpretation which stresses Labour's weak presentation of policy rather than the policies themselves.

Other explanations of the 1983 election result place more weight on the substance of Labour's programme. For example, Crewe (1986) attaches considerable significance to Labour's defence policy. He notes that the campaign polls reported a mounting public rejection of its policies. A large majority opposed the unilateral renunciation of nuclear weapons and, reversing earlier opinion, small majorities supported the government's plans on Cruise and Trident. Miller, by contrast, again stresses the ambiguity in Labour's defence policies. The party failed to heed Lord Melbourne's dictum: 'It matters not much which we say but, mind, we must all say the same.' Assessment of the future impact of the defence issue depends on which of these interpretations is correct. Miller would argue that competent leadership and presentation can defuse the defence issue, even if Labour's policy remains unaltered. Crewe would argue that content as well as presentation is important, implying thereby that defence may yet re-emerge on the campaign agenda if Labour's policies do not change.

In any case, it is clear that public confidence in a party's ability to govern effectively is a crucial influence on its electoral fortunes. Indeed some analysts (Heath *et al.*, 1985, ch. 11) argue that the influence is so strong as to be spurious. After all, voters who decide to vote for a particular party (for whatever reason) are hardly likely to accuse it of incompetence. Certainly, advocates of a pure issue-voting model (which do not include Heath *et al.*) would want to argue that the elector's own policy preferences are at least as important as his or her assessment of the parties' capacity to implement them. But the relationship should not be dismissed just because it is difficult to disentangle the causal processes involved, especially when the impact of perceived competence on votes is what one would expect from the nature of election campaigns themselves.

The Economy

Economic issues such as jobs and prices normally top the pollsters' charts of national problems and deserve special consideration in a discussion of the issue-voting model. At first glance, however, the 1983 election offers little support to an economic interpretation of British voting behaviour. The Conservatives won despite a massive increase in unemployment since 1979, a victory which is usually attributed to political factors such as the Falklands and Labour's weaknesses. Yet for three reasons it would be wrong to dismiss the impact of economic issues in 1983. First, the Conservative recovery in the opinion polls during the Falklands episode coincided with growing public optimism about the economy, so it is difficult to say which was the real force behind the Conservative recovery. Second, Conservative success at reducing inflation after 1980 at the very least reinforced the support of its own sympathisers. As ever, the impact of prices has been underestimated in post-mortems on the 1983 election. Third, although unemployment proved to be the dog that failed to bark during the election campaign itself, the fact is that unemployment was the main economic determinant of variations in government popularity between 1979 and 1983 (Husbands, 1985). The economy provided a backdrop against which the campaign was fought.

The 1983 election exemplifies the findings of mathematical models of the relationship between the economy and government popularity. These studies show some effect of unemployment and inflation on voting intentions, especially when economic conditions are deteriorating towards crisis point, but political factors generally turn out to be more important. These include not just dramatic events such as the Falklands but a regular cycle between elections. In this cycle, a post-election honeymoon is followed by a deep decline in government popularity which is followed by a swing back to the government before the next election. From this perspective, the surprise about the period 1983–5 was not that Conservative support fell but was rather that the fall was so small. Some commentators, however, believe this cycle itself reflects government manipulation of the economy for electoral purposes.

The missing dimension in these mathematical models is how electors themselves assess the economy. In 1983, for instance, the Conservatives performed one of the all-time great electoral escape acts on unemployment. They convinced most voters that the world recession was the real cause of growing unemployment. In consequence many voters were sceptical about whether any party could do much about the problem and the Conservatives successfully defused the time-bomb which had been ticking away under their electoral prospects. What research has been done on economic attitudes points to two important conclusions. First, the electorate is myopic, weighting its judgement of the government towards economic performance in the recent past. 'What have you done for us lately?' is the question voters want answered. This explains pre-election booms. Second, national economic trends reported in the media are more important than the voter's own economic position though the two are obviously related. Mosley (1984) found that the *Daily Mirror*'s own 'shopping clock' was a better predictor of government popularity than the official inflation statistics themselves. In general, this suggests that the electorate's voting intentions are influenced more by the overall tenor of public debate about economic performance – is it boom or is it gloom? – than by a close assessment of the parties' economic policies. The impact of the economy on electoral behaviour thus offers only ambivalent support to the issue-voting model.

Overall this review suggests a mixed verdict on the issue-voting model. Although issue-voting has probably increased, many electors still do not think in ideological terms or have clear views on policy matters. They do, however, judge governments on the basis of their economic performance and these assessments help to account for short-term electoral change. But the issue-voting model underestimates the impact of party identification on political attitudes and tells us little about the roots of party loyalties in the social structure.

Locational and Regional Variations in Voting

The 1983 election served to confirm an increasing geographical variation in voting patterns which began in the 1950s and developed substantially during the 1970s. A Conservative/ Alliance two-party system appeared to be emerging across southern England, where Labour won only three seats outside London. But in the midlands, north of England, Scotland and Wales, the old two-party pattern remained dominant. In Britain as a whole there were only 285 seats in which Labour and Conservative took the top two places; with Conservative/ Alliance contests in 265. The Council elections and European Parliament elections of 1984 apparently solidified this pattern, with Labour support reviving chiefly in its existing 'heartland'. However, in 1985 the county elections showed a different pattern with Labour support recovering most in areas where it had slumped most badly in 1983, a finding confirmed by a large-scale analysis of 1985 Gallup polls conducted by Rose. So the trend to 'dual two-party systems' may have halted temporarily.

As Labour, and to a lesser extent the Conservatives, have found their support concentrating in their areas of strength, so the number of marginal constituencies which change hands with small swings of support between the major parties has been reduced. In the 1960s, there were on average 160 such marginals; by 1983 only half that number. The polarisation of voting patterns protected Labour from the erosion of its vote base in 1983, and increased the difficulties which the Alliance confronts in winning seats in proportion to its vote. In 1983, the Alliance won 23 seats with 26 per cent of the vote; had it gained 6 per cent more votes it would have won only 7 more seats. Disappearing two-party marginals also imply that the 'First Past the Post' electoral system is less likely to deliver clear majority governments and more likely to produce hung parliaments than in the past. The fragmentation of the anti-Conservative vote still provided a Tory landslide in 1983. But in closer elections the electoral system no longer so exaggerates the leading party's ability to win seats.

The widening divergence between how people in different constituencies vote has been explained in three ways: pure geographical effects; social contact mechanisms; and social interest differences. Pure geographical effects would exist if people's territorial locations (their position on the map) was itself a stimulus influencing voting behaviour – perhaps because parties in government favour the parts of the country from which they draw most of their support. Some authors argue that a centre/periphery cleavage (between the Home Counties and the rest) is important (Hechter, 1975: Orridge, 1981). For example, attempts to measure geographical effects try to show that Conservative voting gets less the further a constituency is from London, or that party support depends on which region a seat is in. Naturally, these effects have to be measured after we have controlled for the different class compositions of each area. In practice, most analysis shows that the social character of an area swamps any pure geographical effects (McAllister and Rose, 1984). However, while Labour support can be fully explained by socio-economic influences, Conservative voting does vary significantly by region, being lower in Scotland and northern English constituencies than social influences would predict (Dunleavy and Husbands, 1985, ch. 8).

Social contact mechanisms have been the most commonly cited explanation of geographical variations. The evidence shows that in more-working-class areas all social groups (not just manual workers) are more likely to vote Labour than normal, while in non-manual areas all social groups are more Conservative. Hence, people seem to be pulled towards voting in line with their 'class environment' as well as their individual class position (Miller, 1978; Johnson, R., 1985). Party-identification writers argue that this reflects voters being influenced by their social contacts: 'those who talk together, vote together'. Indeed, for some authors the 'class environmental' effect is a core example of how socialisation influences alignments (Franklin and Page, 1984). However, critics argue that this explanation relies on a mysterious 'contagion' mechanism, and that evidence of the requisite types, levels and political impact of community contacts has not been forthcoming (Dunleavy, 1979).

Social-interest explanations are put forward by the radical model. On this view manual workers in working-class and non-manual areas vote differently because their interests vary substantially. For example, a manual worker in an inner-city area is much more likely to own a run-down house or live in council housing, use public transport and other public services, be unemployed, or belong to an ethnic minority: while a manual worker in an affluent suburb is more likely to own a decent house, drive a car, have a reasonably paid job and be white. In this account there is no convincing reason to suppose that 'class environmental' effects are due to social contact mechanisms rather than to differences in other aspects of people's interests (Dunleavy and Husbands, 1985, ch. 8).

Prospects

How will future social trends alter the partisan balance? There can be no firm answer since a dealigned electorate is a volatile, responsive electorate in which political rather than social trends predominate. None the less, trends in the occupational structure, consumption patterns and demography are most favourable to the Conservatives and least favourable to Labour. The Alliance, as ever, occupies a middle position.

In the *occupational* structure, the decline in manual work will damage Labour, especially as many jobs will continue to be lost in manufacturing. Heath *et al.* estimate that the contraction of the working class may have cost Labour as much as 7 per cent of its support between 1964 and 1983. By the end of the century, a strong position among the working class will no longer be sufficient for electoral success. It will be increasingly important for Labour and the Alliance to capture the sympathetic (and expanding) sections of the middle class – union members, professionals, public-sector employees, graduates and routine office workers. There is a real competition here between Labour and Alliance; on its outcome may hinge the future complexion of Britain's left.

In the *consumption* sector, trends are more uniformly favourable to the Conservatives. Private ownership of homes and shares, the provision of education, health care and pensions through the private sector – all are factors which in the future may reinforce Conservative support in the middle class while cutting across Labour's working-class base. As yet, however, there is little evidence that changes in consumption patterns have altered the party balance, and Labour might conceivably benefit from increased salience of issues such as private health care as long as most people still rely on the public sector.

Demographic trends are the biggest single threat to Labour. As we have seen, electoral change usually proceeds through generational replacement. All voters die; only a minority change their basic political outlook while alive. Until recently, the march of the generations has favoured Labour, as people socialised before Labour became a major governing party in 1945 die off. This factor will continue; the proportion of the population who came of electoral age before Labour's peak in 1945 will continue to fall over the remainder of the century. Unfortunately for Labour, however, this factor will soon be overtaken by the growing dominance in the electorate of voters who do not remember Labour's golden years between 1945 and 1970. Post-1970 cohorts will form a majority of the electorate by the year 2000. These are dealigned generations, unsympathetic to class appeals, for whom the Labour Party has never been an object of strong emotional commitment. Weak partisanship among the young means that a Labour triumph in future elections will not have the long-term, imprinting effect of its 1945 victory. In short, Labour's past is rapidly becoming an electoral liability rather than an asset.

3

The Party System

HENRY DRUCKER AND ANDREW GAMBLE

A Dominant or a Multi-Party System?

Britain currently has four major national parties: Conservative, Labour, Liberal and SDP. The Liberals and SDP though organisationally separate have combined to form the Alliance and operate as one party electorally. These are national parties because they maintain their organisations and contest elections in all regions of Great Britain, and seek to control the government machine at the centre. Northern Ireland has its own party system, once joined to Britain's but now separate. There are a few minor national parties, and two important regionally based parties in Wales and in Scotland, whose aim is not to win power at the centre but to secure devolution of central powers to regionally based governments.

The national political parties compete for votes, for power and over policy. Politically ambitious individuals can climb to power only through one or other of them. They attempt to organise the political bias of forty million people. Each attempts to put together a winning coalition on the basis of a distinctive public philosophy and attempts to appeal to a broad range of interests. Only a tiny proportion of the total electorate joins political parties. Party members help to raise funds and to mobilise votes during election campaigns. Many become active because they wish to influence the policies which the party adopts. In this way parties are one specific channel through which some citizens participate in politics.

Conflict and the co-operation between the parties takes place in several different arenas, each of which is linked to the others. It takes place in parliament and in local government; at local, national and European elections; in the daily press, and radio and television. The way each party behaves is strongly affected by such factors as the number of parties it has to compete with, its chances of forming a government alone, its support from electors and organised pressure groups, the intensity of the conflicts between it and the other parties, the degree to which these differences can be compromised, the nature of the electoral system, and the loyalty of each of the parties to maintaining the authority and integrity of the state. What a party does is affected by the nature of the *party system* in which it operates.

Until quite recently books about British government had no need to emphasise this truth because since 1945 the nature of the system was clearly settled. Britain had a stable two-party political system in which each of the two major parties, Conservative and Labour, had a real chance of forming a government after each election. The two parties competed in virtually all parliamentary seats and nearly all local government wards as well. Each had bastions of local support but these were roughly equal in population over time and each formed governments from time to time. Britain employed the 'plurality' electoral system which reinforced the position of the two major parties and made it all but impossible for any third party to force its way in.

The most important recent development in the party system has been the challenges to the dominance of the two major parties; from the territorially based nationalist parties, Plaid Cymru and the Scottish National Party; and from a nationally based centre party grouping, the Alliance. From 1974 to 1979 Britain began to take on some of the appearance of a multi-party state. The most obvious symptom of this was the failure for a time of any party to gain a secure majority in the House of Commons; for a few months in 1974 and again from 1976 to 1979 the country was ruled by a minority Labour government. In the second period an explicit agreement was made with the Liberals (the Lib–Lab Pact) to keep the government in office. It appeared that subsequent elections would produce parliaments in which no one party would command a majority making coalition government necessary.

But this expectation was not fulfilled in the 1979 election when the Conservative Party won a clear majority of seats in the House, and it was confounded again in 1983, when, on a slightly reduced percentage of the vote the Conservatives succeeded in winning a considerable majority (60 per cent) of the seats in the House. From the standpoint of that election Britain appeared to be moving not to a multi-party system in which there were three major parties, but to a dominant party system in which the Conservative Party could expect to win all but the most exceptional elections.

Which characterisation is applied depends on which part of the system is taken to be most important. At the electoral level there is some confusion, since voters often vote one way in parliamentary general elections, another in by-elections, and differently again in local and Euro elections. Nevertheless, the persistence of Alliance support is unmistakable. What it means is that there is now an emergent three-party system in England and a four-party system in Wales and Scotland.

The territorial challenge from the nationalist parties has been contained for the moment, and it is the national challenge from the Alliance that poses the greatest threat to the continuation of the two-party system. There remain, however, major obstacles to its ultimate success. The two other parties will do everything in their power to frustrate it. The Conservatives in particular have most to lose from the success of the Alliance since they are the party that has been in office most often under the two-party system. Both the existing electoral system and the current facts of political geography are on their side. With Alliance voters concentrated in the South and Labour voters in the North, the opposition is divided in ways that up to now have benefited the Conservatives. Having won 60 per cent of the seats in parliament with 42 per cent of the votes, they are in a very strong position, and it will need a political earthquake to loosen their hold on office at the next election.

At the level of parliamentary competition the old two-party system was, at any rate superficially, intact. Conservative and Labour MPs made up 93 per cent of the House in 1983. In the sense that a two-party system excludes 'third' party MPs, the system was still two-party. But in the sense that a two-party system gives complete control over the administrative machine to the leaders of the majority House of Commons party, the system was less securely two-party than it once was. MPs of both parties were less loyal to their party leaders' wishes than in the past, and governments, even governments with large majorities, did lose occasional votes.

Parliament's actions and the statements of its members still dominate journalists' ideas of what British politics is all about. For this reason the two-party dominance of the House of Commons has tended to be reflected in their dominance of the news. Between elections it has been difficult for the Alliance to make much impact. The structures built up in the old two-party world continued to reproduce the conditions for its perpetuation. But during elections the Alliance got a much nearer equal share of time and journalistic space. Much as the two bigger parties would have liked Party Political Broadcasts to continue in proportion to MPs elected at the previous election, the unfairness of such a distribution was so palpable that the Alliance suceeded in changing it in 1983. The airwaves were usually two-party but during elections they became multi-party.

Size and strength of party organisations were an important bulwark of the two-party system in the past. A good organisation could keep the leaders in touch with public opinion and produce successive generations of new candidates as well as the cash and local knowledge to fight elections. In the post-1945 two-party system the parties in government have been careful not to interfere with each other's organisation or finance. The Conservative bid in the 1983 trade-union act to remove one of the major props supporting the Labour Party – trade-union political funds – is a major departure reflecting a desire by the Conservative leadership of Mrs Thatcher to make her party still more dominant. The Alliance parties, particularly the SDP, cannot match the two major parties in terms of their party organisation and membership. But this may not be of such significance in determining the nature of the party system because of the decreasing importance of party organisations. Party leaders communicate to their publics more frequently and more powerfully by the media than by party meetings, and they learn more frequently and more reliably what the public is thinking through public-opinion polls than through informal local soundings by party members. The success of the SDP since its launch in 1981, despite its skeletal organisation, underlines this point.

It was at the local level that the third party found it hardest to break through but this now seems to be happening. In the English local elections of 1985 the Alliance won about half the seats won by either Conservative or Labour and denied either Conservative or Labour control in 27 of 47 countries. Here one sees multi-party politics in its most developed form. This development is particularly threatening to Labour, for local councils provided an energetic and imaginative generation of local Labour leaders from 1981 to 1985 with a base for experimentation and for opposition to central government policies.

To discuss whether British politics is evolving into a multi-party or a single-party dominant system in the mid-1980s, we need, then, to look in more detail at party competition in several different contexts: parliament; party organisation; thc electoral system; and the policy agenda.

Parliamentary Majorities and Electoral Minorities

The overwhelming Conservative dominance at the parliamentary level in 1983 was much less pronounced at the electoral level. The Conservatives were elected on a minority vote (43 per cent in 1979, 42 per cent in 1983). Electoral support for third parties – the Liberals and the SDP – was both significant and increasing.

At the parliamentary level the British party system is a two-party system marked by increasing Conservative dominance. But at the electoral level the British party system is a multi-party system, with different parties dominant in different regions. For a true dominant party system the Conservative party would need to be dominant in every region. What the electoral map in fact shows is a sharp divergence in the support for the two major parties in different regions. The ability of the Labour Party in 1983 to win 200 parliamentary seats and therefore maintain the appearance of being a major party of the state and a contender for Government, was due to the concentration of the party's support into certain regions – the inner cities, parts of the North, Wales and Scotland. In the south of England outside London, Labour's representation almost disappeared and its support slumped. In Scotland, however, Labour remained the largest party. In South Yorkshire, Labour won fifteen of the sixteen constituencies.

In electoral and regional terms the Conservatives have become an English party. Previously it had broader strength. It used to include within its coalition the Ulster Unionists, and it used to be the leading party in Scotland and in several of the leading English cities including Birmingham and Liverpool. Since the imposition of direct rule in 1972, Ulster politics have been detached from the British party system. The Ulster MPs no longer take the Government or the Opposition whip.

The Conservative Party has a position of great strength in the south of England, but it is less of a national party than at any time in its history as a mass party, and it is also being challenged in its heartland by a new electoral competitor, the Alliance. The scale of Labour's collapse in the south of England in 1983 – the party lost 111 deposits – was a reflection of the success of the Alliance, which came second in two-thirds of the seats held by the Conservatives. In many parts of the south of England the Alliance replaced Labour in 1983 as the main electoral competitor to the Conservatives.

Parties in Parliament

The House No Longer Representative

Traditionally the focus of the party battle between elections is in Parliament. It is here that government and opposition leaders fight weekly to establish the authority of their case. It is on the floor of the House that new backbench MPs can establish their claim to the ear of the nation. The battle on the floor, especially the four weekly question-time periods, has become the central arena of national politics watched over by the political press and reported elaborately. Here reputations of parties and politicians are made and destroyed.

But parliamentary practice has long been based on the assumption that there were two, and only two, parties that mattered: made up of those who support and those who oppose the government. The right of this assembly to represent the nation, and the claim of its debates to be taken seriously as the cockpit of the nation's politics, rested on the further assumption that the citizenry were also divided, without important exception, into those who supported the government and those who supported the opposition. With the recent advent of multi-party politics in the country, these assumptions no longer hold.

The 1983 General Election created a House of Commons which did not fulfil the conditions necessary for the operation of the tradition in two ways: first, there were 44 MPs from parties other than the major two, of which the most substantial were the 23 Liberal/SDP alliance and the 11 Official Unionists: second, the House did not represent the country in that the Conservative Party (397 seats from 42.4 per cent of the votes) and the Labour Party (209 seats from 27.5 per cent of the votes) – respectively the government (Conservative) and the Official Opposition (Labour) – won 93 per cent of the seats with only 69.9 per cent of the votes. In other words, the 25.3 per cent of the votes which went to the Alliance was represented in the House by less than 4 per cent of the seats.

Radical changes in the procedures of the House would be necessary for the Alliance MPs, who won one vote in four, to make that weight felt. Little alteration was made either by the Speaker or by the major party whips (who control the agenda of the House) to correct or compensate for the ill fit between its membership and procedures, and the votes of the British public. The Alliance parties have contested the amount of time their parties should be allowed to have for debates of their choice, the numbers of places they should have on Select Committees, as well as the number of Party Political Broadcasts and Party Election Broadcasts they were entitled to.

Parliamentary Dissent

Strong party discipline was a major feature of the two-party system. Parties enforced loyalty in their own interest. The majority party could govern with little fear of losing votes on major legislation. The pattern began to change under the Heath government of 1970–4. With a majority of over 30, that government lost six votes; three of them on three-line whips. During 1974–9 the Labour government, lacking a secure majority, suffered regular defeats, but did not treat the votes as issues of confidence so did not feel obliged to resign.

The control of the executive over the Commons returned after 1979. In the 1979–83 parliament, dissent by MPs of either party occurred on only 1 per cent of whipped votes, and the government was actually defeated by its own MPs voting with the Opposition on only one occasion (in December 1982 on immigration rules). But MPs' attitude is less deferential and more critical than it once was (Norton, 1985, p. 28). The significance of MPs' change in attitude is felt not so much in lost government bills, but in governments' increased need to consult with and seriously consider the attitudes of their backbenchers.

Private Members Legislation

Some changes are more important, and more generally interesting, than they first appear. This is true of Private Members legislation. This is legislation proposed on the initiative of a backbench MP. During each parliamentary year six or seven such proposals become law. By tradition they do this without either support or interference from any party's whips; equally, Private Members legislation avoids the topics which are divisive between the major parties.

Before the 1966 election most Private Members legislation was on minor matters. Government MPs, in particular, would often ask their whips for suggestions of small things which their Leaders would like enacted. After the 1966 election MPs responded to the strong demand for a loosening of the law on moral matters and used Private Members legislation as a device to push for important changes. Once the habit of cross-party voting was established in the Private Members procedure it could be extended to other areas with the assent of the whips, free votes were sometimes allowed on controversial parts of public legislation. Abortion was legalised, divorce made easier, homosexual relations between consenting adults legalised, hanging was abolished and the theatres were freed of censorship.

The force of this reformist wave which had cross-party support weakened by the mid-1970s, and has been largely reversed; but the practice of non-party voting especially on subjects affecting the liberty of the subject and on sexual matters continued. 'Video nasties' have been outlawed. Front-seat passengers in cars have been required to wear seat belts. The advertising of cigarettes has been restricted. There have been several attempts to limit the age of a foetus that can be legally aborted. There was an attempt to limit research on human embryos.

All these bills have required cross-party support – each of them has been opposed, in turn, by an *ad hoc* cross-party coalition. Collectively they have done much to change the lives of ordinary citizens. The expansion of extra-party voting into morally contentious, often emotive, and usually easily understood and hence much publicised, areas of public policy is an important phenomenon. It has occurred without help from the parties in precisely the arena (Parliament) that the parties unequivocably control, and has denied the electorate any direct influence over which policies are adopted.

Parliamentary Committees

Observers of British politics have often noted the weakness of House of Commons committees compared with committees in other legislatures. The American Congress, in particular, has much more powerful committees than the British House of Commons. The strong unified discipline within the British parliamentary parties was the main explanation offered for the weakness of the British committees. Whatever the formal procedures, so the argument ran, members of the House will vote as the whips suggest.

Recent experience with Select Committees suggests that the

discipline is not as strong as it once was. In 1979, following years of pressure for change and a Select Committee on Procedure Report in 1978 outlining a new system, the Conservative Leader of the House, Norman St John Stevas, brought in a new comprehensive system of Select Committees. The new system of fourteen committees, one for each major ministry, has more authority and powers than its predecessors. The new committees have also been very active. In the 1979–83 session they held more than 2,000 meetings; they called 1,312 civil servants to testify before them 1,799 times; 161 ministers also appeared; and the committees produced 193 reports (Downs, 1985, p. 62).

The work of the Select Committees is circumscribed by their government party majorities. To operate effectively the Committees need to establish their own elan; members must be willing to speak and vote as members of the Committee, not simply as party members. Government party MPs are under the greatest pressure in this respect – opposition MPs have little difficulty voting and speaking against government. The Committees had a number of useful victories in this. The Treasury and Civil Service Select Committee, led in 1979–83 by Edward Du Cann who was also the leader of the Conservative backbenchers' 1922 Committee, criticised some of the central parts of the government economic policy, and the Education Committee influenced the government on several occasions, as did the Energy, Employment, Home Affairs and Foreign Affairs Committees (Downs, pp. 65–6).

No one would yet regard the new Select Committees as equivalent to the American Congressional committees, but the new system has established itself. It does draw information out of government; it does occasionally change public policy; members of committees do their work faithfully and do find in them further platforms for cross-party co-operation. Their success is possible because of the looser texture of party discipline, and their operation will serve to unravel the discipline further.

The Lords

Until quite recently 'parliament' was taken to be a shorthand for the House of Commons. When the House of Lords was discussed at all it was only to argue about how it had survived for so long and how it should be reformed, replaced or abolished (Norton, 1982). With its permanent Conservative majority it gave most trouble to Labour administrations. Under the two Thatcher administrations it has become the only effective parliamentary check – not insuperable but certainly troublesome to the will – of the government.

During the 1979–83 session the government escaped formal defeat in the Commons altogether. Early in the session the Lords defeated the government over the Education (No. 2) Bill which would have allowed local authorities in rural areas to charge for transport to school. Thirty-eight Conservative peers, 28.7 per cent of Conservatives voting, voted against the government (Baldwin, 1985). The Lords also defeated the government when it prevented them from forcing local authorities to sell council houses specially equipped for the elderly; and it forced the government to accept a change in the British Nationality Bill which gave special rights to the citizens of Gibraltar. There were defeats for the government, too, over the Police and Criminal Evidence Bill, the Trade Union Bill, the Housing and Building Control Bill and the Ordnance Factories and Military Services Bill. Between the return of the Conservative Party to power in 1979 and the beginning of the 1985 session, the government were defeated in the Lords on sixty occasions (Baldwin, 1985, pp. 100–101).

But the most important defeats the Thatcher government suffered in the Lords concerned the legislation to abolish the Greater London Council (GLC) and the Metropolitan authorities. The government proposed to deny London and the large English conurbations a directly elected local authority, and to give the powers of the existing bodies to lower-tier authorities (the Districts or Boroughs) or, in some cases, to newly created QGAs (see Chapter 5). The government were forced to abandon their timetable in 1984 when an all-party Lords amendment wrecked it, in the face of government whipping, by 191 votes to 143.

Why has the unelected chamber become more assertive? It has more life-peers and cross-benchers active within it; some peers may calculate that they can defy the government with impunity because the task of reforming them is too difficult; and there is some evidence of tension between the ideological radicalism of the Thatcher leadership and more traditional Conservative attitudes of some of the Tory peers. On this view the Lords may be acting unusually in voting against a Conservative government, but they are voting for a different kind of Conservative Party.

Policy-Making and Factions

The Conservative Party has long possessed significant organisational advantages in comparison with other parties, which has made it extremely resilient and a very tough electoral competitor. Party organisation covers many things, including how policies are formulated, how MPs and leaders are chosen, who controls the party bureaucracy, what role party members play, and how the party raises funds.

In relation to the way they make policy and choose leaders, political parties can be broadly divided into *leaders'* parties and *democratic* parties. In the former, policy-making tends to be the prerogative of the leader, while in the latter the responsibility tends to be shared between the parliamentary party and the mass party. In Britain the Conservatives and the SDP are examples of leaders' parties, though the Conservatives have become noticeably more 'democratic' in the last twenty years, while Labour and the Liberals are democratic parties, with leaders constantly feeling the need to assert themselves.

One of the reasons for the Conservative Party's long success has been its adherence to a form of organisation and policy-making which has given considerable autonomy to the party leadership and very few formal means of making the leaders accountable to party members. Despite her populist campaigning style, Mrs Thatcher has not changed that aspect of the Conservative Party at all. The Conservative Party Conference is not a sovereign body; its votes do not bind the party leadership; it does not elect the party chairman or any party officials. All key appointments in the party, including the policy committees, remain in the hands of the party leader. The formulation of policy, the drawing-up of a manifesto, the choice of campaign styles and messages, are all firmly in the control of the leader. Margaret Thatcher has been a dominant and forceful leader, and the tremendous concentration of powers of patronage and influence at her disposal have been used to stamp her personal style on the party.

Indeed, some addition to the strength of her position derives from the changes in leadership election procedures forced on her immediate predecessors. The leader of the Conservative Party no longer emerges through informal channels of consultation whenever the party feels the need of a new leader. She or he is elected annually by Conservative MPs. This formal procedure has the intended consequence of protecting the incumbent from swift political assassination.

Emphasising the lack of formal constraints on party leaders makes their parties appear monolithic and their leaders all-powerful. This is mistaken. There are many informal channels through which opinion is consulted and assessed and a consensus forged, and this sets limits on the policies leaders can impose on their followers. The real point about leaders' parties is not that they are autocratic but that their leaderships, lacking many formal constitutional restraints, are much freer in devoting all their energies to choosing that set of policies and electoral strategy which will maximise their chances of winning votes. Leaders in democratic parties attempt to do the same, but have to endure much more determined attempts from their mass parties to share in the formulation of the party's programme and to give instructions to the party leadership.

Liberals and SDP

The difference between a leader's party and a democratic party is illustrated by the different policy-making processes of the two parties in the Alliance. The SDP has an elaborate policy-making and formal democratic machinery. But, so far, important policy initiatives have come from the leadership. David Owen committed the SDP to a social market economy in speeches before the issue had been discussed by the SDP Conference. Owen became leader of the party after the 1983 election when Roy Jenkins resigned without election because no other SDP MP would stand against him. This was hardly surprising given that there were only four SDP MPs, but it did negate the party's recently written democratic leadership election procedure. The Liberals, by contrast, have a long-established and thriving party membership. They were the first of the parliamentary parties to give greater powers to their mass party. Since 1977 the party leader has been elected by the whole party, not just by the MPs. David Steel has never been able to impose policy on the Liberals and has occasionally suffered defeats on policies he has recommended to the Conference, particularly on defence.

Labour

The Labour Party is the best example of an internally democratic party. Between 1977 and 1981 it was rent by major struggles over its constitution, which resulted in a major split in its ranks and the formation of the SDP. The Labour Conference is formally sovereign in the party but the parliamentary party has always attempted to assert its autonomy. The rival authority of Party and Conference, was expressed in the 1970s in the constant conflict between the Party leaders (elected by the Parliamentary Labour Party – the PLP) and the National Executive Committee (NEC) (elected by the Conference). The supremacy of the parliamentary party was generally confirmed, due to the alliance between parliamentary leadership and the trade-union leadership, to ensure victory in the crucial policy votes in Conference and control of the NEC.

The campaign for greater democracy in the party was an attempt to make the parliamentary party subordinate to the party conference and the NEC. To succeed, it required the support of trade unions. Success was only possible because of the growing rift between some trade unions and the party leaders, and the opportunity that gave to build up the NEC and the Conference as sources of power against the parliamentary party. The three reforms involved the election of the party leader by the whole party; an automatic re-selection process for all sitting MPs; and responsibility for drawing up the election manifesto to be given to the NEC, rather than shared between the party leader and the NEC. This last proposal was defeated, but the first two were carried in 1980. Subsequently, at the special Wembley Conference in 1981, the electoral college for electing the party leader gave 40 per cent of the vote to the trade unions and 30 per cent each to the PLP and the constituency parties. The pressure for further reforms has abated, but several continue to be advocated by Tony Benn and his allies on the NEC, such as the election of the Shadow Cabinet and the Cabinet by the whole party; while others, such as 'one member, one vote' elections within constituency parties for party leader and MP re-selection, are advocated by Neil Kinnock and his supporters.

The battle over the constitution helped force the split in Labour's ranks and advertised its deep divisions. The Right of the party tenaciously fought to retain the autonomy of the PLP, and when they appeared to have lost, a large group left the party, led by David Owen, William Rodgers and Shirley Williams. The actual impact of the reforms, however, on the party's outlook and performance has so far been very different from what its supporters hoped and its opponents feared. The narrow defeat of Tony Benn in the Deputy Leader election contest in 1981 marked the high point of the Left's struggle to make the party leadership fully accountable to the party Conference. The election of Neil Kinnock and Roy Hattersley by the new electoral college strengthened the authority of the parliamentary leadership. Formerly the leader had only been the leader of the PLP. The hope of its sponsors that the constitutional reforms would lead to a very different kind of Labour leadership have not so far been realised. The new leadership has been as oriented to parliamentary opinion as the old, making its priorities the rebuilding of Labour as an electoral force and as an alternative government, resisting attempts, such as during the 1984 miners' strike, to make the Labour Party into an extra-parliamentary campaigning party.

Kinnock and his team have been showing that Clem Atlee's old dictum that the Labour Party was most easily led from the left (to the right) still holds true. Despite the party's decision to make its internal processes more democratic and more formal, the leadership contrives to appeal to the electorate over the head of the party on the basis of information about public attitudes which come to party leaders from public opinion polls. For the first two years of his leadership, Kinnock consolidated his position quietly, but in October 1985 at the Annual Conference he felt strong enough to take on the hard left in public – much to the delight of the centre and right of the party, and the right in the press beyond.

Conservatives

The Conservatives have avoided the kind of upheavals experienced by Labour, because in their concern to be a party of government rather than a broad political movement, they have always been clear that the purpose of their mass party is to support the parliamentary party, not to direct it. This is one reason why the Conservatives are such a united party. The ethos of loyalty to the party leadership is strong; so too is the dislike of doctrine and ideology and programmatic politics. Because of the hierarchial structure of the party and the lack of formal accountability and a formal democratic policy-making process, there is also a lack of organisational issues that rival factions could contest publicly. Most Conservative disputes go on in private, and there are constant efforts to build and maintain consensus within the party at all levels.

The party has, however, been changing in response to the style of Margaret Thatcher's leadership with its unusual emphasis for Conservatives on doctrine and principle. The party has appeared less united in the last ten years than at any time since 1945. A succession of prominent Conservatives – including two former Leaders, Lord Stockton (Harold Macmillan as was) and Edward Heath, as well as ex-Cabinet Ministers such as Francis Pym, Ian Gilmour, Jim Prior and Geoffrey Rippon – have expressed opposition to the style and some of the policies (particularly economic management) of the Thatcher government. When this criticism culminated in the formation of the Conservative Centre Forward Group in 1985 the former Cabinet 'wets' at last had an organisational focus for their dissent. But their disagreement with the party leadership was not on the kind of issue of fundamental principle such as Home Rule or Free Trade which had split parties in the past and the sting of organised dissent was drawn at the September 1985 Annual Conference when the Chancellor signalled clearly that he was backing off some of the more doctrinaire of his policies. Once again the party had been saved from severe factional dispute by adroit leadership.

The Labour leadership has always had to struggle to keep the internal divisions in the Party from becoming open splits. The factional organisation of the PLP, the trade unions, and the constituency parties on Left/Right lines, is much more pronounced than in the Conservative Party, and has led to continual conflicts over policies, the election of the NEC, and the selection of MPs. The composition of the factions, however, constantly changes. The passage of the constitutional reforms has not led to the domination of the party by the Left. The authority of the parliamentary leadership has strengthened, only a handful of MPs have suffered deselection, and a major split has occurred in the ranks of the Left. Key figures such as David Blunkett and Michael Meacher and Ken Livingstone have distanced themselves from the hard Left and given public backing to Neil Kinnock. The need for party unity which stems from the pressures of electoral competition keeps the factional conflict within bounds.

Membership and Finance

A key factor in the parties' organisation is their ability to attract members and funds. Membership figures are unreliable but all estimates indicate sharp falls for both Labour and Conservatives in recent years. The Conservatives have declined from a peak of 2.8 million in 1953 to 1.5 million now. Membership of the Young Conservatives has slumped. Average constituency party membership is still 2,400. Labour has far fewer individual party members of the party. By the end of the 1970s it had less than 300,000. This decline may now have halted. The imbalance is partly offset by Labour's four million affiliated trade-union members, one million of whom may give help to the party during election campaigns. The SDP and the Liberals are both much smaller (Liberals 200,000 and SDP 65,000), although the Liberals have three times as many members as the SDP and this gives them a powerful edge in joint campaigning and joint selection of parliamentary candidates.

We have very little research on just how much party membership matters to the success of parties. If it did matter a lot we would have to expect the Conservative Party to be able to overwhelm the Alliance in the seats they contest closely, but there is no sign that – in parliamentary elections at least – this is the case. Indeed, in the 1983 election the SDP, from a start barely two years before with only a third of the number of Liberal members and no experienced machinery or well-tested bases, were able, none the less, to capture nearly as many votes as the Liberals. This is one more way in which parties, in this case, the number of party members, seem to matter less than the party activists themselves would like to believe.

The number of individual members of each party is an important aspect, however, of the ability of the parties to raise funds. The Conservative Party has three times as much income as Labour, but the main gulf in funding is at constituency level. The difference between the incomes of the two party HQs is much smaller. The great bulk of Labour funds (close to 90 per cent) comes from the political levy of the trade unions. Conservative funds come from fund-raising activities by the local associations, and contributions from companies. These rarely exceed £50,000 per company. To their embarrassment the largest single recent company donation to a political party was the £110,000 donation from BSM; it went to the Liberals in 1984.

The strength of Conservative local fund-raising enables the party to employ far more local agents than Labour (350 compared with less than 100). They also provide one-third of the income of Conservative Central Office. Labour is almost entirely dependent on financial support from the trade unions. This hinders the development of vigorous local associations, and rules out radical constitutional reform such as reforming the trade-union block vote.

The decline in the number of individual members has created financial problems for both parties, but major difficulties loom for Labour because of new legislation requiring ballots every ten years by trade unions to determine whether their members wish the union to maintain a political fund. If Labour lost the affiliation of some of the big unions, its income would be drastically reduced and its internal balance of power could be changed in unpredictable ways. At the extreme the formal link between the party and the unions might be severed altogether. If this happened it would be a terrible blow to Labour; it would also serve importantly to bolster the Conservatives' status as the dominant political party. On the other hand, the new legislation might easily backfire against the Conservatives. In the first of the ten-yearly votes not a single union disaffiliated from the Labour Party; having reaffirmed their link with the party it will now be tempting to increase their often-miserable weekly subscriptions to it. If that does happen we may say that a Conservative government has resurrected Labour's finances and filled the gap created by a weak organisation.

Short of finance from members, the parties have to consider asking for Treasury subventions: public money thus replaces the money the public refuses. This was recommended by the Houghton Committee (1976), but was opposed by the Conservative Party, and the SNP partly because their local fund-raising was so buoyant. There already exists substantial state funding: £36m. in 1978 (parliamentary purposes grant). The free use of TV time for party political broadcasts was estimated to be worth £10m. The absence of a free market in political advertising on television, in sharp contrast to North American practice, does make the competition between the parties much more equal than it would otherwise be. But state funding is unlikely to be extended unless all-party agreement is obtained.

Electoral Reform

Since the two-party system began to crack in the mid-1970s the electoral system has been on the agenda. The parliament elected in 1983 was 60 per cent Conservative, yet only 42.4 per cent of the voters had voted Conservative. On the other hand, the Alliance won 4 per cent of the seats with 25.3 per cent of the vote. This imbalance kept reform of the electoral system on the agenda. It also shows how a law can thwart the development of a fully multi-party system.

There are hundreds of actual electoral systems. The debate

in Britain has focused on five possibilities. In pressing for one or other of these the parties have emphasised different features of the various alternatives. Four values are repeatedly stressed. The first is proportionality. Elections ought to result in parliaments in which the number of MPs for each party roughly reflects the proportion of votes for that party in the country. This argument has been emphasised by the Liberals. The second is strong government. Elections ought to return governments with majorities large enough to carry out the will of the people, given the opposition that will be put by the civil service, the pressure groups, and other unelected bodies. This argument has been stressed by most Conservative and Labour spokespersons. The third is ombudspersons. An important role of MPs is the representation of their electors' interests to the Whitehall machine, and this can best be done when each citizen has one MP who has an unequivocal responsibility to that citizen. Initially this argument was stressed by Labour and Conservative, but it has been conceded in part by Alliance leaders since 1983. The fourth value stressed is that of 'choice'. Britain is a divided society; moreover there are radically different political solutions to her economic and social problems on offer. The electoral system should not frustrate the ability of the electorate to decide which alternative should rule. This argument is offered by some on the left of the Labour Party as well as some constitutionalists. These values are not easily reconciled. In opting for one you tend to lose another.

'First Past the Post'

The present system is known to the press as the 'first past the post system'. Since there is no post this horse-racing metaphor is misplaced and the system ought to be called by its technical name: plurality. Each party puts up one candidate. Each voter has one vote. The winner is the candidate with the most votes. This system resulted in alternations in government between the Conservative and Labour Parties during the 1950s and 1960s because the two parties had roughly equal numbers of supporters who were spread more or less equally across the United Kingdom. It currently works to the special advantage of the Conservative Party. So long as few people voted for any other party, and this was the case until 1974, the system fulfilled most of the tests of a good voting system. It always failed to give Liberal voters a fair number of seats. Since 1974 the unfairness has been glaring. This is because the Liberal support has been thinly spread, more or less evenly across the country, while Conservative and Labour supporters have congregated in distinctive areas. Thus the Conservatives won some seats and Labour others, leaving the Liberals a good third, since 1983 a good second, in many. The parliament elected in February 1974 had no overall majority and the one elected in October 1974 only a tiny one – this made the legislative work of the subsequent government difficult.

List System

If the main thing one wants from an electoral system is proportionality, the system to opt for is the List System. In this system each party puts up as many candidates as there are seats; the candidates are listed in order of party preference. The voter has one vote which she/he casts for a party (not a member) of her/his choice. Votes are distributed to parties and the candidate elected from party lists in strict proportion to votes. This system is operated in Israel.

MPs represent parties, not definite areas, so no one is politically responsible for all the interests of an area. This system ensures that even small minorities get some seats. The most powerful objection to it is that it makes it almost impossible for large parties to form a majority government without the support of some small parties. Governments are therefore formed after elections by coalitions of parties. Proportionality is purchased at the price of strong government, electorate choice, and MPs acting as local ombudspersons.

Alternative Member System

The remaining three systems try to combine some of the advantages of the plurality system and of the list system without the disadvantages of either. The first of these is the Alternative Member system. This system, a version of which is employed in West Germany, was proposed for Britain by a commission established by the Hansard Society, meeting under the chairmanship of Lord Blake in 1976. Under this system the country is divided into constituencies of roughly equal but rather larger size than at present. Each voter votes for the candidate of his or her choice. MPs for three-quarters of the seats are elected as at present – thus ensuring local representation. The remaining seats are filled by candidates from the parties who have won less than a proportionate share from the first round. In this way a degree of proportionality is added to the system.

The Alternative Member system has not found much favour in Britain, because it would create two classes of MPs: those elected on the first round, and what would inevitably be known as the second-bests. Its rejection on this basis alone shows the strength of feeling for constituency MPs. To the extent that it increases minority party representation in the House it would decrease the chances that single-party majority governments could be formed. The significance of this effect is impossible to judge.

Single Transferable Vote and Fairer Voting in Natural Constituencies

The Single Transferable Vote (STV) system (sometimes misleadingly known as Proportional Representation) was long the favourite of the Liberal Party. Under STV the country would be divided into multi-member constituencies, most probably having six or seven seats. The parties would put up lists of candidates. The voters would be able to distinguish between the candidate of their first preference, second preference, third preference and so on. The seats are filled by fixing a quota and redistributing any excess votes from the most-popular candidates as well as all the votes of the least-popular candidates. A distinctive feature of this system is that voters would be able to choose between the candidates of their preferred party. STV's proponents argue that, in this way, moderate candidates of all the parties would have an advantage.

Six or seven member seats would have to be quite large, and in rural areas massive. To meet this point, as well as the objection that these rural areas are in special need of definite representation, the Liberal SDP Alliance switched their preference from STV to a variant of it called 'Fairer Voting in Natural Constituencies' before the 1983 election. In this variant, constituencies of different sizes are produced. The largest, in large city centres, would have eight seats; the smallest, in rural areas, would have one. Most areas would have an intermediate number. The plan was worked out in considerable detail. From the detail it was clear that, since most Liberals happen to represent rural seats, this variant gave them 'first past the post' advantages in their own seats while producing STV for the rest of the country!

STV and its variant would not distribute seats with quite the mathematical fairness of the List System nor would it give so many parties seats in the House of Commons. On the other hand, it would be more proportional than the plurality system. Its opponents find disadvantages in it, in that it would break the tie between members and their areas and would increase the likelihood of coalition government. In contemporary conditions in Britain it can be assumed that it would give the Alliance more seats and take them away from Conservatives (especially) and Labour. It might produce a succession of elections after which the Alliance, or part of it, was the governing partner of Conservative, then Labour, governments. Some would call this stability; others would call it the denial of choice.

Second Ballot System

The final system proposed for Britain is the Second Ballot system. A version of this system has been recently employed in France. Another variant is in use in Australia. In the Second Ballot system there are single member constituencies with a two-stage ballot. The voter has a single vote. If any candidate wins a majority of the votes on the first ballot he is the winner. If not, there is a run-off between the most popular two candidates; each voter having, again, a single vote.

The Second Ballot system is a compromise; it has some of the advantages and disadvantages of its parents. Next to the plurality system it is the second most likely to produce a majority single-party government. It gives an element of proportionality and would increase the numbers of Alliance MPs. It has some singular features, however; it would produce a House of Commons of MPs all of whom had at least some support from a majority of the voters in their seats. This system is most in the interest of the Labour Party. It makes single-party majority government possible (an interest Labour shares with the Conservatives); it adds to the authority of the democratically elected chamber by giving an element of proportionality and by ensuring that every MP has some support from a majority of his voters; and it reduces the present Conservative bias of the plurality system.

British discussions of the electoral system usually refer entirely to its operation in parliamentary elections, as ours has done. But the plurality system is in use in Britain for local elections also. In local government where no 'Cabinet' is formed, where no laws are passed and where the social consequences of one-party rule are less, the arguments for proportionality are much stronger because the other arguments against it are not so clearly relevant.

Changing the system requires a majority in the House of Commons. It would thus require the support of at least one, and probably more than one, major party. It is unlikely that the Conservative Party would support change. Labour is also not inclined to support change at the moment, but a third successive Conservative victory in 1987/8 or a Labour administration formed with Alliance support might alter its view. The Alliance is in favour of the 'Fairer Voting in Natural Constituencies' variant of STV. Since the Alliance is most unlikely to form a government on its own, it could only get change by agreeing with Conservative or Labour – probably Labour. The terms of such an agreement cannot yet be seen, but it is clear that there is plenty of choice.

Setting the Policy Agenda

The Thatcher government is the first government to win re-election after serving a full term in office since 1959. The regular alternation in government of the two main parties between 1959 and 1979 and their failure to remedy Britain's poor economic performance led to a major change in attitudes to the political parties. In the 1950s the British two-party system was regarded as a major factor in promoting political stability, because it discouraged political fragmentation and political extremism and offered the voter a clear choice between two alternative sets of leaders and two distinctive programmes for government within a framework of broad consensus on constitutional rules and the priorities of public policy. By the mid-1970s, however, proponents of the adversary politics thesis were arguing that the two-party system was producing increasing ideological polarisation between the two main parties and producing marked discontinuities in policy between different party administrations. Each party when in opposition came under the control of its ideological militants who were unrepresentative of party members and party voters. Having secured a parliamentary majority each party would immediately overturn many of the policies of its predecessors, and proceed to implement its 'extremist' manifesto, until practical obstacles and the threat of electoral disaster forced a U-turn in mid-term to more orthodox policies. Before the new policies could bear fruit, however, the party of government had to face another election, which it lost, giving up office to the opposition party eager to begin implementing its own new 'extreme' manifesto.

Doubt has been cast on the existence of such an adversary cycle. Continuity rather than discontinuity marks most areas of government policy. It also seemed to have lost relevance to British politics when the Thatcher government did not succumb to the cycle in 1983. Yet the thesis does direct attention to an important aspect of the British party politics – its potential for producing sharp discontinuities in policy and a progressive breakdown in the conditions for stable constitutional government. This potential may never be realised. A stable consensus-building two-party system might be restored; a series of hung parliaments might institutionalise a multi-party system through the introduction of some form of proportional representation; or one party might assume a dominant unchallengeable position, excluding all other parties from government for a long period. How near is the Conservative Party to achieving this?

A true dominant party system would require the Conservative Party to be dominant not only in parliament, in electoral support, and in organisation, but also in its ability to win the key political arguments, to set the agenda for policy, and to command the support, or at least the acquiescence, of the key organised interests. Much discussion of Thatcherism suggests that this has indeed occurred, and that the Conservatives are constructing a new consensus based upon a much smaller role for government in the economy, the favouring of private over public provision, and reduced influence for trade unions.

The notion of a major shift in the political agenda which would underpin a new dominant party system has been strongly contested. The success of the Conservative Party in winning the ideological argument was reflected in the 1970s and 1980s by the unpopularity of many of Labour's policy positions, even with their own voters. But support for many Conservative policies, particularly on reducing collective provision of welfare, is often ambiguous. Many commentators have also pointed out how frequently the Thatcher government has failed to translate its ideological rhetoric into lasting political achievements. The government has succeeded in launching a major privatisation programme and weakening trade-union power; but the radical restructuring of welfare and the revival of the British economy have proved much more difficult. It will probably not be possible to assess the scale of the change that Thatcherism has made to British politics until it is seen how many of the Thatcher government's policies prove irreversible by the government that succeeds it.

The question of the significance and the durability of Thatcherism is keenly debated within all the opposition parties, since it has a vital bearing on political calculation and strategy. Within the Labour Party there has been a lengthy debate on how far Labour should attempt to alter its traditional trade-union, white, male, manual working-class image, and seek to organise a new popular progressive coalition, articulating an alternative populism to Thatcherism. Opponents of this view argue that Labour should emphasise and further strengthen the class and socialist character of the party, and campaign to change popular attitudes, so as to win an electoral majority for a socialist programme as the only possible solution for Britain's long economic and political crisis. Under Kinnock, however, the party leadership has moved no closer to the class perspective than before, and has launched initiatives on a range of issues, such as council house sales, that imply accommodation to several of the changes associated with the Thatcher government and the abandonment of some aspects of Labour's radical programme developed in the 1970s. The extent to which the party could transform itself into a social democratic party of the Spanish or French kind is severely limited, however, by its continuing dependence on the trade unions.

Under Kinnock, Labour is seeking to re-establish itself as a national party, a contender for office at the centre, and this has to involve, given the present electoral geography of Britain, some dilution of the party's radicalism. The Conservatives are seeking to maintain the dominant position which they secured in 1983, and therefore have to hope that the opposition to them remains divided. The Alliance parties seek to change the electoral system, but to do that they must first win enough seats under the existing system to hold the balance of power in a hung parliament. The Alliance may well hold the key to future developments in the British party system, for it is their continuing strength which makes a return to either a stable two-party system or the consolidation of a new dominant party system unlikely. If Alliance strength persists and a hung parliament or a series of hung parliaments results, then a key question will be whether a majority can be assembled in the House of Commons for changing the electoral system. The character of the coalition that emerges will determine the kind of electoral system that is adopted, since different electoral systems are likely to have very different effects on the fortunes of the existing parties.

Section II

Government

4

Inside Whitehall

GEOFFREY K. FRY

Rhetoric Versus Reality

There are three main reasons why Margaret Thatcher's Conservative government tends to be portrayed as tough. The first is the political personality of the Prime Minister. The second reason is because the main framework of referencc within which British politics is conducted and analysed continues to be that of the Keynesian consensus (for example, Riddell, 1983). The third reason is that the Thatcher government's immediate predecessors, whether Labour or Conservative, were markedly pusillanimous. So, the contrast which the Thatcher government provides seems to be a stark one. The fact of the matter is, though, that the Conservative government's performance has not matched its radical rhetoric. After six years of Thatcher government, the radical Right was able to complain, with some justice, that 'the Government is not proposing deep cuts in public spending'. In other words, the Prime Minister and the two men who have held the post of Chancellor of the Exchequer since 1979, Sir Geoffrey Howe and then Nigel Lawson, have continued to preach economic liberalism of market economics, as have their allies in the government and in the Tory Party. As might be expected, though, the Conservative government has not always, and in some areas not often or not at all, been able to practise what many of its most prominent members preached.

This is not to say that, for example, the restrictive 1981 Budget was other than politically brave especially at a time of rising unemployment. In marked contrast, too, with the behaviour of its immediate predecessors, the Conservative government has proved relatively courageous in its various confrontations with the public sector unions. It has to be conceded too that the Thatcher government's privatisation programme has been ambitiously prosecuted, even if at least some of the ambition can be explained by the insistent pressure on the Chancellor to raise still more revenue. Nevertheless, that the Conservative government has managed to restore marginally more exacting standards of financial discipline in public administration, and that it has presided over, while not being entirely responsible for, the diminution of inflation, hardly adds up to a Thatcherite Counter-Revolution. For, on current trends, even at the end of the present Parliament, public expenditure and taxation seem unlikely to account for a lower proportion of the Gross Domestic Product than they did in 1979.

During the Thatcher years the Conservative government has not been engaged in an 'unreasonable' and remorseless pursuit of radical Right objectives. The economic liberal beliefs of the Prime Minister and her allies have been important factors in the government's behaviour, as can be seen, for instance, in its dealings with the Civil Service. The facts show, though, that, in general, the Conservative government has been more pragmatic than it normally suits both friend and foe to recognise. For critics both inside and outside Tory ranks to admit pragmatism would blunt such sharpness as their criticisms have. Further, the Prime Minister and her economic liberal allies in and around government have no incentive to concede their pragmatism in advance of having to practise it, because to do so would rob their creed of such cutting edge as it possesses.

Second, Margaret Thatcher is *sui generis* among recent British prime ministers. Her open and genuine adherence to economic liberalism certainly marks her off from her post-1945 predecessors. To take an even longer view, she is not in the Stanley Baldwin mould. She is not of The Establishment. Mrs Thatcher is an 'outsider'.

Third, there is no substantial evidence that the Thatcher

years have witnessed the displacement of Cabinet government by Prime Ministerial government. Certainly, institutional changes have taken place at the centre, but a notable absentee has been anything resembling a Prime Minister's Department, the establishment of which is a necessary pre-condition for Prime Ministerial government.

Fourth, the Conservative government has made a sustained attack on the heavily unionised career Civil Service as an institution. This attack on the constitutional pretensions, the policy preferences, and the organisation and rewards of the Civil Service has met with opposition. The defeated Civil Service strike of 1981 was one example of conflict. Other examples have included the series of 'leaks' of confidential information by civil servants, spectacularly so in the case of Clive Ponting (Ponting, 1985; Norton-Taylor, 1985), designed to harm the Tory government (Pyper, 1985). The antagonistic relationship which seems to have developed between the Thatcher government and the Civil Service has made the future of the career Service as such a matter for debate.

Fifth, not only has British politics reverted, broadly speaking, to the form that it had in the 1920s, so has British economics. What marks Mrs Thatcher off from so many of her contemporaries, especially in her political generation, is that she seems to relish this.

The Political Style of Thatcher

'I think your woman's gone crazy. Please God she'll never be Prime Minister.' This was, according to one of her aides, the reaction of a higher civil servant at the Foreign and Commonwealth Office to Margaret Thatcher's Dorking speech in July 1976. The then Conservative Leader of the Opposition chose in that speech to mark the first anniversary of the Helsinki Agreement by an attack on the policy of Detente which it embodied and on the Soviet system in general. To do this was to deliberately flout not only the conventional political wisdom of the time but also the known policy preferences of the FCO. What may well have alarmed the FCO official was that such behaviour made it certain that when she became Prime

Minister, Mrs Thatcher would disturb the British political system. She did not crave to be 'accepted'. She did not wish to 'belong'. She was never going to be a creature of the Establishment and, in this sense, as one of her former Conservative ministers later put it, Margaret Thatcher was the first 'outsider' to be Prime Minister since Andrew Bonar Law (Bruce-Gardyne, 1984). This 'outsider' status is the first distinguishing characteristic of Mrs Thatcher's tenure of the Prime Ministership that we need to recognise.

The second distinguishing characteristic, a concomitant of the first, is that Margaret Thatcher has departed from what can be called the somnolent model of British political leadership. In terms of the British political game as it has come to be played in peacetime over the previous half century and more, Mrs Thatcher has proved to be a disruptive 'dynamic force', much as Baldwin once accused that other 'outsider', Lloyd George of being.

The third distinguishing characteristic of Margaret Thatcher as a Prime Minister, and one which ensures that she is outside the Baldwin tradition, is that she appears to be a radical ideologue, certainly in post-1945 terms. Mrs Thatcher's commitment to economic liberalism seems to be of an order which *in principle* threatens the continuance of the Keynesian welfare state which was established in the 1940s by the wartime Coalition and its immediate Labour successor, for which there was a political agreement or consensus on the part of both the Conservative and Labour Parties to conserve thereafter. Sir Ian Gilmour (1977, 1983) suggests that Mrs Thatcher's behaviour means that she is not really a Conservative. But British Conservatism always has been an amalgam of paternalism or statism on the one hand, and economic liberalism or market economics on the other. Baldwin's governments (for example, that of 1924–9), like every other modern government before the Keynesian revolution, practised a form of economic liberalism; and since Keynesianism was officially pronounced to be dead by the then Labour Prime Minister, James Callaghan in 1976, a form of economic liberalism will have to be practised in the future, to the extent that inherited state commitments make this practicable, which seems likely to be not very much. Baldwin, though, kept quiet about the supposed truths of economic liberalism, and Callaghan had every incentive to do the same. Mrs Thatcher, in contrast, glories in the advocacy of market economics. Thus, the social democratic creed which coloured the Keynesian consensus is treated by Mrs Thatcher as a wimpish variant of socialism, to be accorded a similar status of moral inferiority. Margaret Thatcher's attacks on socialism, its variants, and all its works, made with zest, laced with quotations from Alexander Solzhenitsyn, have proved to be not just the rhetoric that she indulged in to help to make her mark as Leader of the Opposition. The rhetoric has remained much the same since Mrs Thatcher reached Downing Street. For many it seems to have effectively disguised what we have already shown to have been the reality of what the Conservative government has actually done since 1979.

The Persistence of Cabinet Government?

That in several respects Mrs Thatcher has proved to be a distinctive Prime Minister has predictably encouraged some commentators to believe that the Thatcher years have either witnessed, or that they provide fresh evidence of, the displacement of Cabinet government by Prime Ministerial government. In fact, the absence of evidence, even of the partial kind provided by Richard Crossman (1975, 1976, 1977) and by Barbara Castle (1980, 1984) about Harold Wilson's Labour governments of 1964–70 and 1974–6, makes this latest manifestation of the Prime Ministerial government thesis difficult to examine. The view tentatively advanced here is that the Cabinet was certainly at the centre between May 1979 and the September 1981 reshuffle, and very probably enjoyed a similar role down to the June 1983 Election victory. Since then, there have been rumours about the Prime Minister being more dominant than before. Mrs Thatcher's failure, though, to obtain preferred policies even in this post-1983 period (for example, early reform in the rating system, and speedy trade-union reform) suggests that the Prime Ministerial government thesis in the case of this Conservative government remains to be proven.

The fact that, when she became Prime Minister, Margaret Thatcher was unwilling to offer a post in her government to her predecessor as Conservative Leader, Edward Heath, obscured the point that in the Cabinet she actually did form, 'there were, at best, nine voices to be counted on, and at least a similar number of potential rebels, several of whom had made no secret of their distaste both for her opinions and her style in Opposition'. As one admirer of the Prime Minister put it, 'the distribution of portfolios in 1979 was taken to surprising lengths of catholicism'. Edward Heath had not composed his Cabinet on the basis of the Conservative Party being a Broad Church, but Mrs Thatcher did (Bruce-Gardyne, 1984). Certainly down to the reshuffle of September 1981, and after the fall of Heseltine and Brittan in 1986, the presence of so many 'Wets' in the Cabinet was suggestive of Cabinet government rather than Prime Ministerial government.

In the first two years of the Thatcher government, the decision to commit Britain to the Trident programme, and the manner in which the 1981 Budget was presented to the Cabinet, tended to be cited as the major examples of the manner in which the Prime Minister wished to circumvent her Cabinet. *The Economist*, though, pointed out that 'neither defence procurement decisions, however costly, nor the Budget preliminaries have ever been matters for full Cabinet debate'. Subsequently, the Prime Minister, as a result of its pressure, conceded to her Cabinet a greater say in Budget decisions. Warming to its theme, *The Economist* (30 May 1981, p. 23) concluded:

> Indeed, Mrs Thatcher has quite deliberately loaded the dice against herself in her supposed bid for presidentialism. Not for her the Wilsonian 'kitchen cabinets' . . . She has kept her Downing Street staff small and relatively weak on the explicit grounds that 'my Cabinet are my political advisers'. As a direct result, the various organs of Cabinet Government – the committee system, inter-departmental groups, bilaterals and 'meetings with the Prime Minister' – are probably seething with as much collective argument and dissent as they have been for years. Contact between Mrs Thatcher and her colleagues may often be peremptory, bossy, even

downright rude, but officials are constantly aware of its frequency. 'Madame' may love the sound of her own voice and be convinced of the initial rectitude of her prejudices. But even that does not mean she gets her own way. Cabinet Government is alive and kicking. You can tell by the kicking.

Successful Prime Ministers seem to find it essential to have an able second-in-command in the manner that Herbert Morrison acted for Clement Attlee in the Labour government of 1945–51, and William Whitelaw, initially as Home Secretary, seemed content to play this valuable supporting role in the Thatcher government. The other 'Wets', though, tended to be more troublesome. Norman St John Stevas was eased out of the Cabinet in January 1981, and in the following September two more 'Wets', Sir Ian Gilmour and Lord Soames, were dismissed. At the same time, James Prior, who had made haste slowly with trade-union reform, was transferred to the Northern Ireland Office from the Department of Employment, where his replacement as Secretary of State was the Prime Minister's ally, Norman Tebbit. Another ally, Nigel Lawson, was made Secretary of State for Energy. A further ally, Sir Geoffrey Howe, Chancellor of the Exchequer since 1979, was provided with a ministerial team of economic liberals at the Treasury. To Leon Brittan, who had been made Chief Secretary in January 1981, was added the junior ministerial support of Nicholas Ridley and Jock Bruce-Gardyne. The last named later described the September 1981 reshuffle as constituting 'a sea change' in the Thatcher government of 1979–83 (Bruce-Gardyne, 1984), which was the case in terms of Cabinet personnel. In terms of the government's strategy, Sir Geoffrey Howe's March 1981 Budget eschewing explicit 'reflation' had been the decisive turning-point. The 'Wets' did not resign in protest. When Sir Ian Gilmour was later dismissed, he described the Conservative government as 'heading for the rocks'. It was re-elected. The Cabinet lost another 'Wet' when Lord Carrington resigned as Foreign and Commonwealth Secretary for his part in not foreseeing the Argentinian invasion of the Falklands. Another 'Wet', Francis Pym, replaced him, to be dismissed after the June 1983 election victory, it was said because he was too much a creature of the

Foreign and Commonwealth Office, a department whose appeasing outlook had little attraction for the Prime Minister. James Prior voluntarily resigned from the Cabinet in September 1984. The political position of Michael Heseltine being difficult to establish before his resignation during the Westland affair in 1986, Prior's departure left Peter Walker as the only prominent 'Wet' to have survived in the Thatcher Cabinet since 1979.

Whatever political convictions the Conservative ministers had, it seems likely that they acted in much the same manner as Richard Crossman had described the Labour ministers of the 1960s: 'we come briefed by our departments to fight for departmental budgets, not as Cabinet Ministers with a Cabinet view' (Crossman, 1975, p. 275). The Plowden Committee on the Control of Public Expenditure in a seminal report published in 1961 (Cmnd 1432, 1961) had recommended that ministerial proposals for public expenditure should be explicitly related to the projected growth in national income and ordered in relation to each other. As a consequence of the Plowden Report, the traditional mechanisms of Treasury control of expenditure of an annual money-based kind came to be overridden by procedures related to 'real resources', under which commitments to public spending were made over a number of years ahead.

The machinery for these exercises was provided by the Public Expenditure Survey Committee (PESC) comprising senior officials with a chairman drawn from the Treasury. The PESC machinery was overseen by ministers. The Heath government tried to supplement PESC with control in detail applied by Programme and Analysis Review (PAR) procedures but without satisfactory results (Fry, 1981, pp. 35, 76–93; Gray and Jenkins, 1982); and its Labour successor emphasised cash limits in an attempt to contain the growth of public spending (Cmnd 6440, 1976). The Thatcher government, attempting to restore the economic liberalism under which old-style Treasury control had evolved, entirely abandoned the 'real resources' approach, which bore the hallmark of Keynesianism, and replaced it with cash planning in 1981 (Ward, 1983). The PESC machinery survived, but cash planning at least in principle restricted the basis on which ministers could appeal beyond bilateral dealings between the Treasury and their departments to the Prime Minister and the full Cabinet.

In October 1981, Mrs Thatcher interposed between departmental ministers and herself and the full Cabinet a formal Cabinet sub-committee, designated Misc. 62, to resolve the 'hard cases' which persisted beyond the bilateral dealings state. Misc. 62 became known as the 'Star Chamber'. William Whitelaw (now Viscount Whitelaw) has been chairman of the 'Star Chamber' from the outset, and in 1984 it also comprised the Home Secretary, the Leader of the House of Commons, and the Scottish Secretary. The 'Star Chamber' is serviced by a small unit drawn from the Cabinet Office whose members are usually the only officials present. Appearances before the 'Star Chamber' normally take the form of the Chief Secretary to the Treasury confronting the relevant spending Minister. Four out of the eight contentious spending programmes submitted to it seem to have had disputes about them settled at the 'Star Chamber' stage in 1984, and only one remained after the next stage of trilateral negotiation with the Prime Minister. This programme, for housing, survived intact the scrutiny of the full Cabinet, primarily because, it was reported, the Prime Minister was unwilling to countenance the resignation of the Minister concerned, Ian Gow, her former Parliamentary Private Secretary, who had threatened such action if cuts ensued (Jenkins, 1985). That Gow, a supposed economic liberal, behaved in this way illustrated the Thatcher government's continuing difficulties in containing the growth of departmentally based public spending.

The Prime Minister's Policy Unit and the Machinery for Central Direction

That the Conservative government has had such difficulty in realising its economic liberal goals has naturally led to calls from fellow believers such as Sir John Hoskyns and Norman Strauss, who were at one time Mrs Thatcher's advisers, for the Prime Minister to have greater institutional backing in the form of a Prime Minister's Department, from which base, they believe, she could wage more effective war against the interests ranged against the government, notably those of the Whitehall bureaucracy. The foremost academic critic of this proposal, Professor G. W. Jones, has argued that:

> the establishment of a Prime Minister's Department in Britain would be a significant constitutional change. It would set the Prime Minister more apart from the Cabinet, frustrate Cabinet cohesion, and symbolize a shift from Cabinet to Prime Ministerial Government. The Prime Minister would still confront other constraints on his/her power, but a major one would have been removed. Prime Ministerial intervention would have been legitimated and the Cabinet undermined. A Prime Minister's Department is not compatible with collective responsibility and Cabinet Government. (Jones, 1983, p. 84)

The issue, as Sir Kenneth Berrill, the former Director of the Central Policy Review Staff (CPRS), put it, when advocating the establishment of a Prime Minister's Department, is 'whether a Prime Minister should have a support system with time to work on problems in some depth across the width of government activities. At present the advice is given and very presentably too, but the depth is invariably patchy' (see Weller, 1983, p. 61). So, three well-placed advisers to the Prime Minister all shared the opinion that the existing machinery at the centre did not allow the effective practice of Prime Ministerial government. They wanted this situation remedied.

The most, though, that Mrs Thatcher has been prepared to envisage has been a modest reorganisation of her personal office, and offsetting this has been the abolition of two central institutions dating from the Wilson–Heath years. First, as we shall see later, the Civil Service Department was abolished in 1981, to be survived only by a rump organisation in the form of the small Management and Personnel Office. Second, the CPRS was abolished in 1983.

The Central Policy Review Staff had been originally set up by the Heath government in 1971. Later on, the CPRS engaged in a series of studies of particular policy areas and organisational questions, but initially its task had been to provide the Cabinet with a strategic overview. That Edward

Heath's Conservative government should assign such a task to a non-political body, originally with a Labour peer, Lord Rothschild, as its Director, was a curious act. For the point that was almost universally missed about the CPRS was that in its role as government strategist, it was not just supportive of the Cabinet. It was doing work which the Cabinet should do, but which ministers lacked either the time, will, or ability to do (Fry, 1981, pp. 52–4). There was no evidence that Mrs Thatcher's Cabinet Ministers were superior to their predecessors: but the Prime Minister and her allies at least were always less likely to feel the need for supposedly non-political advice, especially from the CPRS which was, rightly or not, associated with Mr Heath's fateful 'U-turn' in economic policy in 1972. That the CPRS survived as long as it did under Mrs Thatcher was more noteworthy than its demise.

The Prime Minister's Policy Unit, an institution first established by Harold Wilson when Labour returned to office in March 1974 (Wilson, 1976, pp. 98–9, 202–5) continues to exist under the Thatcher government. At the beginning of the 1984–5 Parliamentary Session, the Prime Minister's Policy Unit, which is located in the Prime Minister's Office, had nine members, including its head, John Redwood (since succeeded by Brian Griffiths). Previous heads under the Conservatives had been Sir John Hoskyns and Ferdinand Mount. For the first three years of Mrs Thatcher's Prime Ministership, Sir John Hoskyns was her leading political adviser and, together with her then economic adviser, Professor Alan Walters, he is commonly credited with the Policy Unit's greatest coup to date, the 1981 Budget. Hoskyns and Walters are said to have told the Prime Minister that without substantial cuts in public expenditure there would be insufficient room for political manoeuvre later. The Prime Minister's initial reaction was that cuts of the order suggested were not practical politics, but, eventually persuaded by the arguments presented to her, she is said to have required the reluctant Sir Geoffrey Howe as Chancellor to make the necessary economies (Gimson, 1985, pp. 9–11). Professor Walters had been brought across from the USA in 1980 to provide economic advice in the Prime Minister's Office, and he was not replaced when he departed in 1983.

Mrs Thatcher's distrust of the Foreign and Commonwealth Office as an institution was indicated when she brought the former British Ambassador to the United Nations, Sir Anthony Parsons, into the Prime Minister's Office as her foreign policy adviser in 1982 (*The Economist*, 27 November 1982, pp. 25–9). Fears that the appointment was a precursor of future duplication of foreign policy advice of the kind that follows from the divisions between the National Security Council and the State Department in Washington, proved to be exaggerated. The Parsons appointment was short-lived, and his successor in 1983, Sir Percy Cradock, remained a high-ranking FCO official. The Parsons appointment seemed to reflect Mrs Thatcher's lack of confidence in her Foreign and Commonwealth Secretary, Francis Pym. When Sir Geoffrey Howe replaced Pym in 1983, relations normalised (Jenkins and Sloman, 1985, pp. 115–16).

The Mini-Reorganisation of Central Government Departments

The re-allocation of duties between government departments, the creation of new departments, and the merger of established ones, activities that occupied so much political activity in the Wilson–Heath era, has not been a major characteristic of the Conservative government since 1979. Admittedly, the merger of the previously separate Departments of Trade and of Industry in 1983 to form the Department of Trade and Industry superficially recalled that era, and in particular (aside from the separating-off of Energy in January 1974) the department which Mr Heath had established in 1970. It can be said also that, like the Wilson and Heath governments before it, the Thatcher government has been frustrated in trying to realise many of its general ambitions, and the machinery of central government and its staffing has been one thing that the government has been able to consistently 'do something' about. Similarly also, the Heath government, at least initially, had talked a great deal about the superiority of 'business methods', and in 1970 it too had brought Sir Derek Rayner into the machinery of government from Marks and Spencer to advise on rationalisations in government organisation and practice.

In stark contrast with the behaviour of the Heath government, however, the political impetus which the Thatcher government gave to the Rayner exercises did not diminish. Indeed, as we shall see, they were universalised in forming part of the Financial Management Initiative. This sustained concern with the detailed operations of government departments, with the aim of keeping their costs under continuing tight control, was one of the things which made the Thatcher government's approach different from that of the Wilson–Heath era. As will be seen too, and which is illustrated in the abolition of the Civil Service Department, the Thatcher government displayed none of the deference to the interests of the Civil Service which characterised its immediate predecessors.

The Thatcher government showed scant respect, too, for the vested interests of the higher ranks of the Armed Forces when, with effect from the beginning of 1985, it made changes in the central organisation for defence. These changes took further the process begun in 1964 when the separate Service Ministries – the Admiralty, the Air Ministry, and the War Office – were absorbed into a unified Ministry of Defence. In 1972, defence supply functions had been brought within the unified structure too, when the Heath government established the Defence Procurement Executive (Fry, 1981, pp. 119–26). Of the resulting Ministry of Defence structure, Lord Hunt of Tamworth, a former Secretary of the Cabinet, suggested to the House of Lords on 13 June that 'anyone who studied its organisation chart closely would think it a miracle that it worked at all'. Down to November 1981, compromises were all that could normally be expected to result from the workings of the Ministry's higher decision-making machinery. John Nott, the then Secretary of State for Defence, and the then Chief of the Defence Staff (CDS), Sir Terence Lewin, introduced changes which emphasised the position of the CDS as the government's principal military adviser in place of the previous status of being co-equal with the Chiefs of Staff. The latter, though, retained their Service-based, separate supporting policy staffs.

In 1984, Michael Heseltine, who had succeeded John Nott as Defence Secretary in January 1983, published, first, a consultative document, and, then, a White Paper, *The Central Organization for Defence* (Cmnd 9315, 1984), which envisaged that in 1985 a new unified Defence Staff, displacing the separate high-level Service staffs, would be established. The Chiefs of Staff, reduced to the status of senior advisers only, exercised their right of access to the Prime Minister to protest about Mr Heseltine's proposals, but the plans were implemented. The government also established an Office of Management and Budget within the Ministry of Defence designed to obtain 'stronger central determination of priorities for expenditure and control of resource allocation'. This resembled American arrangements. As with the various reorganisations of the 1960s and 1970s in Whitehall, of which those relating to defence were among the more worthwhile, the overriding problems remained those of policy and not machinery, especially in the case of defence where there continued to be a marked mis-match between commitments and resources.

The Civil Service and the Drive for 'Efficiency'

The Thatcher government has asked two main things of the Civil Service. The first is that the Service should be leaner and more 'efficient' and that it should shed some of its 'privileges'. The second is that the Civil Service should revert to playing that subordinate role in governing the country which constitutional theory has assigned it. The realisation of both aspirations required the Civil Service to give up what to many in its ranks must have seemed a lot, in terms of rewards and promotion prospects, and, in the case of some leading officials, influence too, without compensating benefits being offered. The Thatcher government's combative attitude was different from that of earlier reformist governments, which had simply wished for a better career Civil Service. So had the Fulton Committee on the Home Civil Service of 1966–8, whose recommendations had not seriously challenged either the role or the 'privileges' of the career Service. The culture and the hierarchy of the Service had emerged substantially unscathed. From the outset, the Thatcher government showed itself to be more threatening. The heavily unionised career Civil Service's self-interest in high levels of public expenditure wedded it to the Keynesian consensus, and made the Service a natural target for the economic liberals in the Conservative government.

As a result, Civil Service numbers were cut from 732,000 to 705,000 in the first year of the Thatcher government, and in May 1980 it was announced that by the beginning of April 1984 this total was to be reduced to about 630,000. At that latter date, the total was actually 624,000 and further plans were made to reduce the number of civil servants still further to 593,000 by April 1988. The main increase in the number of civil servants in the 1960s and 1970s had come in the non-industrial Civil Service, and particularly at middle-management level (Fry, 1985, pp. 1–2, 148). Yet the reductions in the number of civil servants have disproportionately fallen on the highest and the lowest grades, and on the industrial staff in the Home Civil Service (primarily meaning employees in dockyards and royal ordnance factories). Indeed, the programme of cuts down to 1988 largely depends on still further reductions in industrial staff. So when Mrs Thatcher told the American Congress in 1985 that 'our Civil Service is now smaller than at any time since the War', this was not the case with the non-industrial Civil Service which, even with planned reductions, will be larger in 1988 than it was in 1945 (*The Economist*, 2 March 1985, p. 45) (see Chapter 6).

The threat of cuts being made in the main body of the Civil Service, added to knowledge of the government's aim to 'de-privilege' the Service, ensured that, from the outset, there was disharmony in relations between the Conservative government and its civil servants and, in particular, the unions which represented those officials. 'De-privileging' was not without its setbacks. In 1981, the Scott Committee, to the government's public chagrin, not only recommended that civil servants should retain the right given to them by the Heath government ten years before to index-linked pensions, but also that all employees should have such pensions (Cmnd 8147, 1981). How this open-ended commitment was to be financed was not revealed. In fact, the Thatcher government soon found itself unable to meet the cost of the findings of the Civil Service pay machinery. This had been bequeathed by the Priestley Royal Commission on the Civil Service of 1953–5, during the Golden Age of Keynesianism, and, essentially, the Commission had granted civil servants pay based on 'fair comparisons' with outsiders. The economic liberals in the Thatcher government disagreed with the Priestley formula in principle. The government dispensed with it, and defeated the unions in the resulting Civil Service strike of 1981. The Megaw Committee of 1981–2 devised a Civil Service pay system designed to have more relation than the Priestley formula to market considerations and to the taxpayer's capacity to pay (Cmnd 8590, 1982). The government also followed up its victory in the Civil Service strike later in 1981 by disbanding the Civil Service Department (CSD), the existence of which had symbolised the Civil Service being treated as an interest in its own right. Making the Treasury once more the main department responsible for the central management of the Civil Service emphasised the priority which the government attached to financial considerations. Further than this, Mrs Thatcher took the opportunity presented by the abolition of the CSD to require its Permanent Secretary, the Head of the Home Civil Service, Sir Ian Bancroft, and his Second Permanent Secretary, Sir John Herbecq, to take early retirement. These dismissals were unprecedented blows at the hierarchy of the career Civil Service and, moreover, ones that emphasised political primacy over the Service, especially as they were inflicted in the context of overall cuts in the number of its leading posts (Fry, 1985, chs 4, 5, 6).

The Conservative government's attack upon the hierarchy of the career Civil Service was combined with a determined attempt to change its culture. The leading civil servants still saw themselves mainly as policy advisers, but not only their particular policy advice was unwelcome to the Thatcher government. It was the role itself. The government believed that its officials should concentrate more on management. The Fulton Report (Cmnd 3638, 1968) had said much the same. The difference this time was that a sustained political impetus supporting change was present. The Rayner efficiency exercises were persisted with. A particularly important Rayner inquiry conducted by an official, Christopher Joubert, devised an organisational structure which divided the Department of the Environment (DOE) into 120 responsibility units or cost centres, each of which had an annual budget to cover running costs, including those for staff. This was related to the Management Information System for Ministers (MINIS) which Michael Heseltine introduced into the DOE on becoming Secretary of State there in 1979. The Financial Management Initiative (FMI), which the Conservative government launched in May 1982, essentially underlined the MINIS and Joubert systems throughout the main central government departments. The FMI's aim was described as being to promote in each department an organisation and a system in which managers at all levels had:

(a) a clear view of their objectives; and means to assess, and, wherever possible, measure, outputs or performances in relation to those objectives;
(b) well-defined responsibility for making the best use of their resources, including a critical scrutiny of output and value for money; and
(c) the information (particularly about costs), the training, and the access to expert advice which they need to exercise their responsibilities effectively (Cmnd 8616, 1982, para. 13).

The Fulton Committee had talked about 'accountable management' being introduced into the Civil Service, and essentially the CSD had been supposed to achieve the results now expected from the FMI, but it had failed to do so. The Thatcher government was sufficiently committed to the FMI to chart its progress regularly (Cmnd 9058, 1983; Cmnd 9297, 1984).

The then Minister for the Civil Service, Barney Hayhoe, declared in 1982 that the FMI meant 'a push to greater decentralization and delegation down the line, which will represent a highly significant change in the culture of the Civil Service. . . . Recruitment, training, promotion prospects and practice will all be affected' (Fry, 1984a, p. 333). The scope for managerial initiative in the Civil Service was bound to remain slender as long as salary scales continue to be standardised and promotion arrangements are highly formalised. So it was a potentially important development when, with effect from 1st April 1985, the government experimentally introduced merit pay at the levels of Under Secretary down to Principal and equivalent grades. This took the form of a performance bonus, and it was in line with the recommendations of the Megaw Committee on Civil Service pay. Greater standardisation, though, had earlier resulted from the government's extension of unified grading in the Home Civil Service.

In 1968, the Fulton Committee had recommended unified grading from top to bottom of the Service. Subsequently, unified grading had been introduced but only for the very highest grades – those of Permanent Secretary, Second Permanent Secretary, Deputy Secretary, Under Secretary and their equivalents. From the beginning of 1984, these grades and their equivalents were re-named Grades 1, 1A, 2 and 3 respectively, and unified grading was taken down to and including the Senior Principal grade. Thus, the Executive Directing Grades and their equivalents became Grade 4, the Assistant Secretary grade and its equivalents became Grade 5, and the Senior Principal grade and its equivalents became Grade 6. Unified grading was extended down to Principal level in 1986. These steps towards greater conformity at the top of the Home Civil Service do not obviously cohere with the government's wish to encourage individual initiative in the Service. For they effected a form of rationalisation whereas other approaches seemed aimed at obtaining greater differentiation.

The coherence of the Conservative government's policies towards the Civil Service seems to have been exaggerated by friend and foe alike, although these policies have had the unity accorded by antagonism, as well as, latterly, the Financial Management Initiative. As the relationship between the Thatcher government and the Civil Service has been characterised by disharmony, the legend that Labour governments uniquely experience obstruction from the career Civil Service has now to be consigned to the dustbin. The supposed 'radicalism' of Labour governments has always stopped short of damaging the Civil Service as an interest. The

Thatcher government has displayed no such reservation, and even the future of the career Civil Service of the kind that has developed since the 1850s is now a matter for serious debate (Fry, 1984b), with Sir John Hoskyns leading the attack (Hoskyns, 1983, 1984), and Sir Douglas Wass, formerly of the Treasury, prominent among the Service's defenders (Wass, 1983, 1984).

Retrospect and Prospect

Many observers have recognised the extent to which three-party electoral competition makes British government and politics seem similar to those of the 1920s. The comparison can be taken further. For, like at present, the 1920s also witnessed politicians needing to be convinced of the 'efficiency' of the Civil Service. Suspicion of the Civil Service today is not a Conservative monopoly. Without much cause, Labour and Alliance politicians seem to view the Civil Service with distrust too. The most important parallel with the 1920s which affects the machinery of government, though, is that economic liberalism is once more the economic orthodoxy. For the forseeable future, whichever government is in office it will have to talk the language of economic liberalism, not least to keep international financial opinion sweet. The result is a more restrictive climate than in the Keynesian era, and one in which probably only a government led by believers, like that of Margaret Thatcher, offers the scale of reassurance necessary for pragmatic manoeuvre.

5

Government Beyond Whitehall

PATRICK DUNLEAVY AND R. A. W. RHODES

Administering the Extended State

In 1979 when the first Thatcher government assumed office, state expenditure in Britain accounted for 41 per cent of the gross domestic product. Six years later, the public sector share of GDP has climbed to 44 per cent. Surely there could not be a more convincing proof of the failure of the New Right's ambition to roll back the frontiers of the state? In fact there is no satisfactory single index of state intervention (Heald, 1983, ch. 2). The public sector share of GDP varied wildly, depending on how a government allocates spending between purchasing real resources (goods or services) and spending on transfer payments (such as pensions, unemployment pay, or subsidies to farmers). Conservative governments in the 1980s have painfully cut back on consumption of real resources, only to see escalating spending on social security payments more than wipe out these reductions. Meanwhile their monetary and exchange rate policies have damaged private firms more than government agencies and public corporations, especially during the disastrous 1980–82 'shakeout' of manufacturing firms. In 1979 public sector workers formed 31 per cent of employees in employment; by 1985 the figure stood at 30 per cent despite major public corporation sales.

Conservative policy-makers have only slowly come to appreciate the limits on their ability to reduce state activities. Since 1979 some policies in which Thatcher's Cabinet initially placed great faith have visibly withered. The assault on the budgetary and taxing powers of local government, which consumed much ministerial energy and legislative time, rumbles on amidst widespread recognition that it has not held down local spending or property taxes. By contrast, originally minor 'privatisation' policies have become the central elements in the attack on 'big government', as they appeared to effect the lasting changes which have proved so elusive elsewhere.

To understand this chequered record of perceived 'success' and 'failure', it is important to examine the quite complex ways in which British public policy evolves. The government, opposition parties, and the mass media characteristically paint a simple picture of Whitehall control of all aspects of public sector operations. In practice, however, ministers and their civil servants almost invariably have to deal with a number of other kinds of agency before any domestic policy change occurs 'on the ground'. Taken as a whole these agencies form what we term the 'sub-central government' of Britain. Table 5.1 sets out the major different forms of state administration in Britain, pointing out their widely varying patterns of control by ministers, openness to scrutiny by elected representatives, and organisational, financial and staff autonomy from central control.

What happens in sub-central agencies is always extensively influenced by developments in Westminster and Whitehall. In a unitary state 'the power to delegate or revoke delegated power remains in the hands of the central authority' (Rose, 1980, p. 50). If a government commands a Commons majority then new legislation can extend central departments' capacity to regulate, provide finance, set service standards, and to proscribe or mandate activities. But the agencies beyond Whitehall are in no sense completely or directly controlled by ministers or civil servants. In dealing with sub-central governments the Whitehall machine is always limited by their incomplete authority to reorder these agencies' day-to-day policies, and by their dependence on sub-central agencies to go on producing services needed by citizens.

British Central Government

British central government is primarily a 'non-executant' tier of administration. Only one in ten public sector employees is a civil servant, and only a fraction of these are in Whitehall. Two-thirds of civil servants work in just three policy areas: collecting income tax and VAT; paying out social security benefits and pensions; and providing administrative support and equipment for the armed forces. Only two ministerial departments, the DHSS and Defence, have large staffs; the two key taxing agencies, Inland Revenue and Customs and Excise, are non-ministerial departments run by boards of civil servants. Even inside the Defence Department, the key function of arms procurement is organised in a departmental agency. Departmental agencies (such as the Property Services Agency in the Department of Environment) or effectively separate units (such as the Prison Service or the Immigration Service inside the Home Office) account for virtually all other sizeable groups of civil servants. Especially following the hiving-off operations with which most departments responded to the Conservative's 1979 demand for manpower cuts, the key to understanding Whitehall ministries is to appreciate that they directly administer very little.

Policy-level civil servants overwhelmingly move money around between subordinate agencies or private sector organisations; appoint key personnel to some sub-central governments; regulate service standards; process new legislation; and monitor policy performance and the emergence of new problems. Ministers and their departments thereby retain a substantial capability to involve themselves flexibly in key issues, and apparently dramatic powers to remould legislation and reorganise state administration. But being divorced from policy implementation also severely constrains their ability to achieve quick or lasting changes. Over the post-war period Whitehall has consistently moved to generalise its relations with sub-central agencies. Perhaps more than any other liberal democracy, Britain is characterised by 'hands off' control systems which lay down general rules for dealing with whole classes of sub-central agencies in the same way.

TABLE 5.1 *The major types of agencies in British government*

	Central departments	Sub-central government	
		Quasi-government agencies (QGA)	
		Single-issue bodies	Public corporations
Basic features:	Includes: ministerial departments, non-ministerial departments, departmental agencies, security agencies.	Executive agencies (e.g. Manpower Services Commission, Atomic Energy Authority); advisory bodies.	Includes nationalised industries and single corporations, but not state shareholdings in private firms.
	Manpower (including some QGAs): 786,000		Manpower: 1,611,000
Day to day policy control rests with:	Ministers and senior civil servants; boards of civil servants plus ministers for non-ministerial departments and security agencies.	Politically appointed committee or board plus senior managers.	Corporation chairmen and directors, senior management.
Accountability to politicians or public:	Direct ministerial control and Parliamentary accountability – except security services. Public consultation on policy initiatives; some special purpose inquiries; maladministration redress; legal redress.	Indirect ministerial control and scrutiny by Parliament. Some public consultation and complaints procedures; legal redress.	Indirect ministerial scrutiny and some Parliamentary overview. Very weak consumer consultation procedures.
Financed by:	Parliamentary vote of Exchequer funds; very minor charges.	Exchequer grants; variable other sources.	Trading income; subsidies for operating deficits and part of investment from Exchequer.
Ministerial control capability:	Direct ministerial responsibility.	Appoints controlling committee, approves budget and influences major policy decisions.	Appoints chairman and board; approves price increases; approves investment programme funding; sets financial targets.

Source: *Economic Trends*, September 1985.

Note: Manpower figures are 1984 total personnel.

TABLE 5.1 – *continued*

Sub-central government		
National Health Service	Quasi-elected local government organisations	Local government
Regional and District health authorities. Manpower: 1,223,000	Includes: Police Authorities (manpower: 187,000), joint boards, joint finance programme, Partnership agencies.	County and District councils (England/Wales); Region and District councils (Scotland). 513 authorities in all. Manpower: 2,697,000
Professional staffs; general managers; appointed Health Authorities.	Indirectly elected or part-appointed boards, usually councillors in a majority.	Council committees and professional staffs, with full Council decisions on budgets and major issues.
Accountability to Health Authority, with ministerial overview. Weak community consultation. Complaints and maladministration procedures; some legal redress.	Accountability to boards. Variable locality consultation mechanisms, mostly very weak; some complaints procedures.	Full accountability to elected councillors. Well-developed public consultation procedures, not always applied; maladministration procedures; legal redress.
Exchequer grants; minor own funds.	Rate precepts on member local authorities; grants from Exchequer.	Local property taxes (the rates); trading income and charges; Exchequer grants.
Appoints Regional Authorities and allocates funds to them; reviews Regional plans and annual progress. Supervises Districts via Regions and circulars. Controls capital spending, sets standards.	Controls capital spending; sets limits on current spending; allocates grant; sets standards. May regulate personnel choices as well.	Allocates grants and current spending levels; approves capital spending; mandates functions via legislation; sets standards; audit powers and direct budgetary control of ratecapped councils.

For example, departments allocate grants and now spending limits to local authorities via computerised formulae. In most cases they refuse to negotiate with individual councils about their grant entitlements, and virtually never allocate funds in a discretionary way to localities. This system means that the gulf between central ministries and municipal governments in terms of lack of personal contacts or administrative involvements is far wider in Britain than in countries which have supposedly been more centralist, such as France. Whitehall's relations with local councils have consistently been characterised by a concern to stay out of those aspects of policy questions which are not important to national political elites (Bulpitt, 1983). A very similar approach has also been adopted with health authorities. This pattern of disengagement leaves Whitehall departments free to embrace sweeping changes in national policy with few misgivings. But it also means that ministers and policy-level civil servants cannot easily overcome systematic resistance amongst subordinate agencies, or acquire any fine-grain control over the quality of policy outcomes or the direction of policy development. In addition, central government rarely initiates new policy changes. Instead ministers frequently pick up pioneering ideas developed elsewhere and generalise them as government policy for a whole area of the public sector.

Central departments vary considerably in their attitudes to sub-central governments. Normally the struggle to maintain a ministry's share of public expenditure implies some defence of its 'client' sub-central agencies. By preserving overall funding levels in their policy area, ministers protect their place in the Cabinet pecking order, maintain their patronage powers, and ensure that they can respond to problems for which they are held publicly responsible. But it seems plausible that ministries which retain direct responsibility for policy implementation will fight harder for their policy area's overall funding, than those departments where much of the global budget total is passed on to other agencies to administer (Dunleavy, 1985). For example, the Defence Department spends almost all the defence budget on its own activities, the armed forces, and their equipment. But the Department of Education and Science spends only 2 per cent of the national education budget on its own activities – with all the remainder going to local authorities and the universities.

Whitehall has a strongly developed code which protects departmental autonomy and is rigorously policed by senior officials and ministers anxious to remain masters of their own 'turf'. Only the Treasury, the Prime Minister's office, and Cabinet Office act as a counterbalance. They are also the departments most completely cut off from links to sub-central agencies. Prime Ministers regularly have to draft in 'star chamber' groups of non-departmental ministers to cut spending departments' claims down to the Chancellor's requirements (see Chapter 4). There is still a basic 'weakness at the centre of British government' because every minister 'knows they serve themselves by serving their department'. Although the central agencies fight 'manfully' (sic), this characteristic 'is a debility all the same' (Heclo and Wildavsky, 1974, pp. 369, 371).

Quasi-Government Agencies

Quasi-government agencies at national level (including the public corporations) account for over 30 per cent of public sector manpower. These bodies, such as The National Coal Board, the United Kingdom Atomic Energy Authority, the Arts Council and the Commission for Racial Equality, are characteristically single-issue agencies, controlled by a Whitehall-appointed board or committee, funded wholly or partly from the Exchequer, subject at best to indirect Parliamentary scrutiny, retaining day-to-day control of their own activities, and accountable to ministers only for capital spending and a number of precisely specified control indices.

Public corporations are the most financially independent QGAs, since they raise most of their revenues by trading. Ministers' controls over their current operations are limited to specifying financial targets which must be achieved, and approving price increases. However, the corporations' investment programmes are partly funded out of the public sector borrowing requirement (PSBR), so that they are vetted in detail by their sponsor ministry and the Treasury. Of course, ministers use their control over investment as a weapon to extract concessions from corporation managements on other issues – such as holding the line on wage increases, maintaining politically sensitive but loss-making operations, or buying key equipment in Britain. This influence is rather episodic however; it can rarely be maintained for long against the concerted pressure of corporation managers. Where the corporations are trading at a surplus they can better resist external control, except the threat that under-investment will hold back their further development. By contrast, corporations running deficits seem more open to ministerial direction. In practice, however, the biggest loss-making corporations (British Steel, the National Coal Board and British Leyland) have been such a drain on the PSBR, and their revitalisation has posed such intractable problems that the government is very dependent on the managements they appoint to rescue them. Ministers' most effective influence is often exerted by changing the leading personnel in the corporation boards or, in the long term, by radically restructuring their activities and organisation.

Other quasi-governmental agencies which depend on Whitehall for current grants to finance all their activities are more financially dependent. But bodies such as the Arts Council, University Grants Committee, or the Commission for Racial Equality, have generally been established as semi-independent agencies for reasons other than financial convenience. Frequently ministers have to appoint a mix of controlling personnel to maintain the agencies' credibility with other institutions in their policy area and with the public at large.

The range of quasi-governmental agencies shades into private enterprise and voluntary organisations in literally hundreds of QUANGOs or quasi-NON-governmental organisations. These bodies are not formally part of the public sector; they retain their own systems of control, recruit their staff outside public sector guidelines, and cannot be directly controlled by the central departments or local councils which fund their activities. They are none the less funded largely or partly by the public sector, and their involvement may be crucial in the pursuit of key public policy objectives. Many of the non-governmental bodies most involved in public policy development are voluntary associations, particularly in the government's current push towards 'care in the community' in the social services.

The National Health Service

The National Health Service is run by decentralised Health Regions and Districts, whose members are partly nominated by local authorities. As links between the NHS and local government services are close and have grown progressively stronger, the role of selected councillors provides a vestige of 'political' control over local health authorities. The DHSS keeps tighter control over appointments to the thirteen NHS regions. The pattern of health service policy-making is quite separate from that in local government. First, DHSS financial control is virtually complete on capital spending for buildings and equipment, and it also sets cash limits on day-to-day spending, despite the efforts of a few NHS districts to generate small sums of money of their own. Second, DHSS and regional controls over district health authorities substantive policies are very detailed, with annual reviews of plans and performance. Third, the medical profession inside the NHS occupy much stronger roles in determining the shape of substantive policy than almost any other group of staff in the public sector. To some extent this professional role offsets central restrictiveness, since what counts as 'good practice' in medicine or nursing is constantly being redefined by countless innovations introduced by grassroots professionals. But it also binds local health authorities to operate within the restrictive limits of current 'good practice'.

Quasi-Elected Local Government Organisations

QUELGOs are consortiums or joint committees of local authorities, such as statutory police authorities spanning several council areas (for example, the Thames Valley Police Authority), and voluntary groupings of councils for tourism, infrastructure, or economic development purposes. Each local council has only a limited number of places on these bodies, and their delegates are only indirectly accountable to voters.

Many QUELGOs also include in their controlling bodies people who are not elected, usually representing interest groups. All local police authorities qualify as QUELGOs because a third of their membership consists of appointed magistrates (JPs), as do Joint Consultative Committees of councils and health authorities administering 'care in the community policies'.

Local Authorities

Local authorities constitute the only form of direct representative government in Britain apart from Parliament. Councils are uniformly composed of 20 to 90 local ward representatives elected for a fixed four-year term, usually organised completely on national party lines, and with a system of majority control akin to the Westminster model. There are no local 'Cabinets', but instead a 'submerged executive' composed of the leaders of the majority party or parties, and the chairpersons of the council committees which administer individual policy functions. Local departments are run by powerful local government professions in each of their key service areas. Councils select and appoint their own staff, and raise their own local property tax (the rates), but also receive just under half their funding from a central government block grant, and are tightly controlled by Whitehall on their current spending levels and capital investments. Councils administer a wide range of domestic policies under more or less rigorous central supervision. However, they continue to retain a substantial autonomy about how they allocate both their tax and grant funds between services, and in the development of new policy initiatives. Because they also retain operational control over policy implementation, and have a local monopoly of professional staffs, expertise in service delivery, and other resources, their active co-operation with central directives is usually necessary for any major success in local policy-making.

Regional Variations

Regional variations in the structure of government primarily affect Scotland, Wales and Northern Ireland, each of which has a central department devoted to its affairs and headed by a Cabinet minister. Scotland and Northern Ireland in addition have separate structures for health service administration and for local government. Scottish local government is dominated by the nine large regional authorities, while councils in Northern Ireland have lost key functions (such as housing and education) to quasi-governmental executives at Provincial level designed to prevent sectarian patterns of service administration. All three regions display much more direct patterns of civil service involvement in the affairs of sub-central agencies. With restricted numbers of local authorities to supervise, the Scottish, Welsh and Northern Ireland Offices have relied much less than their English counterparts on 'hands off' formulae for setting grants or spending levels.

Since 1979 the two Thatcher governments have embarked on the most ambitious attempt to control and redirect the state apparatus of any post-war government. Policy towards sub-central agencies has had four key components: a massive change in central–local relations; the increasing use made of quasi-government agencies; the rationalisation and asset sales of public corporations; and the attempt to impose tighter central control over the public sector as a whole.

Central–Local Relations

Until the 1980s central governments relied on grant funding, standard-setting and controls over investment to shape local government policy. By cutting back on grants, Whitehall pushes more of the burden of council spending onto local ratepayers, and forces them to bear the entire cost of any service improvements. Since tax increases are always unpopular, the theory was that councils would be reluctant to impose extra costs on ratepayers, and hence would cut back their spending plans. However, this relatively simple and partial control was deemed quite inadequate by Conservative ministers after 1979 for three reasons. First, they were convinced that local government was over-spending not in a marginal but in a dramatic way. Councils' wage bills were particularly inflated by unnecessary hiring, inefficient administration, and restrictive union practices.

Second, the Conservatives increasingly queried the electoral legitimacy of local government in Labour areas. Turnout at local elections is often very low (under 30 per cent) in inner-city areas, so that large groups of local government employees (again especially numerous in inner urban areas) can often influence local voting disproportionately. Many voters do not directly pay local property taxes (because they are not ratepayers). A large minority of households have all or part of their rates met from social security payments (again especially in inner-city areas). And businesses bear much of the costs of local property taxes – especially in central city areas where there are large concentrations of factories, offices or shops – but have no voice in local elections. Consequently ministers concluded that under existing arrangements many residents had an incentive to vote for improved services at no cost to themselves. This analysis was confirmed in ministers' eyes when Labour councils reacted to central government grant reductions by pushing through major increases in their local property taxes, without apparently worrying about adverse political consequences.

A third element in ministers' hostility to local government autonomy stemmed from a 1979 manifesto pledge to abolish rates, a promise with which Thatcher was personally identified. The Prime Minister's desire to fulfil this commitment was constantly thwarted in her first term by Treasury and Department of Environment opposition. But it contributed to ministers' preoccupation with curbing large rate increases, a stance strengthened by continuous pressure from business pressure groups, and the furore over 1985 property revaluations in Scotland which sent rates bills spiralling there.

Financial Relations

Much of the government's energy was directed into strengthening Whitehall's financial controls over local authorities. The Department of Environment pays most of the central support for local services in a single block grant to each local authority. Councils can then deploy funds as they like across different services. The Treasury created this system in the 1950s and 1960s to curtail overall local spending; in a period of major growth it removed the risk of piecemeal central government commitments increasing out of control. But in a non-growth period block grants have severely constrained ministers' ability to prune some services of which they disapprove (such as council housing) while preserving funding in politically salient areas (such as school standards). There have been three phases in the development of financial controls.

1979–83

From 1979 to 1983 the government pursued a strategy of tightening general financial controls on all councils. The most successful part of this effort was a series of staged traditional cuts in the percentage of local spending met by Exchequer grants. In 1979 Whitehall picked up the bill for 61 per cent of general local spending, but by 1985 this proportion dropped to 48 per cent. Amidst the controversy over other Conservative measures, this longer-run determination to make local taxpayers foot more of the bill for services has often gone unnoticed. In addition, in 1980 and 1982, the government legislated itself new powers to specify an overall spending limit which local councils could not exceed, even by raising their own property taxes. Councils could still legally spend more than the DOE limit, but if they did they lost government grants equivalent to between two and four times the overspending. Marginally breaching government target figures thus became prohibitively expensive, since an extra £1 million spent on services could add up to £5 million to rates bills (that is, £1 million in extra spending plus £4 million lost in grants through penalties).

In practice, things did not work out quite as ministers planned. The biggest hole in the policy was made by 'new left' Labour councils such as Sheffield and the metropolitan counties and GLC which swung to Labour in the May 1981 elections. They realised that once major overspending of DOE targets had taken place the choices of complying with or defying government instructions were almost equally unattractive. Consequently the GLC and the Inner London Education Authority decided to overspend by so much that they no longer qualified for any central government grants at all. From then on they could add another £1 million to services at any time without having to worry about government penalties, simply paying all the extra cost from rates. Many other local authorities (including some Conservative shire counties) reacted against targets by overspending marginally, since they feared (correctly) that complying fully with ministers' demands would simply produce decreases in their permitted targets in future years. The Liberal–Tory administration in Liverpool from 1979 to 1982 complied faithfully with government demands, only to see their targets cutback radically as a bizarre reward for 'good behaviour'.

The vast majority of councils whose spending levels came near to government target levels reacted in three ways. First they radically changed their investment behaviour and accounting arrangements, transferring many items (such as vehicles or small building repairs) previously purchased out of current spending to their capital accounts, which were not so continuously controlled. Since their revenue spending was being squeezed hard, many councils also cut back on longer-term investment, since there was little point in erecting buildings which they could not afford to maintain or staff. By the winter of 1982–3 councils were underinvesting on Treasury limits to the tune of £1,600 million, and much of the spending still financed by borrowing would previously have been paid for from revenue funds. In the run-up to the 1983 election, the slump in local authority investment implied more unemployment in the construction industry, triggering a direct intervention by Thatcher to get councils to take up their capital spending allocations.

Second, councils' used 'creative accounting' to set up special reserves of unallocated funds, where the money had already been raised in taxes and counted as expenditure, and could be deployed to reduce nominally new spending and avoid the full weight of penalties. In the mid-1980s some councils, such as Labour Liverpool, even developed schemes to sell some of their assets (such as mortgages on council houses sold to tenants) to banks in order to maintain their spending.

Third, Conservative and Labour councils alike raised extra property taxes which they put into their bank accounts as a safeguard against increasingly unpredictable Whitehall demands. Even authorities which complied with government spending limits regularly found themselves short of cash when the government 'clawed back' grant from local government as a whole in reprisals for total spending beyond Whitehall's limits. With a new piece of legislation virtually every year, and two or three ministerial interventions in each budgetary cycle, very few treasurers or council leaders were confident that they fully understood the operations of Whitehall's controls. Consequently in the vast majority of local authorities the ratio of reserves to current spending rose sharply between 1979 and 1983. In a very critical 1984 report, the Audit Commission, set up by the Thatcher government to promote greater efficiency in local government, concluded that councils had raised around £400 million extra reserves in each of the previous three years, implying that residents had borne £1,200 million in extra taxes to cushion councils against possible spending restrictions or penalties. Clearly this kind of perverse effect severely damaged government claims that they were curtailing rises in local property taxes, or that general increases in rates stemmed from the actions of a few profligate councils.

The 1983 election to 1985

By the 1983 election, the Conservatives had decided on further action specifically targeted against Labour councils still defying Whitehall control. In a two-pronged offensive the Tory manifesto pledged abolition of the GLC and the six metropolitan counties, and new legislation to 'cap' the local property taxes in councils most prominently overspending the DOE limits. Both measures it was claimed would create savings, abolition by cutting out a basically redundant tier of administration in the conurbations, and rate-capping by directly reducing property tax increases. Although these proposals were subsequently enacted, the GLC and metro counties fought an unusually effective public relations rearguard action. Public opinion in London swung solidly against GLC abolition, severely denting local support for the Conservatives (Husbands, 1985). In the other conurbations, reorganisation may be carried through more smoothly, since only six or seven district councils have to co-operate on the new QUELGOs created to take control of the metropolitan services. In each metro area there is also one large 'core' city authority which might inherit some county co-ordinating functions. In London, however, the 32 small boroughs are unlikely to be able to substitute for the GLC's demise, and a worsening of administration will probably result (Clegg *et al.*, 1985). Unless there is extensive 'policy termination' of previous services, the reorganisation is unlikely to generate any major savings to offset the substantial transition costs involved. Government estimates of cost reductions suggested that at most £100 million would be 'streamlined' but more detailed calculations by opponents of reorganisation suggested that the new metropolitan governments could be more expensive by a similar margin (Coopers and Lybrand, 1984).

The government's rate-capping legislation conferred two new powers on ministers in England and Wales, along lines pioneered two years earlier in Scotland. The first allowed the DOE to draw up a blacklist of 'overspending' councils, and then to fix a rate for their property tax in the coming year. Councils could then negotiate about this decision, but if they did the DOE could examine their budgets in great detail. All but one of the 18 councils blacklisted in 1984–5, and the 10 reselected (with 2 new additions) in 1985–6, were Labour authorities, who refused to negotiate with DOE. The requirements imposed on them varied from small increases only in their rates, to major reductions. Councillors who defied the government were legally liable for any overspending, while ratepayers were free not to pay taxes levied above the permitted level. Left Labour councils reacted by pledging a concerted campaign of defiance, centring on setting illegal budgets where spending and revenues did not match up. In the event their unity broke down under the threat of legal liability actions, and prominent rebels, such as the GLC and Sheffield, caved into government pressure in mid-1985.

The strongest left challenge to the government in fact came from a council which was not ratecapped, Liverpool, where Labour won control in 1982 and a District Labour Party dominated by the Militant tendency achieved tight control over councillors. In 1984 the Labour group threatened to bankrupt the city over government grant cuts and its clearly inadequate 1984–5 spending level. They stage-managed subsequent negotiations with ministers so effectively that they seemed to win major concessions. In 1985 Liverpool repeated its threat and levied a rate which was insufficient to balance its planned spending during 1985–6. In theory this move was illegal and could have been challenged in the courts, but in practice the necessary writ against the council was not moved. By the autumn of 1985, with the DOE now adopting a firm stance against any negotiations, the Labour group decided to sack 'notionally' their entire workforce and borrow money to pay wages as redundancy costs until a new budget in the spring of 1986. But at the 1985 Labour Party Conference in Bournemouth, Neil Kinnock seized on the Liverpool debacle as a chance to attack Militant, supported by the major unions dismayed at the jeopardising of their members' jobs. In the early winter of 1985 Liverpool's resistance weakened with a backlash among the non-Militant members of the Labour group and local party, and the suspension of the District Party by Labour's National Executive.

Although the government claimed that climbdowns by Labour authorities were a famous victory for the rate-capping legislation, their practical impact on overall local government spending was minimal. Even in Scotland where the Scottish Office both picked off two high-spending authorities (Stirling and Edinburgh) and controlled rates in all the other 63 councils, rate-capping produced few selective cuts in spending. In England and Wales the DOE can only target a few of the worst offenders out of its 450 councils. Since some of the blacklisted authorities are quite small districts, the effect of selective rate-capping on local spending totals is completely negligible by contrast with the comprehensive Scottish powers. The government did include a general rate-capping clause in its English and Welsh legislation, but had to promise not to activate the power in order to overcome Conservative local authority resistance and hostile pressure from the House of Lords.

High-spending local authorities not subject to selective rate-capping but, worried that they might run up against government restrictions in future years, have again reacted by raising more local property taxes and increasing their reserves. They reason that if they are at all likely to be rate-capped it is advisable to have their rates cut back from the highest feasible level, and to have some extra money in the bank. As a result, it is difficult to say whether selective rate-capping has marginally decreased total local government taxation (by capping the worst 'over-spenders' in Whitehall's eyes), or substantially increased it (by encouraging many more councils to raise taxes as an insurance).

Finally, the 1983–5 period saw the government attempting to dismantle some of the counter-productive controls over councils introduced during the first Thatcher term. The so-called 'volume spending targets' which limited spending by reference to 1979 budgets were phased out in 1985 in favour of the DOE computer-assessments of each authority's need to spend. This measure was designed to appease many Conservative county councils in the south-east with growing populations. Its most immediate result, however, was a successful legal action by Labour counties claiming that the DOE's transition rules discriminated between authorities rather than following general principles, and raising the prospect of multi-million repayments of previous grants penalties exacted by ministers.

After 1985

From 1985 onwards it has become clear to ministers both that their existing plethora of controls is not working, and that any real resolution of Conservative discontent over rates requires a wholesale reorganisation of local government finance. Three measures emerged as possible options in fierce Cabinet discussions. First, the local property tax on businesses could be levied at a single national rate, and then distributed to councils in proportion to their population. This measure would greatly diminish the financial resources and freedom of manoeuvre of

(Labour) inner-city authorities, making them much more dependent on central grants or taxes on their residents than ever before. By contrast, (Tory) residential suburbs and shire counties would receive increased resources as of right. Second, the local property tax on residents could be scrapped and replaced by two new taxes, one a flat-rate 'residents' tax' payable by all adults whether they own property or not, the other a much reduced property tax. People on supplementary benefits would be obliged to pay a fraction of the residents' tax themselves, instead of the present situation where their local rates are automatically covered by welfare payments. If both proposals were implemented, then current DOE spending assessments and grant penalties might be dismantled. In ministers' eyes the new finance system would make councils fully accountable to their electors for the consequences of spending decisions, and allow some trust to be placed once again in local democracy. Thatcher is particularly keen to reduce radically the rates, but controversy in Whitehall continues over the workability of this plan. Whether these proposals are ever implemented hinges on the results of the general election due in 1987–8. It seems doubtful that widening tax nets so drastically, and altering the incidence of residents' taxes so dramatically, can contribute to Conservative popularity.

Policy Relations

A key, unspoken premise of the Conservative's push for more financial controls over councils was the assumption that the Thatcher government was not dependent on local authorities for the attainment of its key policy goals. Previous post-war governments have wanted councils to do things, and have depended on them for policy outputs such as increased public housing or large school building programmes. Consequently they trod carefully around 'local autonomy' issues, recognising that getting local government to perform effectively required a genuine partnership and self-restraint in the exercise of Whitehall's massive control powers (Dunleavy and Rhodes, 1983). By contrast, ministers after 1979 constantly gave the impression that local government activities were not just wasteful but in large part irrelevant to their concerns. The number of schoolchildren was dropping, in the government's view creating plenty of scope for any improvements in education standards within existing budgets. The gross shortage of fit housing in the UK had been eliminated, so that public housing construction was now a dispensable luxury. The previous decades had seen major investments in general local authority buildings and service development; now Whitehall saw only a period of consolidation and retrenchment. (For more detail on various policy areas, see Chapter 8.)

Yet although some past policy priorities could indeed be written off, the Thatcher governments have time and again confronted problems raised by their inability to secure the active co-operation of local government in meeting new priorities which have arisen in the 1980s. Central–local relations on policy issues have been marked by three key developments.

Direct Centralisation of Power

Direct centralisation of power was the Conservative strategy from 1979–82. Local authorities were compelled by new legislation in 1980 to sell their housing to any tenants who wished to buy, with heavy discounts. A policy of non-compliance by Labour councils and local government unions delayed full implementation of the policy only until 1981, but probably reduced the 570,000 sales total achieved quite appreciably, since real interest rates rose sharply during the delay, choking off a good deal of tenant demand (Ascher, 1986). Government spokesmen, such as Ferdinand Mount, the head of Thatcher's Downing Street Policy Unit in 1983, saw council-house sales as a paradigm case of how the government was 'centralizing in order to decentralize'. Local authorities undoubtedly lost powers to dispose of their housing stocks to Whitehall, but so that ultimate control of the dwellings should devolve from councils to the new owners. Ministers found it difficult to cite any more examples of this dialectical process, however, and subsequent Whitehall interventions simply increased civil service powers without any subsequent decentralisation. For example, in their anxiety to stop new public housing construction, ministers have stopped local authorities from spending much of the money they received from council-house sales on new dwellings or much-needed repairs to the remaining stock. Similarly, after several years' fruitless efforts to shift curriculum development in state schools towards more 'relevant' subjects (Ranson, 1982), the Department of Education and Science finally assumed direct control over how a small amount of councils' block grant funds should be spent on school education projects.

Indirect Centralisation

Indirect centralisation denotes a variety of attempts by central government to promote greater 'efficiency' in local authorities. Early efforts included pressure on councils to use private auditors, and the setting-up of an Audit Commission of highly paid accountants to promote a local government equivalent to the Financial Management Initiative in Whitehall (see Chapter 4). But from 1983 the government has placed its chief hopes on contracting out some local services to private firms as a way of reducing costs and breaking what they see as the 'stranglehold' of local government unions in preserving inefficient work practices. In the first Thatcher term, the key push for contracting out focused on the National Health Service, and the DOE's efforts in local government focused mainly on getting council building works departments to function on more commercial criteria. However, Conservative councillors in several 'dry' local authorities seized on contracting-out primarily as a means of renegotiating wage and conditions deals with some previously powerful groups of manual workers. In the 1970s, dustmen organised by the National Union of Public Employees and Transport and General Workers Union had considerable success in pushing through rather favourable wages and hours deals with individual councils. Contracting-out became a means of pushing back these deals, with many councils going to tender to force their existing workforce to make concessions rather than because they really wanted to bring in contractors (Ascher, 1986). Only a very few right-wing councils, such as

Wandsworth, embarked on a more ambitious programme of bringing in contractors.

By 1985 the government seemed to be about to turn its attention to local government, bringing in regulations which would order councils to seek private tenders as a matter of course in five areas, including cleaning, grounds maintenance, and catering. In a similar move, local bus services were deregulated, forcing councils to stop paying subsidies primarily to their own bus companies or a single public transport operator. Conservative ministers expect these requirements for competition between public and private suppliers of basically commercial services to have a major impact in reducing costs and improving work practices. However, there are very substantial decision costs involved in a uniform use of tendering procedures, both for councils and for contractors themselves, and substantial transition costs in closing down an in-house service (Ascher, 1986). Experience from the NHS suggested that in its first phase a centrally initiated 'privatisation boom' of this kind produces artificially low pricing by private firms anxious to achieve a market share in any 'bonanza', but that its longer-term impact on costs is fairly marginal. When Conservative as well as Labour and Alliance councils responded with hostility to the DOE compulsory tendering proposals, their implementation seemed to be postponed beyond 1986.

By-passing Local Government

This has been the Conservative's third and perhaps most important response. As central–local relations have worsened cumulatively, so the level of trust between the two tiers has rapidly declined, particularly as between Labour councils and Whitehall. Central government has presided rather helplessly over a steady worsening of the social problems of Britain's inner cities, with serious riots in thirty-three locations during the summer of 1981, and a sudden and savage recurrence of rioting in the autumn of 1985. Spending limits, grant penalties and rate-capping have cumulatively diverted government cash away from the areas of greatest need and produced objectively perverse outcomes. For example, in 1982 one of the most affluent shire counties in England, Buckinghamshire, received virtually the same central block grant as the city of Liverpool, which had the same population, a wider range of services to administer, and some of the worst social problems in the country. The DOE has been unable or unwilling to break out of the straitjacket of its overall financial controls so as to help local authorities tackle inner-city deprivation in any effective way, concentrating instead on a purely cosmetic system of special grants, the Urban Programme. Having thus disabled the local authorities by blanket financial controls, freezing up the development of what has previously been the dominant mechanisms for implementing most domestic policies, the government has had to by-pass this tier altogether.

The Growth of Quasi-Governmental Agencies

In their 1979 manifesto the Conservatives were fiercely critical of the growth of non-governmental bodies, all of which they inaccurately termed 'quangos'. The first Thatcher government announced a wide-ranging review of existing bodies, but expectations of any substantial programme of 'quangocide' petered out in the scrapping of barely 100 out of over 2,000 non-governmental bodies, with savings of under £12 million (Pliatsky Report, 1980). Despite this initial hostility the Thatcher governments have in practice become considerable enthusiasts for single-function quasi-governmental agencies. The bodies created or expanded since 1979 have been among the largest and most powerful of such bodies, chiefly because the government has been unable to entrust local government with key functions while maintaining a very tight squeeze on revenue spending.

A major beneficiary of this desire to by-pass local authorities has been the Manpower Services Commission (MSC), whose role in providing employment training mushroomed in response to spiralling unemployment, especially among 16–19-year-olds. Set up in 1976 by the last Labour government to provide a co-ordinated public sector response to growing joblessness, the MSC expanded to a budget of £727 million by 1979. Under the Conservatives this rapid growth continued, since the youth unemployment problem rapidly became a key political controversy. The government redirected the MSC from a basically public sector operation with its own full-time training programmes, into more of a broker role, fixing up placements for youth trainees with firms and businesses. While local authorities provide 16–19-year-olds with training in schools or colleges without any payment, the MSC schemes include a weekly payment to attract young people off supplementary benefits. The first MSC schemes attracted a good deal of trade-union criticism as indiscriminate subsidies to employers, and a revised Youth Training Scheme had to be formulated in 1982. The MSC budget has continued to grow, to £1,906 million in 1983–4. After the 1983 election the MSC has increasingly broadened its role (and that of the Department of Employment) into policy areas previously dominated by local authorities (and the Department of Education and Science). It effectively assumed control over vocational training courses provided in colleges of further education, and under the £150 million Technical and Vocational Education Initiative the MSC even began restructuring the curriculum in secondary schools on more work- and business-related lines.

A very similar pattern has been followed in the economic and social regeneration of inner cities. The 1980 budget created a couple of dozen 'enterprise zones' funded directly by the Treasury, where local taxes are reduced and extra support is available for businesses moving into the area. In 1985 tax-based regeneration was extended to include five 'freeports'. In 1981 the government established Urban Development Corporations in Merseyside and the London Docklands, imitating new town corporations in their sweeping planning powers and use of Treasury-funded physical improvements to stimulate business and private sector housing investment. Criticised by the mainly Labour local authorities in their areas of activity, the UDCs have concentrated on attracting major firms in line with government regional policy. A further move to by-pass inner-city councils was the 1985 replacement of some joint central–local 'partnerships' set up by the Labour government in 1977 by 'taskforces' composed entirely of representatives of Whitehall departments. More radical initiatives still are under discussion inside the DOE, including schemes to entrust the rehabilitation and privatisation of some inner-city public housing to new agencies like development corporations. Finally the government has encouraged the emergence of new forms of public/private QUANGOs to promote more private-sector-orientated economic development.

The abolition of the GLC and metropolitan councils in 1986 has also produced a whole new crop of QGAs. One of the largest is the London Regional Transport Board set up in 1984 to take over the running of London Transport. The new board resembles a public corporation, but is unusual in levying property taxes from Londoners with the approval of ministers. Other new bodies include QUELGOs for fire, police and waste disposal services, and numerous other bodies for co-ordination of planning, aid to voluntary groups, etc. In London, responsibility for around two-thirds of GLC spending is transferred to non-governmental bodies or central departments (Clegg *et al.*, 1985, ch. 1). In the metro county areas, perhaps half of all spending goes to similar bodies. In addition Whitehall has taken special powers to regulate the budgets and staffing of the successor bodies to the metro authorities.

A less conspicuous but equally important by-passing of local councils has developed at the inter-face between the National Health Service and local government. The government's policy of 'care in the community' involves moving long-stay elderly and handicapped patients out of expensive hospital care and into more appropriate decentralised facilities run by the social service departments of local authorities. However, with the government squeezing their spending and resources so severely, councils have been reluctant to take on any extra responsibilities at all without cast-iron government funding. A system of special grants, called 'joint finance' and administered by committees of local councils and health service authorities, has progressively developed to overcome this opposition, but progress remains very slow. The squeeze on local spending thus continues to imply higher costs within the NHS. None the less, local councils and the members of health authorities have lost a great deal of control over the one buoyant area of welfare state spending to a very professionally dominated planning system responding most directly to DHSS wishes.

Contracting out has also played a prominent role in the government's drive to improve what it sees as the lagging performance of the biggest group of QGAs, the National Health Service. The origins of contracting out go back a long way (Ascher, 1986). In 1969 the Wilson Labour government decided to reduce the numbers of civil servants by contracting out much of the cleaning work in central departments to private firms of office cleaners. The Thatcher government pressured departments to do more of the same in 1979 when demanding further civil service manpower cuts. One department which responded particularly vigorously was the Ministry of Defence, who subsequently switched a great deal of cleaning and catering work for the armed forces to private firms, with 20 per cent cost savings in some cases. A vocal lobby of Conservative backbenchers, some of them financially linked to major contract cleaning firms, pressed ministers to do the same thing in the National Health Service. Although a DHSS study group found little potential for similar savings in the health service, ministers pressed ahead in 1983 with circulars compelling NHS district authorities to put cleaning, catering and laundry work out to tender. Ministers directly laid down the procedures to be followed, and on occasion intervened to direct local health authorities to accept private firms' tenders rather than in-house bids. At first, contractors seemed to be making major inroads into NHS work, but as the health service unions became better organised, and as defects in some of the early contracts emerged, so in-house bids by existing NHS staff units began to be successful in a majority of cases.

Finally the Conservatives have consistently encouraged the development of strong corporate management techniques within those local and regional QGAs susceptible to local government or professional influence. NHS authorities have been internally reorganised. The previous system of collegial control by representatives of all the medical and nursing professions has been replaced by powerful general managers in each district and hospital, akin to chief executives in private firms. Similarly, regional water-authority boards used to be large and unwieldy committees, partly 'colonised' by councillors from local authorities in the area and farming interests with special interests in land drainage. In 1984 they were reorganised as small, business-style boards with only a very weak local consultation apparatus representing consumer interests. The more tightly controlled boards still resisted government instructions to raise water charges 'artificially' as a kind of extra tax in 1985, but are now more self-sufficient in their finances. Similarly the government has accelerated moves to switch the water authorities' finances away from their current property tax assessment towards the metering of water supplies. As a logical culmination of these moves the Cabinet decided in 1985 (against civil service advice) to convert one or all the water authorities into limited companies, shares in which can then be sold on the stock exchange in 1987–8.

Public Corporations and 'Privatisation'

Conservative attitudes to public corporations have seen the most rapid and complete transformation during their tenure of power. The 1979 Tory manifesto made little or no reference to the privatisation of state-owned companies. Instead the problems of the nationalised industries were ascribed to the constant interference of Labour ministers, often in cahoots with the overweening power of trade unions in the industry, obstructing necessary change, hampering management's freedom to manage, and preserving uneconomic operations for 'social' reasons that had little or nothing to do with the individual corporations. Yet within two years the Thatcher government developed a passionate aversion to nationalized industries, partly through frustration' (Riddell, 1983, p. 173).

In their efforts to restore economic viability as the touchstone for public corporations' activities, the government interfered more rather than less in the day-to-day operations. Ministers repeatedly claimed that wage increases and job losses were matters for management and unions to resolve between themselves. But, in practice, ministerial intervention was even more visible and extensive than under any previous government. The government clearly did not apply economic cost/benefit criteria to its own policy decisions on public corporations, but instead took a series of principled stands, many of which were overtly 'political'. For example, in 1984–5 the Cabinet completely suspended the ultimate discipline of the market place on corporation managements – the threat of bankruptcy – by granting the National Coal Board a blank cheque to outface the strike by over two-thirds of its workforce. In addition, the restrictions on government borrowing implied by monetarist macro-economic policy, enormously increased the importance of Whitehall support for corporation managements trying to secure their investment programmes against Treasury pressure for a reduced PSBR. Similarly, although the Conservatives initially believed in enhancing the commercial autonomy of corporation managers, they rapidly began differentiating between those managements which shared their vision of the route to viability, and those who had 'gone native', becoming over-identified with the survival of their particular industries in their current form. Finally, while ministers' initial enthusiasms were all for restoring loss-making corporations' profitability even at a high social cost, their willingness to take political risks to enforce shrinkage of 'uneconomic' activities visibly diminished by 1982, as unemployment spiralled past the 3 million mark.

The first and greatest casualty of ministers' determination to make nationalised industries 'stand on their own feet' was the steel industry. Reorganised at great cost in 1973 by the Heath government, the British Steel Corporation was lumbered with far more plant than necessary to meet the greatly reduced 1980s demand for steel and was faced with crippling interest payments on its modern but partly redundant stock. The government appointed an American banker, Ian McGregor, to put through a new corporate plan, which involved firing half the workforce, and was finally implemented after a steelworkers' strike over pay collapsed in 1980. Major job losses were also incurred at British Leyland in 1979–80, but the government propped up the loss-making Belfast ship-builders Harland and Wolfe. In 1981 the Cabinet also backed off from confronting the National Union of Mineworkers over pit closures. Although ministers opted for another McGregor job-shrinking plan for the coal industry after the 1983 election, it was noticeably more incremental than that for steel; for example, it implied no compulsory redundancies for miners. Other loss-making industries in the late 1970s, such as the railways or the Post Office, scraped back into a bare profitability via gradual job-shedding rather than any massive decrease in operations.

Successful public corporations experienced ministerial interventions in three ways. First, the government was abnormally restrictive in approving major new investments, such as British Gas's North Sea pipeline proposals because of their impact on the overall public sector borrowing requirement. By 1981 Britain had the smallest ratio of state borrowing to GDP of any Western democracy, a dubious distinction maintained thereafter into 1986. Second, the government increased the financial targets (that is, levels of profit) which these successful companies were expected to earn in order to pay back interest to the Exchequer and to finance their own investment programmes. Third, despite their formal attachment to commercial viability considerations, successive Conservative Chancellors of the Exchequer woke up to the potential of using a premium on successful nationalised industries' prices as a means of collecting 'hidden taxes' on their consumers. Corporation managements in gas, electricity, the telephone industry and the water authorities protested unavailingly against additional or unnecessary price increases which they were required to make in 1982–4 in order to pull more funds into the Exchequer.

The revenue-raising potential of at least the profit-making public corporations largely accounts, too, for the emergence of 'privatisation' of public corporation asset sales as a central pillar of Conservative policy towards the mixed economy. The approach was begun ironically enough by the Labour government during the 1976 sterling crisis. Callaghan's Cabinet decided to avert some of the spending cuts demanded by the International Monetary Fund by selling some of the 40 per cent stake which British governments have held in the BP oil company since its inception in 1911. The device was not lost on leading Tories and the first Thatcher Cabinet initiated a policy of selling off surplus public agency properties and government stakes in private limited companies. In the six years 1979–85, land and building sales (including leasing motorway service areas to private firms) raised nearly £290 million (Table 5.2). Sales of state equities in basically limited

TABLE 5.2 *Asset sales by the Conservative governments, 1979–85*

1979–80		*1980–81*	
*British Petroleum (5%)**	£276m	North Sea oil licenses	£195m
*National Enterprise Board shares**	37	Public land and properties	84
Public land and properties	34	*National Enterprise Board shares**	83
Shares in other companies	23	**British Aerospace (52%)***	43
Total	£377m	Total	£405m
1981–82		*1982–83*	
*Cable and Wireless (49%)**	£182m	**Britoil (51%)***	£334m
Public land and properties	107	*International Aeradio (100%)*	60
*Amersham International (100%)**	64	**Associated British Ports (52%)***	46
Oil stockpile sales	63	Public land and properties	35
*British Sugar Corporation (24%)**	44	**British Rail Hotels (91%)**	34
Other sales	34	Oil stockpile sales	33
		Other sales	40
Total	£494m	Total	£582m
1983–84		*1984–85*	
*British Petroleum (7%)**	£543m	**British Telecom (50%)***	£1500m
Britoil (49%)*	293	**Enterprise Oil (100%)***	380
*Cable and Wireless (22%)**	263	*Jaguar Cars (100%)**	297
Public land and properties	28	*Inmos (75%)**	95
North Sea oil licensing	19	**British Gas onshore oil wells**	82
Other sales	22	**Associated British Ports (49%)***	50
		Sealink	40
Total	£1168m	Total	£2444m

Key:
Public corporation, whole or part – in **bold** print. State shareholding in private company – in *italic* print. * Stock market flotation.

Sources: OECD Economic Survey of the UK, January 1985; Heald (1984), pp. 30–31.

companies (that is, not forming part of a public corporation or a nationalised industry) raised over £2,300 million in the same period. The major disposals here have included more BP shares; stakes built up by the Labour government's National Enterprise Board in the computer firms ICL and Inmos; Amersham International, a specialist radioactive isotopes company; a historic stake in the British Sugar Corporation; Cable and Wireless, an international communications company built up under state ownership partly for security and intelligence reasons during the days of the British Far East Empire; and Jaguar Cars, the luxury car operation of British Leyland.

From 1980 the programme was extended, so that some profitable parts of full public corporations that competed with or were attractive to private sector companies were floated off as separate companies and then sold on the stock exchange or by tender to individual private bidders. Majority control of British Aerospace, nationalised by Labour only in 1977 was the first to be sold off, followed by British Rail's hotel operations and Sealink ferries, and British Gas's oil interests onshore (at Wytch Farm) and offshore (floated as Enterprise Oil). This programme accelerated in 1982 with the decision to begin disposing of entire public corporations by floating them on the stock market. The pioneering disposal focused on the assets of the British National Oil Corporation built up under Labour in the late 1970s as the government's arm in North Sea oil exploration and extraction. The government-owned Associated British Ports continued this trend, but it was the disposal of half the equity in British Telecom in 1984 which marked a watershed. This sale was the first major public utility monopoly to be transferred to private investors, and the government made strenuous efforts to encourage small private investors to subscribe for shares, linking asset sales to a new ideal of encouraging much wider share-ownership.

Table 5.2 demonstrates the dramatic acceleration of asset sales after the 1983 election. In 1984–5 they realised over four times as much revenue for the Exchequer as in any year during the first Thatcher government. Since these receipts are counted as 'negative borrowing', they are especially valuable to ministers in preserving the form of a restrictive 'medium term financial strategy' (see Chapter 7). As the Prime Minister and Chancellor grapple with the considerable difficulties of delivering any form of tax cuts to voters in the run-up to the 1987–8 general election, the privatisation programme will inevitably expand. The major Conservative target in 1986 is the stock-market flotation of British Gas, the first slice of which should generate £2,500 million for the government. Future sales plans include the current Regional Water Authorities, the first quasi-government agencies not already in a company form to be scheduled for privatisation. In addition to raising finance, asset sales have considerable political spin-offs. By according small investors privileged access to shares at the discounted stock-market prices judged necessary to secure a full take-up of the shares on offer, the government can on occasion transfer remarkable cash benefits to large numbers of voters, as with the British Telecom sale in 1984. This potent vote-catching experiment seems certain to be repeated with the gas sale, and its political impact is of course magnified by saturation government television and press advertising extolling the virtues of privatisation.

The key to pushing through asset sales of major public utilities has been the government's ability to develop alternative forms of enforcing public interest controls on these corporations' managements. The main device has been greater use of regulatory instruments as a substitute for ownership controls – which are highly misleadingly labelled by ministers as 'liberalisation' or 'deregulation'. This example of 'double-talk' flows from the admiration that several Conservative think-tanks have for recent developments in the USA. Most American public utilities (such as the gas, electricity, water, and telephone industries) are run by privately owned companies, but with many legal controls designed to prevent them from exploiting their monopoly control of vital resources. Particularly under President Reagan in the 1980s, the federal government has been trying to reduce its involvement, and to introduce 'free market forces' into public utility operations – a policy known as 'deregulation'. In the British context, most public utilities are not all that closely regulated by the government since they are directly owned by the state already, and ministers can basically just tell managements to come into line with their wishes. None the less the vocabulary of liberalisation/deregulation has been imported from the USA by the Conservatives to describe situations where an existing state monopoly has its undisputed right to undertake a particular activity removed.

Early government initiatives in this direction produced little private-sector interest. Almost no companies took up the option of supplying privately produced electricity to the national grid introduced in 1981. A private-sector competitor to British Telecom, the Mercury company, established in 1981 after determined ministerial pressure, was considerably delayed in its attempt to develop an alternative business-orientated phone network. Bus licensing deregulation introduced in 1983 produced little growth of private services after municipally owned companies aggressively outpriced and outserviced new entrants. Ministers tried again in 1985 with new legislation banning municipal subsidies to their own bus companies, whose basic feasibility remains fiercely disputed. Only in express coach services has the government achieved a modest degree of success in lowering prices and somewhat improving the service choice open to consumers.

The failure of deregulation schemes to achieve change on their own added further impetus to the programme of public corporation asset sales. Since the highly concentrated British public utilities cannot be effectively attacked by private companies, the remaining option was to convert public corporations into private ones. The dangers that natural monopolies will then be exploited for profit by their new shareholders have to be headed off by extending government regulatory powers over the public utilities. In the case of the British Telecom sale, a new quasi-governmental agency, the Office of Telecommunications, was established to police greatly extended powers over Telecom's pricing, the terms on which it connects services to its system, the way it treats bids to 'license' parts of its system from alternative services the equipment purchasing policies it follows, and its provision of 'social' services such as callboxes, etc. In addition the government retained special powers to prevent a foreign takeover in the 49 per cent of equity which it has not yet sold; so-called 'gold shares' give ministers an effective veto power over major policy decisions in special circumstances. This experience is certain to be repeated in the British Gas sale in 1986. Privatisation therefore implies reregulation, not deregulation.

The Centralisation of the State?

Margaret Thatcher's highly personal style of premiership differs from that of almost all her post-war predecessors, not just in the alleged 'presidentialism' of her leadership (see Chapter 4), but also in the breadth of her involvement with sub-central government. It is hard to think of any previous Prime Minister who has become so publicly involved in so many esoteric debates, ranging from how to cure football hooliganism, to local government finance. Media commentators constantly make reference to the individual views and foibles of the Premier in explaining government policy. But these impressions of centralising control of the state apparatus within Whitehall, and even more in the Prime Minister's office, are overstated. We have already indicated many of the areas where government objectives have been achieved, constrained or redirected since 1979. Only in three areas has a standardised policy line apparently been imposed on the whole government machine:

Public Sector Incomes Policy

Public sector incomes policy has been one key element of central control over the state apparatus since 1980. Partly, the Cabinet's firm stance reflects the difficulties of trying to control wages growth via monetary policies alone (see Chapter 7). Partly it is a reaction against the first Thatcher government's weakness in 1979–80 when it accepted the findings of pay inquiries set up by the previous Labour government, and concluded lenient arrangements of its own. Partly it reflects a blanket government commitment to defy trade-union pressure and public-sector strikes. In a number of major disputes, culminating in the 1984–5 coal dispute, the government has funded prolonged resistance to industrial actions. Pay comparabilities between the public and private sectors consistently worsened over this period, and government 'reprisals' have often followed in the aftermaths of major strikes. For example, the 1982 hospital ancillary workers' action spurred the government to advance NHS privatisation plans.

Cash Limits

Cash limits on revenue spending and cuts in public sector investment have generally slowed the expansion of growing public services or public corporations below the levels advocated by managers and professionals in sub-central governments. By creating a climate of restrictiveness, it is probably true that public resources in growth services have been concentrated on higher value-added projects than hitherto.

Privatisation Strategy

By contrast it is still hard to see a coherent 'privatisation' strategy being employed in all areas of the state apparatus, despite the efforts of propagandists to create one (Pirie, 1985). We have already noted that in a British context public utility asset sales and deregulation are not consistent. Neither of these policies is inherently related to the third 'privatisation' effort, contracting out some public service activities to private firms. The fourth 'privatisation' element, encouraging private sector substitutes for public services (such as health care or education), is different again, despite vague fears that it could become linked to contracting out – for example, by using private hospitals for NHS patients. The 'privatisation' label gives a bogus unity to disparate policies connected chiefly by the ideological prejudice that 'private equals good' while 'public equals bad', and the electoral calculation that intensifying public/private conflicts and 'rolling back the state' provides a long-term boost to Conservative electoral support (Dunleavy and Husbands, 1985).

Nor does 'privatisation' necessarily imply a permanent centralisation of power between Whitehall and the rest of the state apparatus. By selling public corporation shares the central state clearly surrenders some powers over areas of economic life previously controlled directly. The Thatcher governments have so far been confident that their priorities and those of the new private sector company managers will not diverge. The first occasion on which a cash-rich British Telecom concedes over-the-odds wage increases to its highly unionised workforce may trigger a re-evaluation of this perception. And, of course, for future governments with quite different values, asset sales unambiguously constrain their behaviour. Similarly, although contracting-out in the NHS has been artificially imposed on reluctant local administrators by central department fiat, one reason for its delayed extension to local government may be the high administrative costs of compulsory tendering where most contracts are anyway won by in-house bids.

The public perception of increasing central control over the wider state sits very oddly with the experience of Conservative policy change in the central departments themselves. With manpower down 15 per cent in six years, and a very considerable shrinkage in the 'open structurc' of senior policy-level officials (Chapter 4), it would be surprising indeed if the policy-making capacity of central departments had markedly expanded since 1979. The record of local government legislation in 1979–83, the ham-fisted transfer of housing benefits administration from the DHSS to local councils in 1982, severe doubts about the basic workability of the 1986 metro government reorganisation, plus the string of legal defeats for ministers on social security, transport and local government finance issues (Chapter 11), all suggest that the effectiveness of civil service and ministerial decision-making may be declining.

The problem for the Conservatives is that the New Right suggestions for pruning 'big government' seem to have only a one-off impact on the welfare state. Equally they do not seem to be well adapted to the realities of a non-executant central government, where very few areas of domestic policy are run directly from Whitehall. The euphoria for privatisation stems directly from ministers' frustration at only being able to influence the framework of sub-central government's policies, and by the inexact targeting of policy change that can be achieved across organisational boundaries. The government's attempts to renounce 'corporatist' negotiations with local government, NHS professions or public corporation unions, has pushed back the limits of what it is feasible for any central government to achieve against sub-central agencies' resistance in one area after another. Yet these gains have often been temporary, triggering an adaptation by other agencies to evade ministers' controls, especially in the case of local government.

'Central intervention', the sheer number of times that Whitehall interferes with sub-central agency policies, has clearly increased. But the picture is less clear for 'central control', measured by the extent to which 'interventions achieve the purposes of central government' (Stewart, 1983, p. 61). The record of the 1980s appears to be one of declining public policy effectiveness – in areas such as worsening social tensions and strains in public order, economic stagnation and continuously high levels of unemployment. A rearrangement of formal powers between central departments and subordinate agencies which is associated with such a broad front of policy decline cannot properly be termed a centralisation of decision-making capacity. Instead central government control has coarsened and narrowed, focusing on a dwindling range of financial aggregates and politically salient *idées fixes*, while the capacity for effective long-run policy co-ordination has simply ebbed away from the state sector as a whole.

6

The State and Civil Liberties

GILLIAN PEELE

British politics in the 1970s and 1980s were increasingly marked by ideological tension as the two major political parties developed their own ways of coping with the crumbling of the Keynesian consensus. Mrs Thatcher's two administrations reflected this change of mood and explicitly sought to promote policies which would stimulate individual initiative, give wider scope for the free market and reduce public expenditure. Although much of the commentary on her government's policies correctly focuses on the success of her economic strategies, it should be recognised that the philosophy of the Thatcherite brand of conservatism had an impact beyond the economic sphere and even beyond the discussion of such cognate themes as the proper level of public spending and the welfare state. In particular, the years of the Thatcher governments saw important constitutional issues arise – sometimes, though by no means in every case, as a result of initiatives and actions taken by the government itself. And the period also saw a heightening of social conflict, which was most dramatically evidenced by a series of riots and racial disturbances in Britain's cities including London. Of course, the events of Toxteth, Brixton and Tottenham could be explained in a number of different ways. Whatever the causes of the violence, it was clear by the end of 1985 that many of the familiar assumptions about policing and law and order in Britain no longer held good, and that these issues had been thrust into the centre of political debate. In this chapter, therefore, the related themes of constitutional reform, civil liberties and law enforcement will be discussed. Not merely are these issues intrinsically important but they provide clues to the less well-aired aspects of Thatcherite conservatism – the approach to the political relationship between the state and the individual.

It might have been expected that the Conservative government elected in 1979 would have displayed a strongly libertarian approach to questions of constitutional reform. After all, part of the justification for reducing the economic functions of the state was to provide increased choice and enhanced personal freedom. Moreover, some of the intellectual advocates of the neo-liberalism on which Mrs Thatcher based her policies – notably Samuel Brittan (Brittan, 1983) – were also libertarian on non-economic issues. However, Mrs Thatcher's governments displayed little interest in promoting measures which would give the individual citizen greater political power within the state. Unlike the Alliance, for example, which emphasised the need for a range of constitutional reforms, the Conservatives treated calls for open government with scepticism and ignored the extent to which they had themselves contributed to the debate about bills of rights and written constitutions while in opposition.

Thatcher's Constitution

The stance taken by Mrs Thatcher's government on questions of constitutional significance seems to have two aspects. On the one hand it is apparent that the Prime Minister is personally rather bored by discussions of institutional reform. Thus, unlike Lord Hailsham, whose writings led many observers to believe that the Conservatives might seek ways of strengthening constitutional protections available to individuals, Mrs Thatcher has put her emphasis on policies, and has made major institutional changes – for example, abolishing the Central Policy Review Staff (CPRS) and the

Civil Service Department – only when policy considerations have pointed in that direction. Open government obtains little sympathy among Thatcherite Conservatives, partly because of this lack of interest in institutional and machinery-of-government questions and partly because of the fear that such a reform would add to the costs of administration.

Yet there is another aspect to Conservative impatience with constitutional and civil libertarian concerns. The brand of Conservatism which proved so popular in 1979 is an amalgam of economic neo-liberalism and populism; and the latter is difficult to reconcile with a high degree of sensitivity to civil liberties. The most visible advocates of civil liberties have not done much, it is true, to make their cause an all-party one either, as was revealed by the debate within the National Council for Civil Liberties (NCCL) about the investigation (which the NCCL announced in August 1984) of police tactics during the miners' strike. It is therefore hardly surprising that those concerned with electoral support should be more anxious to emphasise their hostility to crime and its perpetrators – and support for the police – than to look for ways of giving to those in conflict with the forces of law and order additional protections. Put simply, civil liberties is frequently seen as an issue monopolised by the Left, and one in which there are no votes. Indeed, far from there being any votes in civil liberties issues, it came increasingly to seem that the Conservatives had much to gain from taking a hard stance on law and order. Thus a *Sunday Times* poll published in November 1985 revealed that despite the riots the Conservatives were seen as the party best able to cope with law-and-order issues (see Chapter 2). With the increased salience of these issues on the voters' agenda (in the same poll it had leapt to second place), it seemed very likely that the Conservatives would seek to emphasise their 'hard' image in this policy area. This they duly did by promising in November 1985 to bring in new measures to cope with such problems as soccer hooliganism, drugs, and public order generally. And in early December 1985 it seemed that Mrs Thatcher was considering measures to curb the level of violence in television programmes.

The increased political salience of law-and-order issues was not welcome everywhere. The police themselves expressed concern at the extent to which their position had become subject to political controversy, and there was little doubt that on the Left they had become seen as the instruments of state oppression. Yet, as commentators were quick to point out, the police had to some extent entered the public arena themselves, by their increased willingness to campaign on behalf of a tougher war against crime (which was bound to be identified with the Conservative position), and by their cultivation of new links with the media (Reiner, 1985). And even if they regretted the developments, it was, by 1985, difficult to see how the police could prevent themselves from being caught up in a range of social, industrial and political conflicts when those conflicts erupted into violence.

Conflict, albeit of a more subtle kind than the riots of Toxteth, Brixton and Tottenham, was at the heart of debates about official secrecy and open government. That conflict was a conflict of values – a clash between those who thought the traditional practices of British government should remain as they were and those who felt that a new balance needed to be acknowledged between the government's and the citizen's rights. What was perhaps unusual about the period of the two Thatcher governments was that this conflict manifested itself in the heart of Whitehall among the higher civil servants, whose adherence to traditional constitutional norms had previously been taken as axiomatic.

The deterioration of the relationship between the civil service and the government had many contributing causes. But it led to a series of incidents that in turn further soured a hitherto sacrosanct understanding, as well as generating some spectacular, if controversial, prosecutions under the Official Secrets Act. Coupled with prosecutions that were initiated for espionage, the result was to subject the widely criticised Official Secrets legislation to further hostile scrutiny. And they served to stamp the Thatcher government with a mark of authoritarianism which it did not altogether deserve, as well as to give it a certain mark of incompetence, since at least one of the prosecutions resulted in a 'not guilty' verdict.

The Constitutional Context

Cabinet Government

In trying to understand why the relationships between the government and its civil servants should have been subject to strain in the period after 1979, it is as well to remember that there are few periods in which the orthodox theories of cabinet government and of the constitutional relationships between ministers and officials have worked perfectly. The exigencies of war, the particular pressures of events such as the Suez invasion, and even the special friendships between individual civil servants and their ministers, have as often underlined the flexibility of cabinet government in Britain as they have served to illuminate its character. Nevertheless, two general ideas have customarily been acknowledged to be central to cabinet government. The first idea is that members of the cabinet collectively are responsible to parliament for the government's actions. This in turn means that the central strategic decisions on policy will generally be taken with their knowledge, although the extent to which any particular decision will be discussed in cabinet will obviously vary. Conversely, once a decision in any particular policy area has been taken, it is the decision of the government as a whole, and ministers who cannot accept it and defend it publicly must resign. As the experience of the period of Harold Wilson's premiership showed, the manner in which this convention is interpreted may vary, and perhaps some commentators are correct to emphasise the extent to which Labour cabinets must inevitably be more flexible in this respect than Conservative ones (Peele, 1978). Nevertheless, for the most part, collective responsibility works and stands in sharp contrast to the single executives modelled on the American pattern.

The second idea associated with cabinet government on the Westminster model, is that ministers, not civil servants, are answerable for policy judgements. Any advice given by the civil servant to a minister is therefore confidential, and civil servants may not themselves make public the nature of their conversations with ministers or indeed speak on political topics. When an administration changes – normally as the result of a general election – the ministers of an incoming government may not see the files of its predecessor or know the nature of any advice given to them, although in exceptional circumstances such as the Falklands Inquiry some slight concessions may be made by the prime minister of the day.

The institutional form of cabinet government has, of course, changed markedly since the eighteenth century. Perhaps the most important change in the nature of cabinet government occurred in the middle of the First World War when, in order to prosecute the war more effectively, the old informal methods of conducting business without an agenda or any minutes of decisions were discarded and the cabinet was given a permanent secretariat. Although much of the centralisation associated with Lloyd George fell into disfavour after the war, the structure of cabinet government he established was retained. Thereafter the cabinet secretary not merely serviced the cabinet and its various committees, but also became the custodian of the *mores* of cabinet government and, in particular, the very strong convention of secrecy surrounding cabinet deliberations.

A second major change in the structure of cabinet government which was accelerated after the Second World War was the growing number of committees attached to the cabinet. These committees could be either standing or *ad hoc*, and in some observers' eyes they strengthened prime-ministerial power (Crossman, 1975). In these committees much of the preparatory work for full cabinet was done, and although many of the major committees were composed entirely of government ministers, there were committees which recognised the central role of senior civil servants by combining ministers and civil servants on a single committee. In addition, a number of the more important cabinet committees were 'shadowed' by parallel official committees enabling officials to brief their own ministers on the business in hand.

In addition to these structural changes in the way cabinet government worked, there were, however, less formal ones. Politicians became increasingly prone to reveal their disagreements, despite the image of agreement conveyed by cabinet minutes and required by the traditional doctrine of collective responsibility. The extent to which the Labour governments of 1964 to 1970 saw open disagreements has already been noted, although it is perhaps true to say that the extent to which party factionalism could be accommodated within the cabinet varied with the issues on the agenda. However, it was not merely Labour governments, deeply divided though they were, which needed to accommodate public disagreements between their members. Although Edward Heath seems not to have needed to exert much effort to contain his dissidents (St John Stevas, 1974), Mrs Thatcher has not enjoyed the same luxury. Her early period as prime minister was unusual. Initially she seemed in a minority position in cabinet and was forced to use cabinet committees to circumvent her weakness in full cabinet. In addition, however, she also developed a penchant for taking decisions outside the formal cabinet structure. The personal authority of Thatcher did not, however, prevent her Conservative governments behaving much as their Labour predecessors had done, and it was notable that, by whatever means, the period after 1979 saw much evidence reaching the public of cabinet disagreements over policy, especially policy relating to public expenditure and the incidence of cuts.

'Leaks'

Perhaps it was inevitable that Mrs Thatcher's style of cabinet government should have become more personal as her command over the party grew. However, one reason that was sometimes given for the increasing tendency to take major decisions in small non-official groups outside the cabinet or its committees, was the tendency for these committees and the full cabinet to 'leak'. It must be remembered, of course, that British cabinets have on many previous occasions been liable to find their discussions circulating freely. The details of cabinet proceedings were frequently the subject of instant newspaper reporting and common gossip, even though the cabinet was frequently concerned with such sensitive subjects as military matters (Naylor, 1984). And although Venetia Stanley may not have been the precise equivalent of Sarah Keays, it is obvious that individual ministers will frequently share cabinet matters with their close companions. Indeed the whole question of what constitutes a leak is a difficult one. Even in such obviously undesirable instances as those surrounding Cecil Parkinson's alleged revelation of cabinet discussions to Sarah Keays, the situation is not clear-cut (Keays, 1985). Individual ministers, as the 1978 White Paper, *Reform of the Official Secrets Act*, pointed out, are effectively self-authorising, so that efforts to have them prosecuted for unauthorised disclosures of information are usually useless, although they may of course be subject to political censure for their indiscretions.

Moreover, there is sometimes a strong incentive for ministers to leak information about cabinet discussions. Not merely do ministers often find it convenient to allow the press to have details of how they defended their position if they are in a minority; but it is also the case that all ministers, including the prime minister, may find it useful to manipulate the press by giving information in advance in order to obtain favourable coverage (Cockerell *et al.*, 1984). The conventions of authorised leaking – or 'briefing', as it is more politely known – have become public, and the behaviour of the 'sources close to the prime minister' (a euphemism for the Downing Street press office) or of 'senior government sources' (a euphemism for a cabinet minister) has been further affected by the growing professionalisation of public relations in British politics. To a large extent, therefore, it is not simply that the government of the day has access to information which it is reluctant to share; it is also that it has become increasingly common for that information to be manipulated in order to present the government's position in the most favourable light.

Whatever the distorting effects of the monopoly of some kinds of information by the government and the existence of techniques for exploiting that monopoly, it is, of course, the case that the disclosure of information by ministers, however done, is in a quite different category from that of a disclosure by an official. What the development does perhaps help to explain, though not necessarily to condone, is the increasing pressure that may be put on a civil servant to behave in a manner that is less secretive than in the past, and even the doubts that it may raise in his or her mind about the ethics of a difficult situation. Fundamentally, however, civil servants are in a different position from ministers, because they sign the Official Secrets

Act and risk prosecution as well as their jobs if they disclose information in an unauthorised manner. Yet what was interesting about the period from 1979 was the extent to which information reached the press in a way which suggested it had come not from politicians but from civil servants. Perhaps this tendency was the product of the new style of post-Fulton civil servants – less elitist and drawn from a wider social background, so that he or she was less likely to be wedded to the traditional norms of Whitehall and the culture of secrecy. Perhaps it was the result of the import into Whitehall of more special advisors and temporary civil servants who, being outsiders, could behave in a way which was more overtly political than would be customary for regular officials. Certainly the embarrassment caused by press speculations about policy discussions based on a number of leaks from the so-called 'think-tank' – the Central Policy Review Staff – was suggested as one reason for the abolition of that body after the 1983 election. The number of leaks which occurred at the time of that election prompted the cabinet secretary to circulate all permanent secretaries to warn them of the need for greater vigilance in respect of confidentiality – a warning which may have been prompted by the Prime Minister herself (Pyper, 1985).

It would be a mistake to see concern about breaches of confidentiality as being a feature only of Mrs Thatcher's administrations, however. Indeed, under the Callaghan administration there were some major incidents, as, for example, when *New Society* published cabinet papers relating to the withdrawal of a child benefit. It was clear from the circumstances that either a member of the cabinet or one of the limited number of people with access to cabinet documents had revealed the information to the editor of *New Society*, and the disclosure caused an internal cabinet inquiry and a police investigation, although both of these proved inconclusive. What does seem clear is that the Thatcher brand of Conservatism has been sufficiently radical to seem to challenge very large numbers of vested interests and to threaten a number of existing programmes. It would be surprising therefore if a few officials, who have, after all, spent a period of their careers developing programmes which were suddenly questioned, did not use methods of bureaucratic infighting which they might once have rejected as unprofessional.

Open Government?

By the end of 1985 it seemed difficult to restore the style of confidential decision-making preferred by Whitehall's politicians. Even in the absence of a Freedom of Information Act, civil servants, journalists, lobbyists and even some politicians, as well as the public, had grown accustomed to having greater access to the details of the policy process. Moreover, there was an increasingly powerful intellectual case being made for a more open style of government, and for legislation to allow greater access to the wealth of information held by government and public bodies. What was remarkable about this case after 1979 was the way in which some former civil servants – such as Sir Douglas Wass and Sir Patrick Nairne – joined the campaign. The case of Sir Douglas Wass, a former Permanent Secretary at the Treasury, was especially interesting, since he developed his views into a coherent critique of British government as a whole (Wass, 1984). And when the 1984 campaign for freedom of information was launched, to campaign for a general statutory right of access to official information and the reform of the Official Secrets Act (by restricting its scope to the protection of material which if disclosed might be injurious to national security), Sir Douglas became an adviser to it. Equally significant was the support lent by the civil service unions – including the First Division Association which represents the most senior officials – to that public campaign.

The public identification of retired senior mandarins with a campaign which, strictly speaking, challenged one of the central features of the role of the civil service in the constitutional structure, was itself remarkable. But it complemented an already noticeable trend of civil servants after retirement to comment frankly on the policies with which they had been associated in government. Thus Sir Leo Pliatzky, in his book *Getting and Spending*, was quite outspoken about the conduct of economic policy; and Lord Bancroft (the former Permanent Secretary at the Civil Service Department) and Lord Croham (the former head of the Treasury) all contributed freely to such sensitive debates in the House of Lords as the Ponting prosecution and general civil service matters. Similarly Lord Allen of Abbeydale often speaks in the House of Lords on Home Office policy – frequently in a manner which is critical of it. Perhaps this was a by-product of the breach in the *Crossman Diaries* of the convention about criticising named civil servants; more probably it was simply another reflection of the changed climate of decision-making, a climate which encouraged greater openness and discussion about the internal affairs of government. Television has also played a role in creating this climate, so that not merely are the Lords now televised, but there was even a television filming for a programme, 'Inside No. 10', of the cabinet assembled for a meeting – a programme which revealed the Prime Minister's Parliamentary Private Secretary, Michael Allison, in attendance, a departure from precedent made possible by the fact that Mr Allison is a privy councillor.

The trend towards more open government was one which Parliament had itself participated in, by its first experiments in the late 1960s with specialist select committees and its subsequent determination in 1979 to create a permanent and comprehensive select committee system which could aid Parliament's investigations of the executive. Civil servants had been profoundly affected by these developments. Not merely were they likely to have to appear before select committees to explain matters of departmental structure, they were also likely to be made aware of the greater oversight powers operating in relation to government as a whole. Thus the period 1964 to 1984 saw an expansion of knowledge about the inner recesses of government and its decision-making processes, as a result of a range of reports from select committees and individuals such as the Parliamentary Commissioner for Administration. Taken together they shed light on the decision-making process but they also underline the extent to which civil servants in the 1980s were working in a very different world from that in which Maurice Hankey operated.

There was nevertheless still a very strong disposition towards secrecy in Whitehall, although it would be unfair to attribute it to the attitude of officials. Indeed the course of the debate about open government in the 1970s suggests that the real objections to experiments with freedom-of-information legislation tended to come from ministers. It was not, however, simply that many of them were reluctant to embark on a potentially costly and time-consuming exercise of making government information available to the public; Labour and Conservative governments alike found themselves frustrated in their efforts to reform the Official Secrets Act, which after two decades of criticism remains on the statute book as a blunderbuss for use against unauthorised disclosures of government information, however trivial, and a reminder of the dominant ethos of Whitehall.

The Official Secrets Act and the Difficulty of Reform

The Official Secrets Act has long invited criticism. Apart from the assumptions which it encapsulates about the proper balance between the public's right to know and the government's right to withhold information, the legislation is open to a number of other objections. The section which is used to deal with espionage (section 1) is relatively uncontroversial, although on occasion – as in the 1978 so-called A.B.C. trial – charges may be pressed under this head which later have to be withdrawn (*The Times*, 31 October 1978). It is, however, section 2 which has been the source of most controversy. The clause which was not debated at all when the legislation was passed during the height of the Agadir crisis in 1911 is extremely broad, and one authority has estimated that it effectively created over 2,000 separate offences (Street and Brazier, 1983). Its catch-all nature protects information that is highly sensitive and information that is trivial from 'unauthorised' disclosure. The element of intent which is required to commit an offence under section 2 is not clear, and certainly section 2(2) seems to create a crime of strict liability. Although it is a feature of the English legal system that there should be as much certainty as possible about what kinds of behaviour will lead to criminal prosecution, the Official Secrets Act of 1911 makes such prediction very difficult.

The 1911 legislation was passed to strengthen the 1889

legislation which had typically been used to prevent leaks by civil servants to the press. The Official Secrets Act of 1911 was reinforced by various Defence of the Realm provisions introduced during the First World War and consolidated during the 1920 Official Secrets Act. Apart from a small amendment in 1939 it remains virtually unaltered today, although the Franks Committee in 1972 recommended major reform and a government White Paper in 1978 envisaged legislation to cope with its most restrictive features. An attempt to pass a Protection of Information Bill by the Conservative government elected in 1979 foundered, because that bill seemed to many to be worse than the existing legislation (and was certainly more restrictive in relation to the coverage of security issues), and because at the time of its introduction into the House of Lords the Blunt affair broke as a result of the publication of Andrew Boyle's *Climate of Treason* (James, 1982; Wilson, 1984).

Information within the civil service is routinely classified on the basis of its sensitivity from 'restricted', to 'confidential', 'secret' and 'top secret'. The important point to notice about section 2 of the Act, however, is that no matter how information is classified it is still an offence to disclose it if the disclosure has not been authorised. The nature of authorisation is itself, as has been pointed out, ambiguous and ministers are generally regarded as self-authorising.

Dissatisfaction with the Official Secrets Act increased during the 1960s, but it was perhaps the trial of Jonathan Aitken and other *Sunday Telegraph* journalists in 1971 which marked a watershed in the Act's life. The *Sunday Telegraph* had published information about the Nigerian civil war which not merely irritated the Nigerian government but revealed the extent to which the government had been misleading Parliament about the scale of its aid to the federal government. In the trial, the judge, Mr Justice Caulfield, indicated that that section of the Official Secrets Act ought to be 'pensioned off'; two months later the Franks Committee was set up to examine that section of the legislation. No action was taken on the basis of the Franks Report but there was a 'take note' debate in 1973. In the October 1974 general election, however, Labour came out with a strong commitment to both reform of the Official Secrets Act and open government – commitments which were later decoupled by James Callaghan urging a greater degree of openness on the part of the civil service (through such minimal measures as the Croham directive) and by a White Paper, *Reform of Section 2 of the Official Secrets Act* (1978) which built on the Franks proposals.

The Franks Committee declared that section 2 was a 'mess'. They recommended that it be repealed and replaced by a new Official Information Act which would have narrower and more specific provisions. Briefly, the new act, as recommended by Franks, would have protected classified information in the areas of defence and internal security, foreign relations and matters pertaining to currency and the reserves – issues which became less sensitive once the fixed exchange rate was abandoned. Information, Franks thought, should be given legal protection only if *correctly* classified as 'secret' or 'defence-confidential'. It was to be the responsibility of the minister to ensure before any prosecution that the information in question had been correctly classified and to furnish the court with a certificate to that effect. The certificate would be conclusive. In addition the new legislation envisaged by Franks was to apply to all of certain categories of document, the most important of which was cabinet documents, including cabinet committee papers.

This approach which was broadly followed in the Labour government's White Paper of 1978 was changed quite radically in the Conservative version of the reform which was abandoned in 1979. Although, as many critics have pointed out (James, 1982), the list of categories covered by the newly labelled Protection of Information Bill was in some respects less restrictive than Labour's 1978 recommendations; in other aspects it was more draconian even than the 1911 Act. In particular, all discussion of the security services and domestic surveillance was to be covered by the legislation – a provision which the press was quick to point out would have meant that Boyle's book relating to the Blunt affair could not have been published.

With the withdrawal of the Conservative bill, it became apparent that the discredited Official Secrets Act would remain the only weapon available to the government to control leaks and espionage for some time to come. What was not so apparent then was the extent to which the government would have cause to use it. Although it was plain that the new Tory government would take a strong line in relation to espionage and spying, strictly defined, it was the deteriorating relationship with the civil service which brought the Official Secrets Act to centre stage again.

The Conservative Government and the Civil Service

Abolition of the Civil Service Department

If a number of factors were operating over the 1970s to push the civil service and its views into the open, the Thatcher administration contrived by its policies and attitudes to make officials increasingly resentful of their treatment. When Mrs Thatcher took office in 1979, it was obvious that the goal of reducing public expenditure and of altering the balance between the public and private sectors of the economy would have profound implications for the civil service. At the very least its size would provide the rough measure of how far the Conservatives were succeeding in their goal of reducing the extent of the public sector. Thus to the threat to civil service morale from the explicit criticism of public endeavour was added the tangible threat to civil service jobs, pay and prestige. The change in attitude to the civil service was made even more apparent when after more than a decade's debate about the proper disposition of responsibilities for civil service matters, Mrs Thatcher finally abolished the Civil Service Department, and divided its functions between the Treasury and the Office of Management and Personnel which was absorbed into the cabinet office.

When the Fulton Committee had reported in 1968 it had seemed proper that there should be a department charged with the special task of promoting the interests of the government's personnel and dealing with questions of manpower. In the changed climate of the 1980s, not only was it thought illogical to separate manpower and financial functions but it was also thought undesirable to continue a separate department which could be seen, however unfairly, as a lobby for civil service interests. The abolition of a separate Civil Service Department had a symbolic significance therefore, both in terms of rejecting the Fulton analysis and as a clear indication that the Thatcher administration was impatient with any civil service claims for special treatment. To this was added a new determination to apply outside standards of efficiency to government, which again tended to undermine morale to the extent that it underlined the government's belief that existing practices were inferior to those of the private sector. The Prime Minister also displayed a strong interest in the promotion and deployment of individual civil servants. This she was constitutionally entitled to do, but the rapid promotion of men such as Peter Middleton and Clive Whitmore threatened the established assumptions of the civil service as being a neutral body, where career would develop along well-established lines and where it would be difficult to mark an individual permanent secretary out as the protégé of a particular prime minister. It was not that the individuals selected for rapid promotion by Mrs Thatcher were identifiably partisan; simply that they reflected her preference for a certain style of mandarin – ones who displayed a commitment to government objectives rather than a detached neutrality. Unfortunately the public discussion of these individuals meant that they in a sense became 'marked men' and might prove difficult to use by governments of a different political persuasion (Beloff and Peele, 1985).

In addition to rewarding committed civil servants, Mrs Thatcher antagonised members of the regular civil service by her use of outside appointments. One such appointment was that of Peter Levene in March 1985 to head the Ministry of Defence Arms Procurement Agency. The government was perhaps justified in making this appointment by Levene's ability to bargain with arms contractors; but critics pointed to the fact that his salary was very large and that the appointment was illegal under the 1982 Orders in Council. The argument that someone with an inside knowledge of the arms industry could better assess procurement needs was countered by the argument that Levene himself had a direct interest in the field.

Banning Union Membership at GCHQ

All of these factors would have produced an unusual tension between an elected government and the civil service. However, it was the decision to abolish union membership at GCHQ Cheltenham which caused a storm of hostility and brought a simmering resentment into the open.

The Government Communications Headquarters at Cheltenham is a part of the Foreign Office. It had been established in 1947 but its security role was not generally known about until May 1983 when the Prime Minister acknowledged the importance of its function in answering a Parliamentary Question. (Parliamentary Questions may be put down on security matters but they need not be answered.) On 25 January 1985 it was announced that trade-union membership would henceforth be prohibited to GCHQ employees, and that staff would be given a period in which to decide whether to seek transfer elsewhere (which would not necessarily be possible) or to renounce their union membership and remain at Cheltenham. Those who opted for the latter course were to be given an *ex gratia* payment for the removal of their trade-union membership rights.

The reasons for the move – which was taken without any prior discussion with the civil service unions – were closely related to the dangers of having GCHQ tied to what was perceived to be an increasingly militant union. In 1981 there had been major industrial action by the civil service which had included the GCHQ and which had led to concern being expressed by the United States about the vulnerability of Britain's security establishments to industrial pressure. The government had also come to think it undesirable on security grounds that individuals employed at GCHQ should have access to industrial tribunals. Any organisation to represent the interests of employees, the government felt, should be based solely within GCHQ and its ties with the wider civil service unions should be severed.

Naturally, the civil service unions did not take the same view of the matter as the government. They sought an injunction to restrain the Prime Minister from altering the terms of employment at Cheltenham. Initially they were successful because, although Mr Justice Glidewell in the High Court recognised the right of the Prime Minister to make changes in the civil service, he held that the lack of prior consultation in the matter constituted a breach of natural justice. On appeal, the decisions of the Court of Appeal and of the House of Lords overturned the decision. The Court of Appeal based its decision on the fact that national security was involved, and there, it was argued, government ministers were the sole judge of what should be done. The House of Lords agreed but based its decision on slightly narrower grounds. What was of interest throughout the case was the court's apparent willingness to review the use of prerogative powers as opposed to statutory ones (*The Council of Civil Service Unions & others* v. *The Minister for the Civil Service* [1984] 3 W.L.R. 1174).

The GCHQ case illustrated the tension between the government's attitudes and the assumptions of the representative bodies of the civil service. However, even more controversy was generated by the prosecution of an individual civil servant, Clive Ponting. His case brought into focus the extent to which attitudes within the civil service might have changed, the extent of the alienation of some civil servants from the government, and the weaknesses of the Official Secrets Act.

The Ponting Case

Clive Ponting was not the only civil servant to be disciplined because of his disagreements with government policy. During the 1979–85 period there were at least three major controversies over officials who revealed sensitive information to newspapers or, in Ponting's case, to an MP. All three civil servants were willing to make this public gesture of dissatisfaction with government policy, or with the way the government was handling the conduct of public business, because they believed that the ethics of their situation demanded it. Thus when Ian Willmore, an administrative trainee in the Department of Employment, thought he had discovered an irregular discussion between a senior official and the Master of the Rolls, Lord Donaldson, he leaked a note of the meeting to *Time Out*. Willmore admitted what he had done and resigned; the government did not prosecute him. However, in the more obviously security-related case of Sarah Tisdall, a junior Foreign Office clerk, who sent the *Guardian* details of the timing of the delivery of cruise missiles to Britain and the manner in which the announcement of their arrival would be handled in Parliament, there was a successful prosecution under the Official Secrets Act and Sarah Tisdall was sent to prison.

The background to the Ponting affair was the Falklands war. In that war an Argentine ship – the *General Belgrano* – had been sunk with the loss of many lives. It was argued that the decision to sink the ship had been taken in breach of the announced rules of engagement and was an unnecessary act of violence. The domestic significance of the episode was the allegation that the government had attempted to conceal the full details of the episode from Parliament and the public. Clive Ponting passed documents (which were not in fact classified) to Tam Dalyell, who sent them to the Select Committee on Foreign Affairs which was then investigating the Belgrano Affair. Ponting claimed that the government had deliberately tried to mislead Parliament and the public over the matter. The government's response was to prosecute Ponting under s.2(1)(*a*) of the Official Secrets Act.

The prosecution itself raised a number of interesting questions. To what extent did putting a document in the hands of an MP constitute an offence under the Act? To what extent was the Attorney General (who must consent to a prosecution under the Official Secrets Act) acting independently in this case? There has always been an ambiguity in the Attorney General's position, since he is simultaneously a member of the government, answerable to Parliament for his decisions, and a law officer of the Crown, who is in theory removed from partisan considerations. It has generally been assumed – at least since the Campbell affair of 1924 – that, while an Attorney General may consult with his colleagues, he may not be directed by them when deciding whether to prosecute or to enter a *nolle prosequi* order (Marshall, 1984). Moreover, in reaching a decision as to whether to prosecute, the Attorney General may take into account the public interest but not the interest of the government in a party sense. In the Ponting affair, it was clear that what was primarily at stake was not the national interest but the government's desire not to be embarrassed by what looked very like a cover-up. As critics were quick to point out, the circumstances suggested either that the Attorney General had misused his discretion or that he had been influenced improperly by other members of the Cabinet in his decision to prosecute Ponting (Ponting, 1985; Norton Taylor, 1985).

The distinction between the interests of the government of the day and the public interest loomed large in the Ponting trial. Ponting indeed made it a defence that he conceived a broader constitutional interest to be served by telling the truth, rather than concealing it at the behest of ministers. However, the judge made it absolutely plain that the definition of public interest had to be determined by the government and not by individual civil servants. Given that Ponting admitted his action in sending his documents to Dalyell, that ruling should have meant that Ponting was convicted. However, such was the sympathetic publicity surrounding the case that the jury, rejecting the judge's ruling, returned a 'not guilty' verdict.

The government was, of course, dismayed by the 'not guilty' verdict in Ponting's case. However, its response was to issue through Sir Robert Armstrong, the Cabinet Secretary, a statement which was unyielding in relation to the issues thrown up by the trial (Norton Taylor, 1985). The 'Notes of Guidance on the Duties and Responsibilities of Civil Servants to Ministers' was unequivocal about the fact that civil servants were servants of the crown, and for all practical purposes this meant the government of the day. Sir Robert emphasised that officials had no constitutional responsibilities independently of their ministers, although some commentators have suggested that this is not strictly correct given that permanent secretaries are the accounting officers answerable to Parliament for their departments. Sir Robert's 'Notes' suggested that when a civil servant found himself in the situation faced by Ponting he should either insert a memorandum of dissent to written instructions or speak privately to the cabinet secretary about the matter.

Although in many ways an incident which could be quickly forgotten, the Ponting case tied into a number of general concerns which will continue to figure in British political debate. Obviously it fuelled the fire of those who had expressed doubts all along about the Falklands war, although the committee established under Lord Franks was convinced that the invasion could not have been foreseen and that the government could not have done anything to prevent its occurrence. The affair also underlined the extent to which the difficulties associated with the Official Secrets Act made juries unwilling to convict, especially if they suspected a political motivation in a prosecution. Most of all, however, the Ponting case again demonstrated how unwise it was for a government to ignore the issues of reforming its Official Secrets legislation in the changed climate on open government.

The Control of the Security Services

The use of the discredited section 2 of the Official Secrets Act highlighted the changing attitudes towards open government in Britain. Yet the other aspect of the legislation – its provisions for dealing with what might more normally be thought of as espionage – also attracted attention to this period. Linked to the attention given to the methods of dealing with threats to national security was concern about the efficiency and accountability of the security services themselves. Indeed one reason why some observers thought it unwise to press for reform of any aspect of the Official Secrets Act was the experience of a number of security lapses. The cases of Willmore, Tisdall and Ponting were not of course cases of espionage, although the Tisdall case underlined how easily sensitive material could escape from a government department. There had been sufficient cases over the years to convince the critics of Britain's approach to security issues that there should be some stricter scrutiny of the way these services – especially MI5 and MI6 – worked. Both the counter-intelligence service (MI5) and Britain's overseas intelligence service (MI6) had their origins in the work of a sub-committee of the Committee of Imperial Defence before the First World War, although their true potential had developed enormously during that war (Andrew, 1985). Throughout their history, however, there had been breaches of security, and after the

Second World War there were revelations that they had been infiltrated by communists. The Blunt affair (which contributed to the government's decision to withdraw its reform of the Official Secrets Act in 1979) hit the Thatcher government at the begining of its period of office; the second Thatcher government saw a major scandal when a middle-ranking MI5 officer, Michael Bettaney, was sentenced to 23 years' imprisonment on espionage charges on 18 April 1984. The case raised important issues, because it became clear that Bettaney's heavy drinking had earlier suggested that he was not suitable for employment by the security services and indeed should have caused him to be stripped of his security clearance at a much earlier stage. Bettaney's conviction prompted Mrs Thatcher to refer the case and the defects in the system that it revealed to the Security Commission.

Lack of Parliamentary Control

Reform of security practices is not, however, an easy matter. Even the constitutional position with respect to control of the services is obscure, and there are fluctuations in the degree to which prime ministers will allow discussion of their performance. Direct responsibility for the security services is the Home Secretary's, although MI5 and the Special Branch are in a slightly different position, since MI5 is departmentally accountable to the Home Secretary, and the Special Branch is responsible to the Home Secretary by virtue of his role as police authority for the Metropolitan Police Force. Additionally, prime ministers have claimed a special personal responsibility for security sometimes stating that the formal responsibility is theirs (Marshall, 1984).

Parliament's role in the control of the security services is even more confusing. The Liaison Committee of the House of Commons, in its first report on the new select-committee system, suggested that the budgets of the security services could be reviewed by a select committee, but the point was controversial. In March 1985, Mrs Thatcher again vetoed the appearance of any member of the security services before select committees (*The Times*, 26 March 1985). In the United States, concern with the activities of such agencies as the CIA during the 1970s had caused Congress to establish machinery for securing some oversight of it, although inevitably much of the questioning has had to be in camera. However, it seems unlikely that any such change is likely to occur in Britain.

Instead, Mrs Thatcher, responding to criticism in the wake of the Bettaney affair, concentrated on the management and efficiency aspects of the security services rather than on their accountability. After a report on the services by Lord Bridges, the government announced on 7 May 1985 that there would be a large-scale reorganisation of MI5. This was designed to produce an improvement in the procedures for recruiting and vetting its officers, and to improve morale, which had been lowered by the Bettaney conviction and the subsequent discussion of MI5 in the press. Thus the government hoped that the services would be able to reform themselves from within, and would create a new style of service which would be more open, more self-critical and more aware of personnel issues than in the past.

This approach did not, of course, satisfy everyone. Apart from the fact that it was unclear how more openness could be reconciled with the nature of the security system, a number of critics (some of them from within the services) alleged that the problem of subversion and personal unsuitability for security work was rarely brought into the open, for fear of censure. In the House of Commons even some conservatives expressed a feeling that there was need for greater public scrutiny of security services, and that the traditional approach of ministers to questions on security matters – which is usually neither to confirm nor deny allegations – was inadequate.

The general unhappiness with the parliamentary control of security matters was not, however, without effect. It did, for instance, mean that both the government and the public were more sensitive to the problems surrounding them, and it also meant that when related issues such as telephone-tapping and mail-interception were discussed the government faced a more critical audience than it had in the past.

The Police, Law and Public Order

The controversies surrounding the Official Secrets Act, the duties of the civil servant and the role of Britain's security agencies highlighted some of the tensions in Conservative philosophy between authoritarianism and the protection of civil liberties. A far more dramatic development over the years of Thatcher's premiership was, however, the spread of violence in British society. This violence manifested itself in a number of different ways. There was the eruption of racial tensions in the inner-city riots of 1981, 1982 and 1985. There was the spectacle of vicious violence at football matches at home and abroad. There was violence in the context of strikes and demonstrations. And there was a rise in the incidence of such violent crimes as robbery, murder, muggings and rape.

The Conservatives had come to power with a clear statement of intent to put additional money into the fight against crime. As the Conservative manifesto declared, because the number of crimes in England and Wales had risen half as much again since 1973 when the Tories were last in government, radical measures were required: 'The next Conservative government will spend more on fighting crime *even while we economise elsewhere*' (Conservative Manifesto 1979 (my italics)).

That commitment – which seemed to some observers to be a direct response to the increasingly militant stance by the police themselves on these issues (Reiner, 1985) – nevertheless had its effect. As one informed commentator noted, by the middle of 1981 the government's implementation of the Edmund–Davies Report on policy pay and conditions of service had vastly improved recruitment and morale: 'Wastage of experienced police officers had been stemmed, police recruiting both in quality and quantity was at its highest since the years of the Depression in the 1930s' (Alderson, 1985).

In addition to investing extra money, the Conservatives committed themselves to revising the law on police powers and criminal procedure, an area which many thought reflected a balance which unduly handicapped the police in the fight against crime. The first attempt to achieve a new balance in this area in a Police and Criminal Evidence Bill was lost with the general election of 1983. But the bill as introduced by the government had caused a storm of controversy because of its grant to the police of additional powers of search and seizure. The second version of the bill – which eventually became the Police and Criminal Evidence Act of 1984 – was far less controversial, but for many it was shorn of the added powers desired by the police in the new environment.

To some extent, of course, the emergence of new kinds of threats to the public order had generated their own responses from the police. The recurrence of riots on the mainland of Britain underlined weaknesses of strategy and organisation and led to the consideration of tactics that had hitherto been confined to Northern Ireland. The use of riot shields became standard practice; rubber bullets, CS gas and protective vehicles were made available by the Home Office, although their use was carefully controlled.

The Miners' Strike

The miners' strike of 1984–5 underlined the extent to which the Conservative governments of Mrs Thatcher were very different in character from previous Tory administrations. In the first place they were prepared to be very tough indeed in response to threats of industrial disruption, partly because they believed the miners' leader, Arthur Scargill, was determined to use his industrial muscle for overtly political purposes, and partly because they had learned the lessons of earlier strikes, especially those of the Heath period. The experience of the Saltley episode in 1972 and the 1974 miners' strike had left deep scars on the Conservatives, who were determined not to be caught again in the same trap, and there were careful preparations for coping with any sudden disruption of coal.

Although the miners were ultimately defeated, the strike left a bitter legacy to British policing. Because of the violent nature of much of the picketing, the police inevitably became involved in the attempt to keep the striking miners from intimidating those who wished to return to work. The police – whose operations during the strike were described by one union leader as being 'more controversial than any previously seen in Britain' (McGahey, 1985) – were not seen as the neutral upholders of a just law but rather as active participants in a political struggle. The operations undertaken by the police also gave rise to criticism in some quarters. The role of the National Reporting Centre – from which the police response to the strike was co-ordinated – seemed to suggest that a fresh turn had been taken in British policing (Kettle, 1985). Here, in New Scotland Yard, there was administered an efficient deployment of the aid and assistance which all authorities give to police authorities that find they are in special need of reinforcements. The problem with what, on one view, looked like an eminently sensible way of handling a dispute, which was concentrated in some areas and not spread evenly across the country, was that it conjured up the vision of a national police force rather than a purely local one. Thus it seemed that the National Reporting Centre was the instrument through which local structure of accountability and control could be by-passed and the wishes of the Home Secretary implemented.

The National Reporting Centre

The National Reporting Centre – which had been established in 1972 – and had been used to handle both controversial episodes like strikes and riots and less contentious ones such as the papal visit to Britain – was in theory simply monitoring the availability of police in particular regions and was not a conduit for the political direction of the police from the centre. However, it is difficult in times of stress to maintain such a distinction, especially given another peculiar feature of the situation, the autonomy of chief constables from local control in operational matters. Thus it seems likely that even without any concerted direction from the Home Secretary, chief constables would have taken the view that it was their duty, once the facts were known, to provide forces to aid authorities with manpower difficulties.

The complex pattern of inter-force policing during the miner's strike raised problems beyond the theoretical objections to what looked like the *de facto* emergence of a national police force. Chief constables might take decisions which involved expenditure, and this was especially unpopular in those authorities where sympathy lay with the strikers rather than the miners who wished to return to work. Some money was recouped from the central government but the question of who was to pay the bill for a massive police operation was an awkward one. Although the number of local police authorities that reacted by cutting the police budgets elsewhere was small, the episode to many underlined the weakness of the local authority in the tripartite relationship between local authorities, national authorities and the chief constable. It also underlined the difficult of budgeting in situations where a dispute was essentially national rather than local (Spencer, 1985).

It also became apparent during the miners' strike that Britain employed specialised police units for riot control as well as mobile units which could be deployed around the country at will when needed. The mobile units made their mark in the first days of the miners' strike, when it became apparent that 8,000 men could be sent to the major trouble spots of Leicestershire, Derbyshire, North Wales, Nottingham and Warwickshire and that 1,000 of them had been mobilised within four hours (Lloyd, 1985). In addition, however, it has become clear that the idea encapsulated in the Metropolitan Police Special Patrol Group – a specialised force with riot training – has not been abandoned but has become common currency within the police. Whether or not the experience of inner-city riots justifies such a development remains a moot point. What is certain is that these developments have occurred without public debate and scrutiny and in contravention of the long-standing British hostility to anything analogous to the French CRS, which has its own distinctive ethos.

The police attracted criticism during the strike for their tactics, although it was also true that there was a high level of violence on the strikers' side. Not merely truncheons and riot gear but also horses and dogs were used to control crowds – a tactic which may have been responsible for the South Yorkshire Police Authority's decision to cut its financial costs by selling its horses and dogs (Spencer, 1985). Certainly the image of the police at the end of the miners' strikes in the areas affected by the coal dispute was very different from that of the bobby on the beat, and it could well be argued that the whole episode added militant unionists to the range of groups who had been alienated from the police force.

Even without the miners' strike, however, the foundations had been laid for such a development. The emphasis placed by the Conservatives on law and order, and the series of doubts about the role of the police in a divided society, would probably have prompted the Left to question the existing structures for control of police policy and to demand more democratic accountability in relation to police matters. In fact the ambiguities inherent in the tripartite relationship between the Home Secretary, the local police authority and the chief constable were in need of clarification in the early 1960s, but the 1964 Police Act did not provide any such enlightenment. Moreover, developments such as the standardisation of equipment between forces, and the growing role of police associations such as ACPO (the Association of Chief Police Officers), were changing the reality of policing however much the formal position remained one of essentially local forces. In 1979 Jack Straw, a Labour MP, introduced a private members bill – the Police Authorities (Powers) Bill – into the House of Commons. This was designed to remove the non-elective element in the police committees (which are composed of elected councillors and non-elected magistrates in a ratio of two to one) and to allow these committees to decide the general policing policies for their areas, to supervise complaints procedures and to have the 'power of appointment, promotion and dismissal extended downwards to chief superintendents and superintendents' (Tredinnick, 1985). The Bill failed, but it was an early indication of Labour's new concern with strengthening the local and democratic element in the structure of police accountability. Labour, in its annual conference in 1981, outlined a 'ten point plan' for the police, which involved, *inter alia*, the abolition of magistrates seats on police committees, giving police committees power to approve police policies and to appoint senior officers as well as to control resources, manpower, training and discipline and to assist in the development of police–community relations. This 'ten point plan' became official Labour Policy and was included in the manifesto of 1983.

The Special Patrol Group

The Greater London Council – even though not itself a police authority – was also at the forefront of the debate about how much control there should be of police practices. After Labour returned to power in London in 1981, Ken Livingstone made explicit the significance of police-related issues in consolidating Labour's appeal when he said: 'The task surely is to break the Metropolitan Police Force as at present constituted and disband the SPG [Special Patrol Group]' (London Labour Briefing 'Special Victory Issue' editorial, GLC, June 1981). The GLC established a Police Support Unit to monitor police behaviour and took an active part in campaigning for a reduction in the operational autonomy of chief constables. It also founded a journal, *Policing London*, whose tone was extremely hostile to the police. In part, such action may have been justified by the feelings of frustration engendered by the fact that, because the Home Secretary is the police authority for the London area and answerable to Parliament for his actions in that respect, Londoners could have no local input to their policing arrangements, or find any easy way to express their feelings about a force which seemed biased against such minorities as radicals, homosexuals, the young and ethnic groups. Although there was an effort after the Scarman report to set up local liaison committees within London, the general impression of the police in London for many of its inhabitants was difficult to change – and the impression was hardly helped by a very detailed study of police attitudes which the Policy Studies Institute published in 1983 (Smith, D., 1983).

Emphasis on Law and Order

The source of this demand on the Left for more democratic policing can be explained on a number of levels. One factor was certainly the heightened emphasis on law and order by the Conservatives and by the very visible intervention by the police in the conflicts of the period. Another was the view, especially strong on the hard Left, that the police were the agency of the state, and while the police provided some physical protection for everyone, they were no longer the neutral enforcers of law in a consensual society. Finally there was a belief that it was Conservative policies which were generating the long-term causes of crime and unrest – poverty, unemployment, homelessness, and racial discrimination – and that as these policies produced more crime, the natural response of the government would be more repression. Thus for some on the Left there was a natural correlation between the new Conservatism's reliance on the iron laws of the market and its need for greater reliance on the forces of law and order.

From the Conservative point of view such abstract analysis was hardly convincing. Crime was seen as a major problem which threatened the security of all people and something which demanded tough solutions. It was also sensed to be an area where the attitudes of ordinary voters differed from that of the party elites, at least on the Left. Whether or not Mrs Thatcher sees law and order as a potential 'Falklands factor' – which would enable her to compensate for any loss of Tory popularity between now and the next election – she did think it sufficiently important to give it a central position in the Queen's Speech for the 1985–6 session of Parliament. There then followed a new Public Order Bill (which had been foreshadowed in an earlier review of the 1936 Public Order Act). It was designed to provide new methods of coping with public order, including football hooliganism. The Bill as presented to Parliament in December 1985 introduced not merely a new offence of 'riot' but provided for a maximum sentence of life imprisonment for this offence. Among other provisions the Bill created a new offence of using words which are threatening, insulting, or disorderly where there is a reasonable cause to believe that the words or behaviour are likely to harass, alarm, or distress another. The vagueness of these provisions, and the stiff sentences that breaches of the new Public Order Laws carry, are unlikely to endear them to those concerned with civil liberties: but they reflect a growing concern with violence in British society – a concern which is already being reflected in the sentences being imposed by the Courts under the existing law.

It remains to be seen how this measure will look when it has completed its passage through Parliament. But two things seem clear from Mrs Thatcher's approach. The first is that the strong law-and-order stance of her first administration is going to be even more explicit in her second – as indeed the 1983 manifesto suggested. The second is that whatever happens as a result of the *next* election, the style of British policing and the question of policing generally will have become politically salient in a way which will be difficult to reverse. Ultimately, of course, the debate also underlines the difficulties of those who seek to protect individuals and minorities in a majoritarian system where there are no constraints such as might be imposed by a bill of rights. It is perhaps one of the curiosities of the British political culture that despite the heightened debate about law and order in the period since 1979, the protagonists have failed to address the broader constitutional implications of their concerns.

Section III

Public Policy

7

Economic Policy

PAUL MOSLEY

'All political history', Harold Wilson argued in 1968 after four years' experience as Prime Minister, 'demonstrates that the standing of a Government and its ability to hold the confidence of the electorate at a General Election depends on the success of its economic policy.' This may or may not be true (and some recent surveys suggest that public order and welfare may count for at least as much as economics); what is important is that it is *believed*. Every British prime minister since the war has grasped with enthusiasm the opportunities for state involvement in the economy which Keynesian economics offered, not least Margaret Thatcher, who beneath an appearance of *laissez-faire* economics has placed herself more centrally within the mechanism of economic policy-making than any predecessor. In this chapter we will first consider how the mechanism of economic policy-making works. In subsequent sections we consider the success of attempts of governments to control the economy, and the implications, and probable future, of the 'monetarist experiment'.

The Policy-Making Structure

In Great Britain, the legislature, in the shape of Parliament, has very limited power to initiate tax, public spending, money supply or interest-rate changes. Hence, political influences on macro-economic policy are confined to those that result from the governing party's pressures on members of Cabinet. The governing party will have a hard job decoding the wishes of the people regarding macro-economic policy; the message from the opinion polls is not easy to read aright. And, in any case, there is no compelling reason for a governing party with a secure majority to pay any attention to them. We conclude that, except during the immediate run-up to an election, the executive machinery of government has considerable freedom of action in determining what macro-economic policy should be.

By 'the executive machinery of government' we mean, in Great Britain, agencies operating on two separate levels. On the upper level, there is the Cabinet of selected government ministers, and the Treasury, who between them determine what the broad thrust of macro-economic policy will be in any given time period. On the lower level, are those agencies which manipulate the levers of economic policy with which we are here concerned. The Inland Revenue obtains money from the public through taxes, the Bank of England borrows and prints money on behalf of the government, and the spending ministries – education, health, transport, overseas aid, and so on – spend it.

It is, however, too simple to see these three 'second-level' agencies simply as bodies which carry out the Treasury's instructions. Each of them, in fact, has a different sort of bargaining relationship with the Treasury.

The *Bank of England* (founded in 1694 but only brought under state ownership in 1946) is, among other functions, the government's banker; when the government runs a deficit the Bank borrows on its behalf. The size of this deficit, or public sector borrowing requirement (PSBR) is generally an important element in the growth of the money supply, since not all of it can be borrowed from non-bank sources. The key decision which the Bank has to take within the constraint already imposed by the size of the PSBR, is what level of interest rates to aim at (which means abandoning control over the money supply), or alternatively what level of monetary growth to aim at (which means abandoning control over interest rates). In the 1950s and 1960s, broadly, the Bank followed the first strategy, and since the early 1970s it has followed the second. But it has recently done so within the limits of a medium-term financial strategy, laid down against its wishes by the Treasury, which still further limits its freedom of action. In principle, the Bank, in the words of one of its most independently minded governors, has accepted the role of 'a good wife: to nag, but in the last analysis to obey'; however, this has not prevented bruising clashes between the Bank and Treasury ministers, either because the Bank was suspected of not allowing the money supply to rise enough to accommodate the needs of businessmen (as in March 1966), or because it went to the opposite extreme, as in the summer of 1980, and allowed money supply to grow between June and August by more than twice the maximum permissible under the medium-term financial strategy.

The Bank of England has a tendency to regard as targets of monetary policy variables such as the exchange rate and the stability of interest rates, which Treasury ministers would think of either as instruments or as variables which should be left to the free play of the market; and this difference of approach can lead to clashes too. In principle, it may be 'the Bank who advises, but the Government who decides', but an adviser's power can become considerable at times of financial crisis, such as the autumn of 1976, when the adviser is the only reliable interpreter of what measures of domestic economic policy Britain's overseas creditors are insisting on in return for short-term credit. The fact that Bank of England staff have a completely independent career structure from civil servants increases the independence of their position.

The *spending ministries* (Defence, Education, Health, Social Security, Transport, Environment, and so on) indicate to the Treasury every summer how much money they would like to spend in the following financial year. They ask for as much as they think they can get away with, but what they think they can get away with is limited by both visible and invisible constraints. In full public view are the figures in the Public Expenditure White Paper, which lays down each March targets for each major branch of public spending five years in advance; less blatant is the constraint imposed on civil servants in the spending ministry by their desire to stay on good terms with the Treasury. Unlike Bank of England employees, civil servants in spending ministries have a common career structure with those in the Treasury, and the headship of both the Home Civil Service and the Government Economic Service is located within the Treasury. The implication of this is that they will think twice before doing something of which the Treasury might disapprove, such as putting in a spending bid which might be exposed as irresponsible or extravagant. Ministers in spending ministries may try to exert an upward pressure on what their civil servants demand but they can seldom make much headway except during the later years of an election cycle. For these reasons, spending ministries do not by any means have a free hand when they bargain with the Treasury every autumn – not even as free a hand as the Bank of England, which at least has one policy instrument under its control. At the macro-economic level the Treasury has, in particular, the advantage of being able to 'divide and rule': it negotiates piece-wise with each ministry concerning that ministry's spending, and keeps within its own walls the central 'Budget judgement' concerning what *overall* levels of expenditure should be for the coming year.

The *Inland Revenue*, finally, in conjunction with the Department of Customs and Excise, collects the taxes which the Treasury tells it to collect. Unlike the spending ministries it does not in any meaningful sense 'bargain' with the Treasury, but simply advises on the administrative feasibility and likely yield of prospective tax measures. Like the spending ministries it is excluded from the 'Budget judgement'.

Macro-economic policy in Great Britain, we have argued, is made by a decision-making inner circle and an executive outer circle whose three members possess different degrees of bargaining power *vis-à-vis* the inner circle. Now let us examine the formal processes of policy-making within the inner circle.

Macro-Policy-Making in Normal Times: The Treasury 'Policy Cycle'

Figure 7.1 shows how functions are divided up within the Treasury. For our present purpose, which is to discuss how

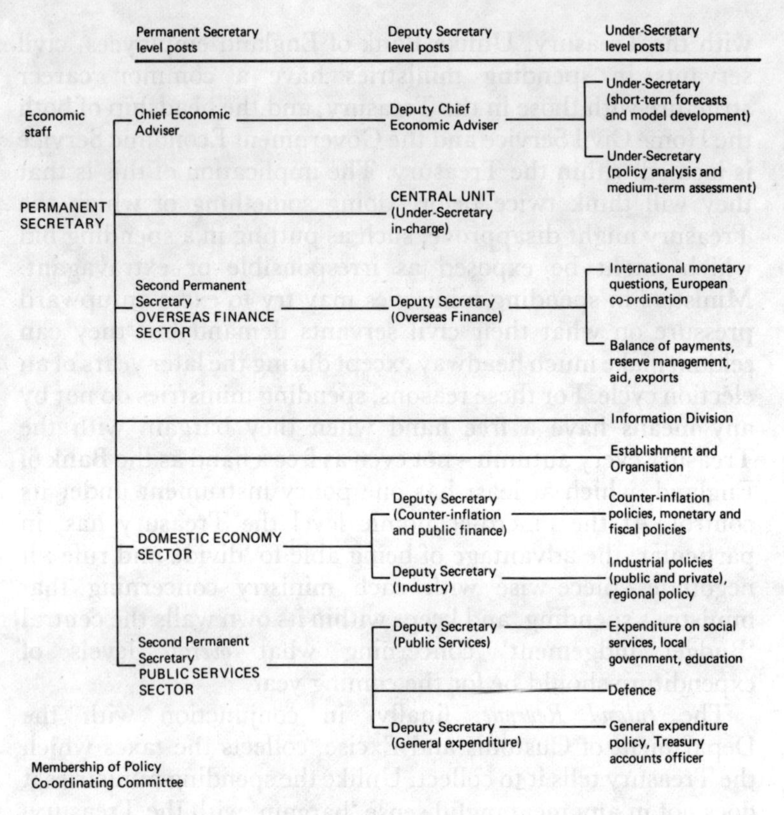

FIGURE 7.1 *United Kingdom: organisation of HM Treasury (since 1975 management review)*

Source: Mosley (1984[a]).

instrument variables such as public expenditure totals, tax rates and interest rates are determined, certain parts of the Treasury which deal with establishments, information and certain sub-divisions of public expenditure are not relevant. The permutation of these instrument variables which Treasury officials offer to the Chancellor as policy options emerges from an interaction between the groups marked off in separate blocks in Figure 7.2: the economics staff, the Central Unit and the Policy Co-ordinating Committee. The Policy Co-ordinating Committee (PCC) consists of the four permanent secretaries, the six deputy secretaries and the head of the

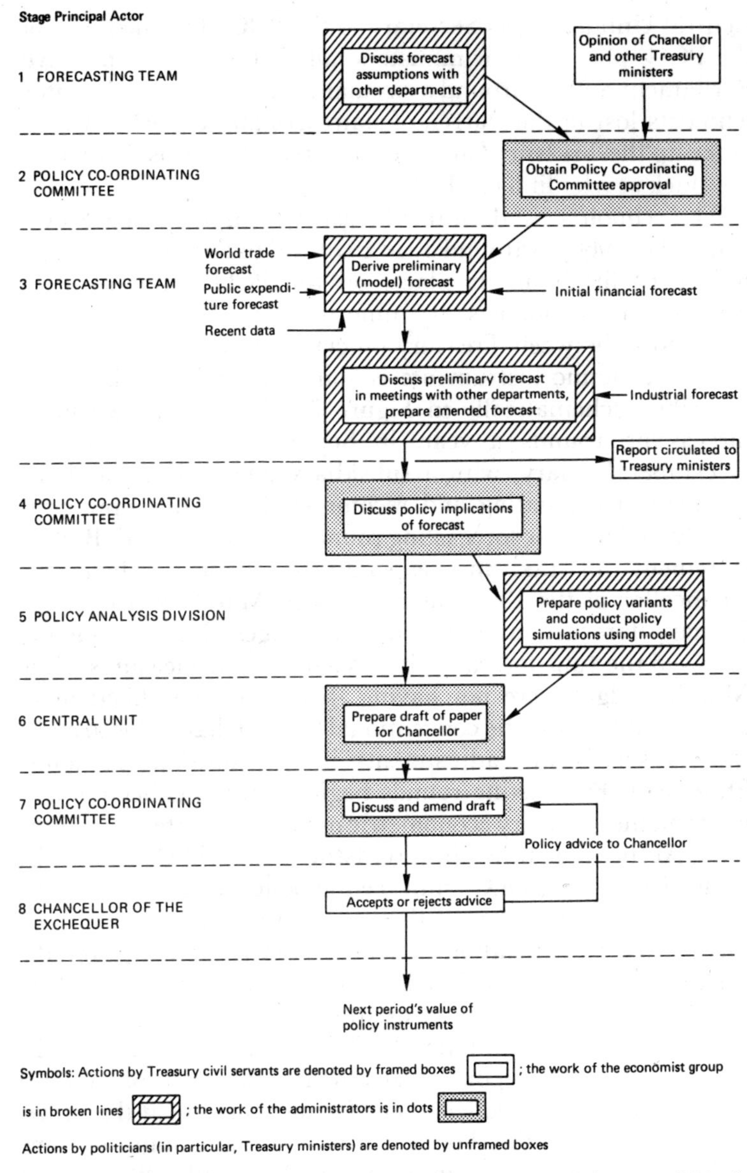

Symbols: Actions by Treasury civil servants are denoted by framed boxes ; the work of the economist group is in broken lines ; the work of the administrators is in dots

Actions by politicians (in particular, Treasury ministers) are denoted by unframed boxes

FIGURE 7.2 *United Kingdom Treasury: simplified diagram of main activities in preparation of forecasts and policy advice*

Source: Mosley (1984[a]).

Central Unit acting as Secretary to the PCC. The members of the PCC, plus the four ministers, all sit two floors up above Parliament Square around a circular inner courtyard. A visitor who gets lost in the warren of staircases and corridors which make up the Treasury Chambers can tell when he is there from the thick scarlet lino on the floor.

The economics staff in the Treasury is large – it has grown from 10 in 1960, to around 50 in 1970 and today nearly 80 – but it is kept firmly in its place: on the PCC, administrators outnumber economists by nine to two. The pattern of interaction between Treasury officials leading to a change in macro-economic instruments is somewhat ritualised, and is described schematically in Figure 7.2. There are two main forecasting rounds a year, in September/October and in December/January, with a subsidiary one in June, and the business of evaluating and recommending policy options is closely tied to these forecasting rounds: the traditional 'Budget statement' of tax and expenditure changes which the Chancellor of the Exchequer makes in March is tied to the January forecast, and 'summer packages' have frequently followed the June forecast if it was felt that the measures of the March budget were not having their intended effect; more recently, it has become conventional for the Chancellor to make these discretionary adjustments in an Autumn Economic Statement. Forecasting rounds begin with the collection and entry on the Treasury econometric model of the latest economic data. At the same time, the forecasting team will clear with the Policy Co-ordinating Committee the policy assumptions to be used in the current forecasting round. One of these will be a 'no-change extrapolation', that is, a forecast made on the assumption that present policies continue unaltered. Others will feature adjustments to existing policy instruments – for example, a 5p cut in the standard rate of income tax – and the introduction of new policy instruments – for example, import controls or incomes policies. The Chancellor can, at this stage, communicate to the Policy Co-ordinating Committee any policy changes whose implications he would like to see explored. Thereafter the forecasting round follows the sequence depicted in Figure 7.2. The forecasting team present their forecast of what will happen over the next year in the event

TABLE 7.1 *Some Treasury 'ready reckoners': estimated policy effects in 1982*

Simulations	Hypothetical policy change	Effect on target variables:	Inflation %	Unemployment '000s	Real personal disposable income % growth	PSBR (£ million) (% of nominal GDP in brackets)
1.	Increase in central government expenditure on goods and services of 0.5 per cent of GDP	after 1 year	0.2	−75	0.4	900 (0.3)
		after 4 years	0.6	−15	0.4	1400 (0.4)
2.	Increase in income tax of 0.5 per cent of GDP	after 1 year	zero	10	−0.7	−1050 (−0.4)
		after 4 years	−0.4	30	−1.0	−1200 (−0.4)
3.	1 per cent increase in rate of VAT	after 1 year	1.0	−48	−1.0	1000 (0.4)
		after 4 years	1.1	−101	−1.4	1250 (0.4)

Note: The table shows the effect, on the assumptions stated, of the change in policy given in the left-hand column on the target variables listed in the top row. Effects are measured in terms of deviations from the 'base run', that is, the value of the target variable which would materialise if policy were left unchanged.

Sources: Rows 1 and 2: United Kingdom (1982), tables 1 and 2, p. 150; Row 3: Rowl (1980), p. 51.

TABLE 7.2 *United Kingdom: consequences of certain hypothetical policy actions proposed for 1982 budget*

Package recommended by:	Proposal in respect of:		Calculated effect after one year (four years) on:			Derivation
	Government revenue	Government spending	Inflation (%)	Unemployment ('000s)	PSBR (£ million)	
1. Chambers of Commerce		5% reduction	−0.8 (−2.4)	300 (60)	−3600 (−6400)	Table 7.1 simulation 1
2. Confederation of British Industry	2% off employers' National Insurance contributions	£250 m increase	0.1 (0.7)	−5 (−40)	1750 (2500)	Table 7.1 simulations 1 and 2, assuming effect of change in N.I. contributions same as change in income tax
3. Conservative Party 'left'	2% off employers' National Insurance contributions	2% increase in unemployment pay	0.2 (1.0)	−7 (−60)	2500 (2900)	Table 7.1 simulations 1 and 2
4. Trades Union Congress	2½% off rate of VAT	£2.1 billion increase in public investment; £1.7 billion increase in expenditure on training	3.0 (4.4)	−320 (−290)	4900 (7300)	Table 7.1 simulations 1 and 3
5. Labour Party	2½% off rate of VAT	£4.5 billion increase evenly divided between: capital projects, training, industrial subsidies, social benefits	3.1 (4.7)	−360 (−300)	5300 (8000)	Table 7.1 simulations 1 and 3

Source: For contents of policy packages, *Sunday Times*, 14 February 1982; for their effects, see final column of this table.

of no change in policy (the 'central case'), which is made in conjunction with experts in other departments, in particular the Inland Revenue, the spending ministries and the Bank of England, who will sometimes tell the Treasury team to override some of the relationships in the model in the light of up-to-date information. The Policy Analysis Division of the economics staff, after receiving a preliminary draft of the forecast, then makes a projection of what would happen to the economy if policy were altered in specific ways. This is presented in the form of a set of 'ready reckoners', that is, forecasts of what would happen if *one policy instrument only* were changed by some handy conventional amount such as 1 per cent. Three such 'ready reckoners', as published by the UK Treasury, are set out in Table 7.1.

These 'ready reckoners' can then be combined by the Policy Analysis Division into hypothetical policy packages which the Chancellor might wish to consider for use in his Budget: for example, '£1 per week on child benefit plus £2 billion increase in public investment'. Five such hypothetical policy packages are presented in Table 7.2; each of them was recommended to the Chancellor in advance of his 1982 Budget by a different British interest group. Their consequences are derived from the ready reckoners set out in Table 7.1. These consequences are the constraint on policy choice; they are spelled out in Table 7.2 and Figure 7.3. In practice, the Policy Analysis Division examine a good many more. The limits to the search process are set on the one hand by the capacity of the Treasury computer, and, more important, by the 'house view' of Treasury civil servants concerning the limits of what is politically feasible.

The limits of what is politically feasible, of course, shift around according to the state of the election cycle. Harsh deflationary measures such as those implemented in 1976 and 1981 can scarcely be contemplated in the run-up to an election; however, these days it is probably counter-productive for a Chancellor to follow the practice of his predecessors in the 1950s and 1960s and cut taxes in the budget immediately before a general election. He must now employ more subtle means: inflationary pay awards, such as those sanctioned throughout the British public service in 1979; postponement of job cuts in

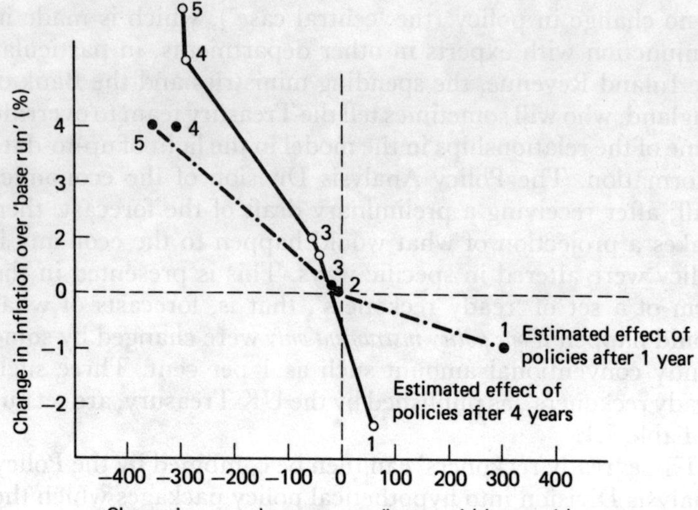

Policies:
1 reduction of 5% in all categories of public expenditure
2 2% reduction in employers' National Insurance surcharge, £250 m. increase in government spending
3 2% reduction in employer's National Insurance surcharge, 2% increase in unemployment pay
4 2½% reduction in rate of value added tax, £3.8 billion increase in government spending
5 2½% reduction in rate of value added tax, £4.5 billion increase in government spending

FIGURE 7.3 *United Kingdom, Spring 1982: estimated consequences for unemployment and inflation of certain hypothetical policy options*

Source: Table 7.2, columns 4 and 5.

the health service and nationalised industries, as employed in 1983; relaxation of monetary targets, as adopted in the 1982 budget. All of these factors will be in the mind of Treasury Ministers as they guide their permanent staff through its annual policy search.

The set of policy options on whose consequences there exists an up-to-date forecast is the portion of the universe which is visible to the Chancellor at that time. He now has to choose, in conjunction with senior Treasury staff (stages 6–8 of the policy cycle in Figure 7.2) where he wants to be within it. This is now done in January of every year at a country house in Kent, at a weekend house-party attended by all members of the Policy Co-ordinating Committee and all Treasury ministers. We shall be considering the decision process in detail later in the chapter. But it may be worth setting down here a not-too-implausible speculation about the way the Chancellor may have seen the decision in January 1982.

The Decision-Making Process in January 1982

(1) The basic objective of the government's strategy is to get inflation gradually down to 5 per cent or less. At present it is 11 per cent, but falling.
(2) Two years ago the Treasury announced a medium-term financial strategy for achieving this basic objective. This involved scaling down the rate of growth of the money supply each year, with proportionate reductions in the public sector borrowing requirement. The target growth rate for the money supply for fiscal 1982/3 is current 5–8 per cent.
(3) Officials advise me that the PSBR which would be compatible with this monetary growth is in the range of £8 billion to £10 billion. If income tax allowances and excise taxes are 'indexed' in line with forecast inflation, but all other policies remain unchanged, the PSBR for 1982/3 is forecast to be about £8 billion.
(4) Hence I can give away at most an additional £2 billion; no more.
(5) Hence all policies which involve an increase in the PSBR of more than £2 billion [for example, those proposed by the Labour party, the TUC and even the Conservative 'left': options 2 to 5 on Figure 7.3] are out.
(6) Within my 'room for manoeuvre' of £2 billion or less, however, I shall give away as much as I can. Output has been falling for two years, the industry lobby is getting dangerously disaffected, and there will be an election in two years' time at most. [In the event, of course, it came the following year.]

The amount eventually given away in the 1982 spring Budget was £2.3 billion, with a further concession in the autumn on the employers' national insurance surcharge. In addition the target range of growth of the money supply (sterling M_3) was allowed to slip upwards – from 5–9 per cent to 8–12 per cent in respect of the 1982/3 fiscal year.

Emergency Procedures

Such is the formal structure of short-term policy-making. Fairly frequently, however, policy and forecasting rounds cannot be allowed to run for their normal period of about six weeks, because of the need to take quick action in response to an emergency. In the 1976 sterling crisis, for example, a run on the reserves developed in July, and a decision on the appropriate policy response had to be taken within a week. Under this stress the search for the appropriate response:

(a) embraced at the first stage only one instrument and two options (a £1 billion cut in public expenditure, or a £2 billion cut);
(b) when extended, concentrated on *speed and certainty of effect*, above all. Hence social security cuts and de-indexing of pensions were quickly rejected as policy tools, because even if feasible they would require new legislation, and local authority expenditure cuts were out because there was no certainty that they would take effect;
(c) was confined to a few people only; even the Cabinet were only brought in when deciding how cuts were to be allocated, not when deciding what the size of the overall cut should be.

In general, economic policy-making, not only in emergencies, but in normal times, can be described as a *satisficing* rather than a *maximising* process. Ministers do not spell out an 'objective function', or consistent set of preferences between alternative states of the world – indeed, as we have seen, it could be described as politically irrational for Treasury ministers to have one; there is only a one-off decision about whether things are good enough or not good enough. What is 'good enough' in normal times is determined by the requirements of the medium-term financial strategy, which is discussed in more detail in the next section; in times of crisis it is characteristically determined by the small overlap between the harshest methods which are thought politically feasible and the mildest methods which are thought capable of resolving the current crisis. There is *no complete specification of the constraint*, that is, the set of alternative feasible states of the world. The Policy Analysis Division, acting on instructions from chancellors, will prefer to explore five or six alternative policy packages thoroughly rather than fifty or sixty at the level of a bare forecast. Changes in policy instruments do not occur simply because changes occur in the external environment; rather, they are generally *prompted by 'crises'* which plunge some aspect of the economy into the unsatisfactory zone, and stimulate search behaviour of a type which depends on the time available. And finally, there is evidence that what is considered 'satisfactory' itself adapts to continuing discrepancies between performance and aspiration. In the 1950s an unemployment rate of half a million was enough to stimulate reflation, whereas at the time of writing anything below two and a half million would be considered a real achievement. The unemployment target was replaced, in 1980, by a medium-term financial strategy setting target ranges for the money supply in each financial year, but this was itself relaxed in both 1981 and 1982.

How has this happened? Let us now consider the evolution of policy-making in the last thirty years in more detail.

The Rise and Fall of 'Keynesian Budgeting', 1945–79

In 1944 the government bound itself, in the *Employment Policy White Paper*, to 'maintain a high and stable level of employment after the war'. This it sought to do through its budgetary policy throughout the 1950s and 1960s: when unemployment appeared to it 'too high' taxes were cut and government capital spending was increased. However, these attempts to maintain 'high and stable' levels of employment were frequently affected by foreign exchange crises which forced it again and again (in 1957, 1960, 1964, 1966 and 1973) to throw its policy instruments into reverse, raising taxes and raising interest rates to protect the balance of payments. These policy reversals were frequently derided as 'stop–go'; but the inference drawn by most observers during the thirty years to 1975 was that the government should try and anticipate rather than respond to crisis, or alternatively bring in other policy instruments such as incomes policy or devaluation. Nobody suggested that the procedure of discretionary government intervention through the budget, to stabilise the economy, as originally recommended by Keynes, was in itself wrong.

Rather suddenly, during the oil crisis of 1974/5, the lamps went out on the accepted pattern of economic policy-making. For the simple rule 'reflate if unemployment is excessive and deflate if the balance of payments is in crisis' was unworkable in an environment where both unemployment and the foreign payments deficit, and, for good measure, inflation as well, stood at post-war record levels and still rising. For several years the British economy, and for that matter those of America and Europe as well, had presented to the onlooker the likeness of a swimmer, frantically swimming from one side of the beach (excessive unemployment) to the other (excessive inflation) but all the time drifting offshore. The only school of thought offering a coherent remedy for this condition of 'stagflation' were the so-called *monetarists*, whose startling point of departure was that the objectives enshrined in the Employment Policy White Paper should be overthrown and that budgeting should concern itself with one objective only: the control of inflation over the medium term. Once this basic objective had been brought down over the medium term, then by the quantity theory of money the rate of growth of the money stock had to be brought down over the medium term. If the growth of the money stock was to come down, then government deficits – which, if met by borrowing from the banks, constituted a major element in the growth of the money stock – had to be reduced, and that precluded them being used for any other purpose, such as, for example, to achieve a target level of employment. And if the growth of the money stock was to serve as an intermediate target of policy, that precluded interest rates – the price at which that money stock could be borrowed – from being used as an instrument. Thus the shift from a 'Keynesian'

TABLE 7.3 *Alternative patterns of economic policy-making*

Characteristics	'Keynesian'	'Monetarist'
Target variable	Employment, with side conditions concerning the level of balance of payments deficit (or inflation) that can be afforded	Inflation
Time period over which targets are to be attained	Short-term (i.e. major part of objectives to be attained within one or at most two years)	Medium-term (i.e. five years or more)
Order of precedence among instruments	Fiscal policy determines the stance of monetary authorities	State of monetary growth in relation to target determines what the budgetary authorities are able to 'give away'
Monetary instrument	Interest rate (plus auxiliary instrument such as controls on bank lending)	Monetary aggregates

to a 'monetarist' pattern for the making of economic policy involved four separable elements, as set out in Table 7.3.

The transition from the first pattern to the second lasted throughout the decade of the 1970s. Monetary targets began to be surreptitiously set by the Bank of England from 1970; the Labour Chancellor, Denis Healey, explicitly refused in his 1975 budget to stimulate demand in spite of rising unemployment; and under the stress of the 1976 sterling crisis, previously mentioned, it was acknowledged (in a letter of intent to the IMF) that henceforward the budget deficit would have to be brought down over the medium term in order to enable the desired cuts in monetary growth to be achieved. Thus the transition to new methods of policy-making was three-quarters complete by the time that the Conservative government under Margaret Thatcher took office in 1979. All that remained for the new government to do was explicitly to abrogate, in the Chancellor's 1979 budget speech, the target of 'high and stable' levels of employment and the instrument of incomes policy as a means of achieving it.

Monetarism and After: Economic Policy in the 1980s

One important aspect of the retreat from 'fine-tuning' methods of economic control was that policy-makers were forced by events into the realisation that macro-economics was a very inexact science: 'more like gardening than operating a computer', as one Chancellor highly trained in economics put it. For especially during the 1970s the basic ratios on whose stability macro-economic policy depends began to jump around in the most alarming way. Let the aforementioned Chancellor, Denis Healey, who had the job of controlling the economy at the time, explain his predicament:

> After the first increase in oil prices the savings ratio, the proportion of income saved, rose remarkably in all countries. But following the second oil crisis in 1979, it jumped up again in Europe – it was 17% last year in Britain – whereas in America just over a year ago, it had fallen below 3%. Nobody has fully explained this fact. Yet the savings ratio is one of the most important factors in explaining how the economy behaves. For example, in Britain, an increase in the savings ratio of 1% is equivalent to a fiscal deflation of £2 billion, add to that the fact that the savings ratio in Britain is now running at 17% of personal disposable income implied a deflation of £16 billion compared with the situation in the sixties . . . Because of this uncertainty about economic rules and facts it is extremely difficult for a Chancellor of the Exchequer in Britain to guess when he frames his budget how much demand he should be putting into the economy or taking out . . . The estimate I was given for the public sector borrowing requirement when I came into office turned out to be £4

billion too low. That was 5.6% of gross domestic product in 1974, a bigger error than any fiscal change in demand brought about by any budget.

The response of the 1979 Conservative government to the instability of the economy was to try and shift the focus of economic policy from the short to the medium term, and in particular to announce a *Medium-Term Financial Strategy* in March 1980, under which targets for monetary growth and the PSBR (Public Sector Borrowing Requirement, or budget deficit) were set out *four years ahead.* Similar targets for a four-year planning horizon have been put forward in every subsequent budget.

Unfortunately for monetarism, throughout the last five years the actual stance of economic policy has differed from its intended stance. Although the government repeatedly stated that its intention was to bring down inflation by means of a gradual reduction in monetary growth, the actual average growth of sterling M_3 has in fact been *higher* under the Conservative government of 1979–85 than under the preceding Labour government. In large measure this failure of monetary control reflects the continuing failure of the Conservative government to reduce the real value of public spending, which was planned to fall by 3½ per cent between financial years 1979–80 and 1983–4. By autumn 1985, the Conservative Party, which was then experiencing 'mid-term blues' and trailing badly in the opinion polls, manifestly did not know whether to boast about or to attempt to conceal this slippage.

That the slippage itself should have occurred speaks volumes about the difficulty any democratically elected government must experience in attempting to control the central instrument of economic policy. Year after year since 1979 the Conservative government's Public Expenditure White Paper has promised a cut in the level of real public expenditure, and year after year the outcome has been a rise, a state of affairs that has been christened the 'weeping willow effect' after the pattern which planned and actual expenditures make if plotted on a graph.

The causes lie deeply embedded within the British social and political system: many elements of public spending, including social security, are still not subject to cash limits and hence are formally uncontrollable; substantial cuts in the real value of certain benefits, including unemployment benefit, tax relief on mortgages and the national health service, are agreed to be politically out of bounds; and, when an election is close, governments of whatever political colour feel constrained to switch from the austerity measures which will please bankers to the reflationary measures requested by the industry lobby. One of the most startling shifts within the economic policy of the Thatcher government has been the transition from the tax increases of the 1981 Budget, which kept the City happy but prompted the chairman of the CBI to promise a 'bare-knuckled fight' with the government, to the autumn economic statement of 1985, which is full of promises of increased government spending and contains no mention whatever of sterling M_3, the original guiding light of the medium-term financial strategy. The transition is reminiscent of that which occurred in the United States three years previously, and of course was largely motivated by the same political imperative.

This unanticipated growth of spending has made tax cuts, long promised as an integral part of Conservative economic strategy, very hard to achieve. If they ever materialise, it will only be possible as a consequence of the Conservative government's policy of privatisation of state assets, which has gathered momentum in recent years with the sale of the British National Oil Corporation, of Jaguar Cars, of parts of British Petroleum and of British Telecom. The sale of British Airways and British Gas is now clearly on the horizon also. Originally conceived as a purely ideological measure, this policy is now yielding financial dividends which may allow tax cuts in 1986. Whether these cuts, if they materialise, will yield the hoped-for political advantage is uncertain: a past Conservative Prime Minister, Lord Stockton, has stigmatised the policy of government asset sales as 'selling off the family silver', and recent opinion polls have suggested that a majority of the electorate would prefer present levels of services and present levels of taxation, to fewer services and lower tax rates. Middle-of-the-road consensus may exist, and always have existed, within the plurality of the British electorate even if it has not been reflected in the pronouncements of British politicians.

Economic policy in Britain therefore stands at an intriguing cross-roads. Having retreated from an attempt to 'fine-tune' the macro-economy to the more modest objective of attempting to control, by means of the Medium-Term Financial Strategy, 'only those things – very few of them – that the Government actually have within their power to control' (Sir Geoffrey Howe, *Hansard*, 26 March 1980, col. 1442), the Government now finds that even those 'very few things', apparently, do not lie within its grasp. The obvious options are to retreat still further and to abandon the attempt to stabilise the macro-economy altogether, to try and make a better job of controlling the 'very few things' – that is, the budget deficit and the money supply – or to go back to full-blown Keynesian attempts to control employment and output. The first of these options would imply so great a loss of face even to Conservatives that no politician would consider it. The second is the official government position: much play is now made in each budget speech with alternative indicators of monetary stance (such as M_0, notes and coins issued by banks), auxiliary instruments of policy (such as discreet management of the exchange rate) and supplementary indicators of economic activity (such as money GDP).

The third is favoured by the opposition parties, with the Liberal/SDP Alliance laying stress on the control of inflation through a long-term, tax-based incomes policy, and the Labour party on redirection of investment by the state. What is striking is that, fifty years after Keynes and fifteen years after Harold Wilson's dictum about the primacy of economic policy, neither government nor opposition is in possession of an economic policy prescription which has in the past demonstrably achieved its stated objectives. Two-thirds of the British electorate, on the evidence of a Gallup poll published in May 1985, feel that unemployment is 'the most important policy issue facing the government', but no party is offering a tried and tested solution to the problem. Economic management, as we have discovered to our cost over the past half-century, is not the same as control engineering, particularly in a democracy where industrial disputes, productivity and wage levels are even less subject to government control than the budget deficit. Whether those in charge of the process will recover their lost morale, or whether the economy will in future years come to be seen as no more subject to 'policy' than the weather, is at present not at all clear.

8

Social Policy

RICHARD PARRY

What Is Social Policy?

Social policy is about both the intervention of the state in individual lives and the setting of norms about social obligations. In Britain, the starting-point for the analysis of social policy is the idea of contingency – that at certain times of their lives, individuals will face circumstances they are unable to cope with by themselves. In the British tradition of teaching and writing about social policy, this is associated with 'need' – the idea of deprivation crying out for treatment in a way to which the average citizen would seek to respond, both individually and through organisations like political parties.

In the nineteenth century, these ills were obvious – slums, epidemics, exploitation of labour, illiteracy. Today, they are less visible but in some ways just as pervasive – unemployment, poverty, disablement. Even though the absolute value of state cash benefits has increased markedly, work by Peter Townsend and others has claimed that upwards of 10 per cent of the population cannot afford the necessities of life (Townsend, 1979, table 7.1; Mack and Lansley, 1985, table 6.3). Television series like London Weekend's 'Breadline Britain' (1983) and Yorkshire Television's 'From the Cradle to the Grave' (1985) have forcefully displayed the persistence of poverty. Particularly when their impact on individuals is considered, these ills seem to be a reproach to the state. From this perspective flows the notion of 'altruism' – that there is a natural and fraternal behaviour different from the acquisitive and selfish behaviour associated with the economic market.

Social policy is a wide-ranging concept. Even nations with ill-developed social services may have strong state control over education, criminal policy, and personal behaviour. Many of the social problems of today – like drug-taking, vandalism and football hooliganism – can in principle be tackled through social policies. Many moral dilemmas – embryo research, removing children from their natural parents, the treatment of the terminally ill – likewise call for social policy solutions. In fact, almost every process or decision of life – forming a household, having children, entering or leaving the labour market, having a healthy or unhealthy lifestyle, caring or not for parents – may be influenced by the state through a variety of means. A survey like the present one which concentrates on government activities should not undervalue the primary importance of legal and social norms, and the considerable role of non-state provision.

Traditionally, social policy has been taken to consist of the main social services – social security, health, education and housing – joined more recently by social work, or the 'personal social services'. These are all human services, delivered to individuals, but they do not necessarily combine in a coherent way. Governments have policies for the social services, but seldom a consistent 'social policy' based on what a 'good society' might be or how it might be achieved. They tend to think in terms of agencies and programmes, rather than of 'client groups' like the unemployed, sick or elderly. Many policies traditionally regarded as economic – especially transport and labour market services – have important social implications, which may be obscured in the policy formulation process.

The objectives of social policy, and the motives for promoting them, are seldom clear-cut. Many commentators have pointed out that, far from merely being an expression of human goodwill:

(a) *Social policy is supporting capitalism* and the smooth running of the economic market through education, public health, and income maintenance benefits. Marxist social policy analysts have traced the way that the 'reproduction' of capitalism (the perpetuation of a trained and compliant workforce) requires the power of the state to provide services and lay down rules that are beyond the scope of the individual employer. Far from challenging capitalism, as might be supposed by those who see social policy as vaguely 'left-wing', it is liable to promote and sustain it.

(b) *Social services have become part of the currency of politics*, especially since the universal franchise was granted and the Labour Party was established by the trade-union movement. It is difficult to deliver specific goods to individuals through economic policy, since variables like economic growth, inflation and employment are outside the direct control of government. But through social policy, the individual can receive a new house, a place at school or university, residential accommodation for an aged relative, a family allowance or a pension.

(c) *Many social policies benefit their employees* as well as those they serve. There is a fundamental difference between services in cash (like retirement pensions or unemployment benefit) and those in kind (health care or education), since the latter consist basically of paying people to provide a service, so providing a benefit to both the consumer and the producer.

(d) *Social policy may become wasteful and even oppressive*, offering services that consumers don't want, monopolising provision of a service without delivering efficiency in return, behaving bureaucratically and unfeelingly towards fellow-citizens, failing to take account of local circumstances, and using people's taxes to underwrite jobs and career opportunities.

In sum, social policy is unavoidably a part of politics, with all the pressures and conflicts this implies. Altruistic behaviour cannot be secured by the state alone, but in Britain it came to be assumed that an effective social policy required vigorous public action. It is the choice between public and private, marketed or informal provision – in terms of both possibility and desirability – that underpins political debate about social policy, and, for decades, the trend seemed to be running in the collective direction. But since the 1970s it seems to have gone into reverse. This chapter charts the change of direction and asks: is the Conservative government really dismantling the welfare state, as is often argued, or protecting its essential principles, as it would claim?

How Is Social Policy Made?

More than in most areas of politics, social policy is about 'winnowing in' issues to the political process. It might seem extraordinary that in the United States a current educational issue should be about compulsory prayer in schools, and a health care issue, the permissibility of abortion; but these are matters at the top of the public agenda, because American social policy is strongly affected by legal norms. In Britain, policy-making reflects an interplay between politicians, experts and interest groups which may be seen to make 'waves' of policy over different periods of time.

In the *long term*, the important thing is the shaping of a political consensus about what the right approach to policy might be. Often the issues involved have little to do with immediate party politics – matters like the education curriculum, public policy towards minority groups, building styles in housing, institutional versus home-based care, possibilities for medical technology, the kind of groups that should receive an income guarantee. Issues often form themselves through a long-term process of adapting public norms – for example, unemployment was not recognised as a social problem until the end of the nineteenth century, and specific public provision for the disabled and single parents has emerged only in the last twenty years. In the mid-1980s, the poor structural state of the housing stock is becoming a subject for debate. The activities of pressure groups are important in defining new areas of public concern; the move to a more generous child benefit is associated with the Child Poverty Action Group, and public action against homelessness with Shelter.

In the *medium term*, the time-horizon tends to be that of the five-year parliamentary term, and the inclusion of promises in a party manifesto; indeed, the phrase 'over the life of a parliament' is often used. Important here are targets for service or benefit provision, managerial arrangements, expenditure plans, and strategic objectives for social policy – such as to encourage work incentives or family care. In the 1950s and 1960s targets for housebuilding were an important medium-term policy, and the provision of jobs or training places is a comparable issue in the 1980s.

In the *short term*, the main elements are the day-to-day pressures of parliamentary politics and of public expenditure control. Social policies are typically not centres of attention for the Cabinet and Prime Minister; they must be forced into their priorities. The active and competitive popular press in Britain plays a major part here. Among the triggers for short-term action are scandals, like hospital malpractices or the ill-treatment of children (like Maria Colwell in 1975 or Jasmine Beckford in 1985), civil disturbances (like the riots in Handsworth and Tottenham in 1985) and the pressures of central–local relations. Indeed, such events can rapidly challenge long-held wisdom about matters like community policing and keeping families together.

Of central importance is the role of professionals. Social policies that are delivered in kind rather than in cash are typically provided by professionals like teachers, doctors, social workers, housing managers and architects. There is a great deal of sociological debate about how professionals operate, but two factors seem important. First, they present an authoritative and expert face to the world, so that tasks that once seemed the job of the individual or the family are reserved for the trained and qualified. Second, they to some degree take control of an occupation and how it is practised, and so make it difficult for political representatives to determine the content of service delivery.

Two professions are especially powerful: doctors – who dominate health and the personal social services because of their powers to diagnose conditions and commit resources; and teachers – numerous, politically aware and in charge of the closed world of the individual classroom. The long-term agenda in education and health is controlled by them, since

TABLE 8.1 *Who implements social policy? (% of expenditure, 1983–4)*

	Elected		Not elected
	Civil service	Local government	Other agencies
Social security	92.5	7.5 (housing benefit)	0
Health	3.9 (DHSS administration)	0	96.1 (health authorities)
Education	0.3 (DES administration)	87.0	12.7 (universities)
Housing	19.1 (rent subsidies)	60.0	20.9 (housing associations)

Source: Calculated from HM Treasury, 1985.

senior posts are held by qualified staff. Short-term policy implementation is also constrained by what is acceptable to professionals. Also important are the architects and planners who can determine the physical context of people's lives. It has become fashionable to blame professionals for the faults of the social services, but this may ignore the importance of proper standards of service and individual responsibility. But a price has to be paid in terms of training costs, higher pay, and the restriction of competition, for the virtues of professional organisation.

As Chapter 5 emphasises, the implementation of social policy is very much a matter of central–local relations. Central government departments like Education and Science (DES), Environment (DOE) which controls housing, and the health side of Health and Social Security (DHSS), administer policy through laws, standards and circulars but generally do not provide services directly. Their financial leverage is also limited by the block grant system of rate support, which makes it difficult for central government to offer authorities direct financial inducements to follow particular policies. Schemes like 'joint finance' (support finance in Scotland), where health authorities fund social-work projects to care for clients outside hospital, are the exception. As Table 8.1 shows, it is only in social security that there is direct civil service administration. The exception here is housing benefit, which was transferred completely to local authorities in 1982 (but is financed directly from the centre). Health is also outside local government, but employees are not civil servants. Education, housing and personal social services are largely administered by local government.

Relatively few social policy tasks have ever been run through civil service departments. Until 1948, local authorities ran hospitals and most public assistance, but there is a reluctance to allow local authorities to administer cash benefits that they might increase; this can be seen most clearly in the Poor Law Amendment Act 1834, which tried to reduce the scope for local generosity to the poor, and is reflected today in the way that, for instance, rates for housing benefit and student grants are set nationally. In addition, professional norms are strong, and directors of education and social work tend to look to practices in their fields rather than the political circumstances of their authorities. Even though there are marked variations in service expenditure between authorities, it is difficult to ascribe them to local political choice.

The present structure promotes irresponsibility on both sides, because neither tier has a clear relationship between revenue-raising, policy responsibility, and accountability. In most fields of social policy, it is possible for central government to renounce concern about day-to-day administration, and for local government to blame inadequacies on central policy. Local rate levels have much more to do with central government manipulation of grant than with local decisions on levels of service, and set-piece central–local battles – like Liverpool in 1985 – tend to be resolved by creative accounting and the arbitrary halting of growth rather than any decision about social priorities.

The Growth and Crisis of Social Policy

It is a commonplace of debate in present-day Britain that the

TABLE 8.2 *The growth of expenditure, employment, and clientele (thousands)*

	Expenditure		Employment		Clientele	
	1950	1980	1950	1980	1950	1980
Social security	2,575	11,727	58	100	1,350	3,247
Health	2,853	5,633	492	1,222	2,788	5,670
Education	2,582	6,783	572	1,551	6,756	10,008
Housing	1,379	3,793	n/a	63	1,660	6,900
Personal social services	107	1,105	116	370	various	
Total	9,496	29,041	1,238	3,306		

Source: Calculated from government statistics.

Note: Expenditure is at 1980 prices, from National Income and Expenditure data deflated for price movements in each economic category; employment and clientele are thousands of persons. Clientele defined as: number of recipients of national assistance/supplementary benefit; number of discharges/deaths from hospitals; number of school pupils; number of public sector dwellings.

welfare state has become too big in relation to what it can offer society. The question needs to be asked because of the increasing share of national resources taken by social policy. As Table 8.2 shows, expenditure has tripled in real terms since 1950, and employment doubled. The number of clients – benefit claimants, patients treated, pupils and students educated, tenants housed – has increased two- or threefold. If resource input alone were the test, post-war social policy would seem to have a been a great success.

To understand why there are still doubts and controversies about the value of social policies, we need to distinguish three reasons why expenditure, employment and clientele grow:

(1) More people consume the same service: there may be more children requiring to be educated, more elderly entitled to a pension, more unemployed claiming benefit. The main variable here is demographic (the balance of age groups in the population), but economic circumstances are also important.
(2) Eligibility is extended: for instance, by raising the school leaving age, making higher education places available to

qualified applicants, changing the conditions for pension eligibility, or creating new types of benefit, such as for the disabled.

(3) Increases in the cost or value of the service: the value of a pension of benefit in real terms, the sophistication of medical equipment, the number of teachers per pupil, or the pay of employees. How far this represents an 'improvement' in the quality of service is not clear.

If cuts in expenditure are to be made, it must usually be in the second or third category, and this is seldom easy to achieve. People must be made worse off, or expectations must be denied. Equally, growth under the first and third categories may be expensive and may not be given political credit by consumers. This is why there tends to be a higher public perception of cuts than of growth.

Until the 1970s, any doubts about the British welfare state were set in a context of growth and optimism. The Beveridge Report of 1942, the Butler Education Act of 1944, and the establishment of the National Health Service in 1948 led to a comprehensive structure of education, health care and income maintenance that was a source of national pride. In the 1950s, there was fairly tight budgetary control by the Conservative government, but it became the orthodoxy in the 1960s that expansion and the injection of resources would solve social problems. High-rise blocks would enable the slums to be cleared; earnings-related pensions would provide an income guarantee for all; education, especially comprehensive and higher, would pay for itself by improving the qualifications of the labour force. A hospital building programme would upgrade the health service; the social-work profession would deal with residual social problems. In First World War terminology, 'one big push' would correct the remaining social problems. The 1960s generation moved smoothly into public sector jobs, and it seemed that an expanding economy would allow a social preference for expanding social services to be realised.

The pattern of growth was not always smooth: we should not imagine that there was ever a 'golden age' when money flowed freely. The Treasury always sought to minimise the cost of improvement in services, and the demands of economic management forced expenditure cutbacks, especially after the pound was devalued in 1967. The social policy plans of the 1964 Labour government, such as an adequate income support for pensioners, were never properly carried through. Nevertheless, for two decades, social expenditure, and especially capital investment in schools, houses and hospitals, seemed to be an essential part of national development, all the more favoured because it was too early to test whether it was achieving its objectives.

The key change was in the conception of public expenditure – from an engine to a brake on national growth. The oil crisis of 1973–4, and the cuts in expenditure agreed with the International Monetary Fund in 1976, set up a connection between financial probity and expenditure restraint. It was no longer possible to assume that future economic growth would finance an increase in public expenditure in real terms. Public sector capital investment and pay came under close scrutiny in the belief that they were 'crowding out' resources from private industry.

The role of the trades unions also changed. During the 1960s and 1970s union membership in the public sector expanded rapidly, and the public social services became strongly unionised at nearly twice the level of the private sector. The 1974 Labour government deferred to the unions (especially by allowing pay to catch up with the private sector), but paid a price for it by a loss of control of public expenditure which forced a pay policy to be imposed in 1975. This was successful in reducing inflation but built up trouble that culminated in the 'winter of discontent' of 1978–9, when industrial disruption hit the health and social services. As a result, the image of social policy employees took on connotations of selfishness and waste.

At this time, there was talk of a 'crisis of the welfare state' – that citizens would demand more of it but refuse to pay for it, leading to political instability. There is little real evidence for this in Britain, but the concept points up the extent of the reversal in expectations (Taylor-Gooby, 1985). This provided an intellectual opening for Conservative thinkers prepared to argue that the state should get out of many areas of policy, and a political opening for the Conservative politicians to promise tax cuts.

Many of the new trends in policy were evident before the Conservatives came into office in 1979, but they were given a strong impetus by the strength of the mandate given to the Thatcher government. The movement may be summed up as one from *socialisation* to *privatisation*. The concept of socialisation is that the public organisation of the social services is both more efficient and morally superior, because it pools risks throughout the community and allows for economies of scale. This owes much to the experience of the Second World War, when the British people put up cheerfully with bureaucratic control and then elected a Labour government which perpetuated it. It also reflects the fact that it is difficult for the private sector to operate hospitals with the latest technology, schools with a wide range of courses, and pension schemes covering all workers. Although the services are provided to individuals, the start-up and overhead costs are high. Private provision in Britain is associated with privilege, privacy and choice, but not necessarily with efficiency or high quality.

However, provision by a multiplicity of producers also has its attractions. There are many areas of life where people are not prepared to be allocated with a service on the basis of 'need' as determined by experts – food and clothing are the obvious examples, as the vitality of the private sector in these services testifies. Over time, the range of these services appears to be increasing, and the private sector is better placed than the public to respond. There is greater awareness of what Fred Hirsch called 'positional goods', where supply is scarce (like specialist education and medical treatment) and there is a temptation to buy your way to the head of the queue (Hirsch, 1977). Employees have also come to expect more than pay from their jobs. Cars, generous pensions and private health insurance became a sign of status and a *de facto* welfare state for the better off, one that was supported by tax incentives from government. Richard Titmuss was the pioneer in delineating these 'occupational' and 'fiscal' welfare states and showing how they reinstated inequalities the overt welfare state was designed to remove.

Under the Conservatives, privatisation has developed in

TABLE 8.3 *The Thatcher record in social policy*

	Real expenditure		Employees		% Private	
	1979–80	1983–4	1979	1984	1979	1983
Pensions	14,549	16,401	101	93	51	51
Income maintenance	13,657	19,990				
Health	13,655	15,686	1,152	1,223	5	8
Education	15,853	16,351	1,539	1,430	5	6
Housing	8,247	4,277	61	70	66	69
Social services	2,413	2,655	344	368	33	40
Total	68,374	75,330	3,197	3,184		

Sources: Expenditure, HM Treasury, 1985, tables 2.5 and 2.6. At 1983–4 prices. Personal social services excludes Northern Ireland.
Employees: *Economic Trends*, March 1985, p. 96; *Annual Abstract of Statistics*, 1985, table 6.7; *Employment Gazette*, June 1985, table 1.7. Education and housing local authority employees only.
Private: pensions percentage of workforce with occupational pension entitlement, from DHSS, 1985, table 1; health percentage of population covered by private health insurance, from *Social Trends*, 1985, chart 7.29; education percentage of pupils in non-maintained schools from *Education Statistics for the United Kingdom*, 1982, table 7; 1984, table 13; housing percentage of dwellings not owned by the public sector from *Annual Abstract of Statistics*, 1985, table 3.7; social services percentage of elderly in residential accommodation in private sector places (England only) from *Health Service in England Annual Report 1984*, table 10.

many fields (see Chapter 5). Nationalised industries have been sold off, especially British Telecommunications; local and health authority support services have been put out to tender; bus transport has been deregulated. In the social field, the most striking policy has been the sale of 850,000 council houses to their tenants. Private health insurance and private residential accommodation have also grown, with government approval and to some extent with government financial support.

But when this is set alongside the record on expenditure and employment (Table 8.3) the limitations of change may be seen. Overall, expenditure has increased by 10 per cent, although the rise in employment has been checked, especially in education. Health (up 15 per cent), pensions (up 13 per cent) and other income maintenance benefits (up 45 per cent) have increased sharply in real terms, the latter reflecting the increase in the number of unemployed and pensioners. The share of total social security benefits going to the unemployed increased from 8 per cent in 1979 to 17 per cent in 1984 (HM Treasury, 1985, table 3.12.5).

Education has been stable, but housing has reduced by half (but only by a quarter if receipts from council-house sales are taken into account). There has been a shift to private provision in most services, encouraged by government policy, but the basic responsibilities of public social policy remain unchanged. The expenditure figures in Table 8.3 do not include 'tax expenditures' like the income tax relief on mortgage interest payments and occupational pension contributions, and many of the people covered by private provision are still heavily dependent on publicly provided income or services.

The limitation of Conservative social policy is that the New Right thinkers who have heavily influenced the Conservative government do not have a coherent policy strategy for the social services, and the actions of the government often fail to correspond to the ideology they profess. The Conservatives are prepared to strike but afraid to wound. The big money savers can only come from positive cutbacks in the range of services provided – de-indexing pensions, closing schools, charging for health care – which affect so many voters that a political price must be paid.

The Conservatives have made many marginal changes – in particular, they have indexed pensions to prices rather than earnings, abolished the earnings-related supplements to unemployment and sickness benefit, and increased health charges and council-house rents. But the basic commitments to maintain coverage and not make employees redundant have been maintained. The paradox is that the Conservatives must bear the political cost of failing to promote the welfare state, but they have failed to secure the economic reward they seek from a real reduction in expenditure.

Strategies for Social Policy

Social-policy-making is usually dominated by the practical

TABLE 8.4 *Strategies for social policy*

	New Right Conservatism	Consolidationalist Conservatism	Technocratic	Unreconstructed Labour	Radical Socialist
Pensions	abolish state pensions; mandatory private provision at minimal level	maintain flat-rate state pension, but shift earnings-related provision to occupational pensions	concentrate state resources on flat-rate pension	SERPS as at present	SERPS, and question tax privileges of occupational schemes
Income maintenance	safety-net through negative income tax	restrict supplementary benefit, especially to young single people	negative income tax	higher scale rates, less means-testing	income guarantee for all, in or out of work
Health	mandatory insurance with competitive hospitals	maintain NHS, with growth money paid for by efficiency savings	improve efficiency in NHS, use private medicine to increase consumer choice	more jobs, higher pay in NHS, control private medicine, phase out charges	abolish private medicine
Education	vouchers given to parents for use at public or private schools	more parental choice; more employment; standards enforced by central government	greater responsiveness to parents, students	more jobs, higher pay, extend opportunities	community schools offering chance of lifelong education
Housing	sell or transfer all publicly-owned houses, abolish rent control	minimise public housebuilding, reduce rent subsidies	end mortgage interest relief, use money to finance repairs; means-tested housing benefit for all	more public housebuilding, more rent subsidies	major public housebuilding, community-based management
Personal Social Services	family responsibilities enforced by law	community care to support families	closer links with NHS to secure more cost-effective care	more resources to replace outdated facilities and extend services	community-based facilities to promote independence of client groups and relieve family carers, especially women

circumstances of the various social services: social security (which may usefully be divided into retirement pensions and the safety-net of income maintenance for the working-age population), health, education, housing and personal social services. But it is possible to identify five ideological tendencies which are influencing current political debate. These are defined by the political strategy they embody rather than by the particular policies they may adopt, and may in practice accept – sometimes without great enthusiasm – similar policy outcomes. The way this happens is illustrated in Table 8.4.

New Right Conservative

At one extreme, the *New Right Conservative* scenario sees the possibility of the state's getting out of welfare altogether except for a minimal safety-net, and transforming social services into private commodities. Patrick Minford, for instance, has argued that pensions, health care, personal social services and even education and unemployment benefit could be provided by the private market, if the state enforced family responsibilities through the law and transferred resources to the poor through the tax system (Minford, 1984). It is a bold vision, but one where the destination is clearer than the method of reaching it. Compulsory private provision may be inefficient; taxpayers from all income groups may value public services; and savings may be swallowed up in other areas of public expenditure.

Consolidationalist Conservative

Next, the *Consolidationalist Conservative* strategy – the one in practice pursued by the Thatcher government – is characterised by a sensitivity towards what is politically possible. This is often determined very crudely – that, for instance, cuts in the basic state pension or in mortgage interest relief are 'not on' – and runs the risk of obtaining only marginal savings and political advantage while bearing disproportionately on the poor and underprivileged. It tends to rely on central government rather than consumer action to improve the efficiency and quality of service. Although its practical effect may seem limited, it has tended to change the public view of what is politically possible.

Technocratic

The *Technocratic* strategy assumes that goals are agreed, and only means need to be perfected. Experience in Britain – notably in terms of organisational change in the National Health Service – suggests that it is difficult to suppress clashes of interest in this way, but the strategy remains influential with the Liberal/SDP Alliance and managerially minded Conservatives. Its best recent expression is the Institute for Fiscal Studies' proposals in 1984 for an integration of tax and benefit systems with greater selectivity (Dilnot, Kay and Morris, 1984). In housing, a notable recent expression of it was the review chaired by the Duke of Edinburgh for the National Federation of Housing Associations, which recommended a phasing-out of mortgage interest relief over ten years.

Unreconstructed Labour

Unreconstructed Labour is the culmination of the post-war feeling that the problems of the welfare state could be solved by the injection of more resources. In practice this has meant that the development of services in kind is secured through improving pay and career opportunities for staff. Its classic expression was the pledge in the 1983 Labour manifesto that expenditure on health would be increased in real terms by 3 per cent a year, and on social services by 4 per cent. This strategy emphasises resource transfers rather than tinkering with institutions; its limitation is that economic constraints may make it as difficult to achieve as were Labour's plans in 1964 and 1974.

Radical Socialist

More interesting is the *Radical Socialist*, which is sceptical about many of the social policies of previous Labour governments and is distinctly less kind to public sector trade unions. It favours area-based approaches, organisation of minority groups, and self-help. It is resistant to what it sees as the oppressive bureaucratic power of state social services. While bold in its opposition to private provision, it hopes that flexibility and choice can be promoted through a socialised system. In its trust of the individual it represents a tantalising convergence with radical Conservative views.

The State and the Social Services

The objectives of the five strategies outlined above have to relate to the differing policy-making contexts of the main areas of social policy. When considering individual services, the usual way of classification is by government agency or clientele served. This sees social policy as a problem of public administration or individual circumstance, but it misses the political essence of the issue – which is whether the state assumes responsibility for providing a service rather than regulating private provision (as it does, for instance, with the motor car). On this basis, we can arrange the social services in descending order of 'dispensability' to government.

Income Maintenance

Most intrinsically public is *income maintenance*. Britain has a national system of income support administered by civil servants, with uniform scale rates. In theory, there is a distinction between national insurance contributory benefits (for retirement, invalidity and unemployment) and others. Politically, it is important that national insurance is 'earned' through contributions paid by employers and employees. But, in practice, the NI system is part of general taxation, and means-tested benefits (supplementary benefit) are paid to a similar level. The Conservatives have aligned the two systems; for instance, the unemployed claim national insurance unemployment benefit (which lasts for up to a year) and means-tested supplementary benefit on the same form. They have also passed the payment of sickness benefit on to employers, allowing them to deduct the cost from their national insurance contributions. The main non-contributory benefit is child benefit, paid direct to mothers: there are no problems of take-up, but it is disliked by government as an expensive and non-selective benefit, and in 1985 the government failed to increase its value in line with inflation.

The social security system is facing a bigger load than ever. A quarter of the population receive an income subsidy benefit, and administration is cumbersome: only recently have plans been completed to computerise case records. Social Services Secretary Norman Fowler led a major review in 1984–5, published as a Green (discussion) Paper, 'The Reform of Social Security', in June 1985, and a White Paper, containing firm proposals, in December 1985. The changes are planned to take effect in 1988. Supplementary benefit would be replaced by an Income Support scheme payable at a higher premium rate to families, pensioners, single parents and the sick and disabled, and Family Income Supplement (a modest scheme for low-paid workers with children) by a Family Credit through the income-tax system. The present 'single payments' for occasional or emergency items would be replaced by a discretionary 'social fund' explicitly intended to promote sound personal budgeting. Any major savings are likely to come from restrictions in housing benefit (help with rent and rates). The review marks a reversal of a long-term policy trend to see income maintenance as a social right, but it does try to maintain the safety-net of income support. In contrast, the New Right would seek to dismantle any state apparatus standing between the taxation system and private charity.

Education

Next most public is *education*, the longest-established of the personal social services. In Britain, it was accepted in the late nineteenth century that the state should provide a basic education service for all. Although private schools (especially the so-called 'public schools') have been important in recruiting and maintaining a social elite, they now provide only 6 per cent of school places. As the school-leaving age was increased (to 15 in 1945 and 16 in 1971) and higher education has expanded, the role of the state has become greater. Private schools have felt threatened by Labour, which abolished direct-grant schools in 1976 and has threatened to withdraw their tax privileges. The New Right has long favoured education vouchers which parents could cash at any school, but consolidationalist Conservatism has not in practice found ways of injecting more private provision, beyond the 'assisted places scheme', which allows schools to subsidise the fees charged to low-income families.

Education expenditure doubled in constant price terms between 1959 and 1973 – but since then it has been static. Part of the reason has been the 20 per cent decline in the school-age population following the sharp downturn in the birthrate after 1964. But that is not the whole story. There has been a loss of confidence in government circles in the purpose of academic education. James Callaghan launched a 'Great Debate' in 1976, and the Thatcher government has emphasised parental rights and the need to conform education to the needs of the economy, a call long associated with a 'technocratic' approach to social policy.

The clearest example of this is the way that the Manpower Services Commission (MSC) has in effect built up an alternative system of post-16 education based on training for jobs. The MSC now finances, and determines the content of, an increasing number of courses at further education colleges. Its Youth Training Scheme (YTS) was introduced in 1983 to provide a semi-compulsory extra year of training and work experience for 350,000 school-leavers who would otherwise be unemployed; it is being extended to two years. Its Technical and Vocational Education Initiative is extending its influence into schools.

This further erodes the autonomy of local education authorities, who run the schools and (in England) non-university higher education. The pressures for uniformity tend to displace the scope for initiative through local politics. These pressures include the examination system, which is influenced by the entrance requirements of universities and itself determines the curriculum; the role of central government inspectors of schools; and the professional norms which protect the position of head and classroom teachers. No political strategy finds it easy to control the content of what is taught by the education system.

Health

The *National Health Service* is a paradox among social services. Unlike education and pensions, it might be expected to work itself out of a job by reducing the incidence of ill-health. At the same time it is a state-of-the-art social service: new equipment and surgical procedures are always being devised. From glamorous surgery like organ transplants and coronary bypass to new operations on the elderly and very young, possibilities of treatment – leading to expectations and pressures on budgets – are increasing. As the NHS is so labour-intensive (55 per cent of budgets go on pay), there are also internal pressures for higher pay, professional status, shorter hours, and backup for skilled workers. Often these have wide public support – as with nurses' pay claims – and are promoted by the 'unreconstructed Labour' strategy; but they do not in themselves improve the level or quality of service.

The NHS is not particularly about 'health', in the wider sense of personal fitness or well-being. Surveys show that an increasing number, now about 30 per cent of the population, report a long-term sickness, and there is wide variation in sickness and death rates according to occupation, residence and class (Townsend and Davidson, 1982). Preventative health measures and improvements in housing and child care might well do more to improve health status than would further investment in hospital-based services. Much NHS activity is for the elderly or mentally handicapped, which have long been neglected in comparison with acute medicine.

The fears of the medical profession and the wish of central government to retain budgetary control have dictated a network of appointed health authorities rather than administration by local government. In England there are 14 regions and 192 districts; Scotland, Wales and Northern Ireland have single-tier structures. The operation of the NHS is based on the general medical practitioner (GP) as the gatekeeper to the system. These are self-employed but receive a per capita fee. Other NHS employees are salaried, but most consultants have a contract that enables them to do some part-time work. The NHS is financed very largely from taxation; charges yield only 4 per cent of the total, but they have been increased by the Conservatives in the few categories where they have become politically acceptable – drug prescriptions and dental charges, which for those not exempt often approach the full cost.

When the NHS was reorganised in 1974, it was claimed that the obvious fault of the original 1948 structure – the separate administration of hospital, practitioner and community services at local level – had been remedied. However, the practical demands of political control had not been met. Recently, a battery of techniques has been devised to try to allocate resources better – territorial reallocation of funds (through the 'RAWP' system), performance indicators, staff limits, efficiency savings, contracting-out of housekeeping services. Above all, general managers (chief executives) are now being appointed to all health authorities following the Griffiths report of 1983 to supplement the consensus management system introduced in 1974, in which doctors, nurses, administrators and treasurers operated as an executive group.

The recent growth of private health care has tempted profit-making American groups into the British market, and consultants into extending their private practices. Private health care may provide a flexibility and choice which many are willing and able to pay for; the danger is that it will weaken the health service and compromise the basically socialised structure of health care in Britain, especially for certain client groups and geographical areas. This is the dilemma facing Conservatives who are not committed to the New Right strategy.

Personal Social Services

The *Personal Social Services* operate in the range of needs between permanent institutionalisation and 'normality'. The principal impulsion of public policy has come from children – adoption, orphans, liability to 'non-accidental injury' and other forms of ill-treatment. From this has developed a wider concern with the family circumstances that put children into difficulties. The desirability of maintaining family structures by appropriate 'interventions' has now extended to the physically and mentally disabled.

Until 1970 in England there were separate local authority childrens' and welfare departments. In that year they were integrated into Social Services Departments on 'generic' lines – that is, area social workers would not work with a single client group. In England (though not Scotland), the probation service, dealing with ex-offenders, remains separate from SSDs.

Social workers have come in for much scepticism and even contempt. Unlike medical professionals, they lack a secure social position based on professional esteem. Their client-centred model of training spawns a jargon which seems strange and far-removed from a practical and intuitive lay approach. The services they provide (like house adaptations and places in residential accommodation) are not as central as cash, a house or health care. But critics often forget that most social-work staff are in residential establishments, and are the 'bottom-line' service, brought in when all else fails; they are plugging holes in the rest of the welfare state.

Pensions

Pensions deserve separate mention, as the main universal social security benefit and the one where the public/private interplay is the most complex. Uniquely, pensions were both an early state benefit (1908 in Britain) and an early occupational benefit. They enjoy high public consent, and do not figure much in the anti-scrounger campaigns which periodically affect those claiming income maintenance benefits. All political strategies see financial provision for old age as a central objective of social policy.

Fifty per cent of workers now pay contributions to an occupational pension – this has not increased in the past twenty years. These are extensively underwritten by the tax system – contributions are tax-deductible, the income of pension funds is not taxed, lump-sums payable on retirement are tax-free. Only the pension payment itself is taxed. Since 1971 public sector pensions have been index-linked, a guarantee that private pension funds cannot give.

This is superimposed on the basic flat-rate state pension; this is low, only 40 per cent of average earnings, and elderly people with no other income will qualify for supplementary benefit. The state has tried to bridge the gap for those outside occupational schemes by two public earnings-related measures – a minimal graduated scheme in 1961, and the much more ambitious state earnings-related pension scheme (SERPS) of 1978. After the Fowler review of social security, the Conservatives planned to end SERPS in 1990 for all aged under 50, but opposition from private employers and pension interests led to that proposal being abandoned in the White Paper of December 1985.

The revised proposal is to reduce the generosity of SERPS benefits to those retiring after 2000, and to make it easier for employers to contract-out their occupational scheme and for individuals to opt for a 'portable pension' that they can take from job to job. No savings are expected until well after 2000. The advantage of SERPS is that it operates on a pay-as-you-go basis, financed out of current taxation, rather than the 'money-purchase' principle of private schemes, where you get out the invested return on what you pay in. It is more efficient to employers than making occupational schemes compulsory, and voters may well prefer to pay higher taxes to ensure proper financial provision, protected against inflation, in their old age.

The Conservatives had pointed out that the number of labour-force members per pensioner is likely to fall from three to two in the first third of the twenty-first century (Ermisch, 1983, table 6.1); but ultimately they have not followed the New Right view that there is no need for state pensions at all. The attempt to abolish SERPS shows that, in Conservative ideology, pensions mark the transition from state handout to an earned right – even though the amount paid by an individual and the net cost to the public purse might be less under a state scheme, or that money-purchase schemes may leave pensioners unprotected against inflation. In the end it was these practical points that prevailed.

Housing

Finally, *housing* is the optional social service – the one where the private market comes nearest to satisfying public demands. Public housing policy may be seen as a temporary interval between the 1920s and 1980s, as owner occupation replaced private renting as the normal tenure for the typical household. In the interim, housing policy was an expression of the same spirit that led to legislation against overcrowding and bad sanitation in the nineteenth century.

Housing expenditure has four components: capital (costing in 1983–4 £2.8bn gross, £1.4bn net); rent subsidies to public sector tenants (£1bn); means-tested housing benefit for rent and rate charges (£2.9m); and tax relief on mortgages (£3.5bn) (HM Treasury, 1985). Only the first two of these enter the housing budget mentioned in Tables 8.2 and 8.3, but the other two are the more important for householders. £2.5bn is also forgone by the exemption from Capital Gains Tax of the sale of a main residence. While under the Conservatives the state is in effect withdrawing from direct housing provision it is accepting long-term commitments to subsidise the housing costs of most of the electorate.

Public housebuilding completions halved between 1980 and 1982. One-third of them are now by housing associations, which are directly financed by central government through the Housing Corporation, rather than by local authorities. Public housebuilding is important politically as the main function of lower-tier local authorities and the most valuable direct gift that local government can make. Public housing policy has been inflexible – it has provided windfall gains to some but has failed to deal sufficiently with the social problems of poor housing conditions. The response of the New Right is to terminate public housing provision – if necessary by transferring houses free of charge to their tenants – and socialists recognise that much greater attention must be paid to the wishes of consumers.

* * * * * *

There is also an outer ring of social policy, encompassing services like employment services, industrial promotion, legal aid, transport, the arts, and to some extent defence and agriculture. The overt objective of most of these is to maintain the framework of law and security, or facilitate economic growth, but the practical effect is to subsidise particular consumers or employees. As these objectives may seem more attractive politically than do explicitly social ones, there is a likelihood that the outer ring may escape any cuts or retrenchment imposed on the core of the welfare state.

The most severe problem of overlap between social policy and other public policies is in the case of the unemployed, whose numbers have increased from 1.5 to 3.5 million under the Conservatives. Traditional British policy assumed that economic policy would make unemployment a declining problem, easily manageable by social policy, but the severity of the collapse in manufacturing employment in Britain since 1980 has made this no longer tenable. The response has been to increase the funds of the Manpower Services Commission, who provide work and training for the unemployed through measures like the Youth Training Scheme and the Community Programme; but their spending in 1983–4 was only £1bn against the £5.7bn that went on cash benefits to the unemployed (HM Treasury, 1985, table 3.12.5).

Social policy has become an intensely political matter, and politics is often not about fairness or efficiency. It is wrong to see social policy as something for marginal groups. It benefits everybody, and the more it benefits the privileged the more concealed it becomes. It is clear that public subsidies for transport, higher education and mortgages disproportionately go to the middle class (Le Grand, 1982). A whole middle-class lifestyle and 'yuppie' culture depend on welfare-state jobs – teachers, doctors, social workers. The future of social policy depends on the consent and preferences of these indirect beneficiaries just as much as on the obviously poor.

What Difference Does It Make?

Has the Thatcher revolution really changed the social services and people's attitudes towards them? It might seem to have done – Victorian values should have displaced reliance on the state, and the growth of services halted if not reversed. But opinion-poll evidence makes it clear that a core of the welfare state still enjoys solid public support. This includes pensions, hospitals and schools. In contrast, council housing, unemployment and child benefit are more vulnerable. On the basic taxes-versus-benefit question there is a political warning to the Conservatives, since the preference for tax cuts over improved services seems to have been a temporary phenomenon of 1979 (Taylor-Gooby, 1985).

The main impact of the Conservatives has been to desensitise the political system to the social ills that inspired the welfare state in the first place. The principal ill is unemployment, but poor housing and chronic ill-health are also important. Increasingly, poverty seems to be concentrated in the major conurbations, especially in the North of England. The poor are socially marginal, and also politically marginal in that they cannot build an election-winning coalition. They, and not the employed and affluent population, bear the brunt of the restrictions in public services the Conservatives have imposed.

Ultimately, consumers of the social services are in a position of weakness. They may be sick, disabled, very young or very old, lacking shelter or any kind of income, distressed, violent or alienated. In a commonsense way their problems tend to be quite simple – but the means of dealing with them are complex. Simple diagnoses of personal problems are replaced by more ambitious ideas of individual well-being which allow the production of both political promises and professional opportunities. Social policy allows government to become a benefactor to individuals; it is less clear that the services produced truly meet the needs of those who receive them or the political preferences of society as a whole.

9

Foreign and Defence Policy

PETER NAILOR

The Example of the Falklands Campaign

The Falklands campaign, from April to June 1982, illustrates how complex the interaction between foreign and defence policy and the domestic base of politics can be. The security of the United Kingdom itself was not threatened, but the ability of the United Kingdom to defend its interests in a crisis, and to maintain objectives and principles that are important to the country, was put under a scrutiny that was tense and urgent, because the Argentinian invasion of the Islands was unexpected.

The invasion was initially a defeat for British foreign policy, that was signalled by the resignation of the Foreign and Commonwealth Secretary and two of his senior colleagues. But it was mitigated by the success of the British defence forces, which improvised a campaign that repossessed the Islands within four months, and was supported by a diplomatic campaign that mobilised political support and military assistance in crucial *fora* and areas. The crisis showed both that Britain could act effectively in support of its interests, and also that such action depended significantly upon the approval and support of its allies. The political circumstances were such that this support could be mobilised; the Argentinian invasion was an effrontery that not even anti-colonialists could easily excuse, and the nature of the Argentinian regime diluted regional or ideological considerations that – for the United States and the Soviet Union, for example – might otherwise have complicated the issue. The need for Britain to act, rather than to rely upon the moral force of international opinion, arose out of a number of considerations, some of them of domestic rather than of foreign policy concern. It was necessary to show, in foreign policy terms, that Britain was not prepared to accept a forceful *fait accompli*, but this was as much a matter of the intangible issues of reputation, style and credibility as it was of any direct read-across to other cases of potentially similar crisis like Belize or Gibraltar – or Hong Kong.

Even though it is right and prudent, as well as fashionable, to talk of arbitration and conciliation as ruling principles of international conduct, the pattern of international politics in action shows that the ability to use force is still an indispensable tool of policy. Without it, even the offer of conciliation may not be credible. Domestically, however, the pressures on the British government were crucial. An explosion of public opinion, reflected as much as orchestrated by the popular press, demanded satisfaction, both for what was seen as an unjustifiable act and for the failure of policy. In spite of attempts, well after the event, to speculate about what might have been the outcome of diplomatic efforts at mediation, it was difficult at the time, and it still seems difficult, to see what might have satisfied public opinion other than a total Argentinian withdrawal. The possibility of a graceful British concession had vanished with the invasion, and the implacable element of honour was introduced into an issue which, in truth, had not previously been seen as significant. The Falklands dispute with Argentina had in fact dragged on for many years, because it was not of much importance in the domestic British environment: not important enough to risk the parliamentary row which the few enthusiasts might have engineered, about a 'sell-out to a bunch of fascists'.

Domestically, the results of the campaign brought credit to the defence forces, which showed high professionalism and skills, and benefit to the government, whose determination was personified by the Prime Minister, with important consequential effects upon her electoral appeal and distinctive leadership position in her party and administration. Militarily, it exposed some problems, in matters of equipment, organisation and support (such as the much diminished merchant shipping resources that could be called upon); diplomatically it provided something of a fillip to the country's general reputation, but it left the Falklands issue itself as a more intractable, and much more expensive, problem. The outcome was therefore both successful and inconclusive; but, as an example, it helps to show how foreign policy and defence are closely interrelated. It also showed that, in spite of a sophisticated governmental machine, the unexpected can throw policy into confusion. The ability to react is something as useful as the capacity to plan ahead.

The Policy Context

'The security of the realm' is, in fact, one of the traditional and most fundamental tasks of any system of government. In the case of a major and prosperous state, it is usually organised, conceptually and administratively, in ways that distinguish between diplomacy and force. *Foreign policy* is not only about treaties and negotiations on major political issues; much of it is concerned with day-to-day exchanges of information which facilitate the international movement of goods, services and resources. It is often as much interested in the interpretation of accumulated evidence and opinion, about what the external world thinks of us, and what we want to express to the external world. *Defence policy* in some states is weighted towards the maintenance of domestic stability, and it is true that all states are concerned with the 'bottom line' problem of the defence of their own territories and political systems. But for states like the United Kingdom, with a history of active international involvement and widespread commercial and political interests, that in our case still extend to the residue of a great empire, defence policy tends to be thought of primarily as an instrument and attribute of external policies.

This tendency is reinforced by historical experience. The United Kingdom's geographical position has been strategically advantageous in struggles with enemies like the Dutch, the French and the Germans, but it has always called for the political and military assistance of allies. Britain is a small metropolitan base with limited manpower resources with which to exercise power and influence. Although for a long time her economic power and overseas possessions enabled her to make great efforts to sustain successful alliances, Britain's dependence upon alliances has been increasing. In both of the World Wars, the resources of the Empire and of her continental allies were not enough to ensure victory against determined adversaries; and the intervention of the United States was a conclusive factor. The massive effort called for by the demands of industrialised mass armies needed more men and more equipment than Britain could herself provide; and in the post-war world after 1945, Britain's relative economic decline has driven her to economic alliance with her neighbours, in the European Community, as well as to a security partnership with the other countries of NATO, particularly with the United States.

Thus, while the United Kingdom is still, by any standard of comparison, a powerful economic, political and military entity, her continued security now requires a much more careful balance between the pursuit of her own interests and a judgement about what will advance the collective interest of the groups to which she belongs.

The Policy Machinery

This predicament shows up in a number of ways. Politically and administratively, it means that the problems of deciding what policy to follow, and whether we can afford it, are sometimes long drawn-out. Institutionally it has led to both formal and informal changes in the way in which British policy is worked out and implemented. The two great departments of state, the Foreign and Commonwealth Office (FCO) and Ministry of Defence (MOD), are still the core organisations which have a formal responsibility for managing affairs. In the last twenty years or so, they have gone through a series of re-structurings: foreign, Commonwealth and, to a certain extent, European affairs have been brought together in the FCO, and the three historically separate Service departments and the organisation for equipment procurement have been melded together in the MOD. Each department is politically directed by no more than five or six ministers; each of them has a range of parliamentary and departmental responsibilities that make them 'heavy jobs', even at a time when all office-holding imposes great demands on competence and resilience. The two Secretaries of State are, naturally, senior appointments in the administration. The Foreign Secretary is historically a very important political figure, and the resources which are now regularly allocated to defence make the Defence Secretary hardly less significant to the fortunes of the government of the day. The political significance of the appointments can still, however, be distinguished, using the difference that Clement Attlee noted. The FCO is primarily a 'policy' department, giving great political visibility; Defence is an 'administrative' department where, although high policy is made, a minister must have good managerial skills as well. The personal relationships between these senior members of the government and the Prime Minister of the day are rather more important than they used to be.

The number of Cabinet-level ministers concerned directly with overseas affairs has declined, and the pattern of consultation with allies, in the EC as well as NATO, formally and directly involves the Prime Minister as well as departmental ministers. The fashion for personal diplomacy, notably in 'summits', and a constant exposure to the various forms of news and publicity (which tends, on television at least, to emphasise personality at least as much as it deals with issues), reinforces this trend. And the pressure to respond, which is one major effect of the enormously improved communication network of the electronic age, puts a renewed emphasis upon personal consultations among senior personnel, like Cabinet members, especially when there seems to be a need, under pressure, to provide a solution or an answer in 'real time'. The same sort of pressure puts a growing burden, too, upon the supporting hierarchy of advisers and officials who may have access to a lot more data, but also may have rather less time to assimilate and reflect upon what it all may mean.

The Cabinet and Cabinet Committees

Foreign and defence problems frequently come to the full Cabinet, and there is usually a routine brief by the Foreign Secretary at the start of the meeting. There is a sub-committee of the Cabinet, chaired by the Prime Minister, which deals with much of the FCO and MOD business that needs ministerial approval; and there is also a ministerial sub-committee which superintends European issues. The growth of the large conglomerate departments has put back into the departmental process some contentious issues that earlier might have come to Cabinet, but there are still many problems about which, because they have sensitive political aspects or simply because they fall between the departmental 'boxes' in which government business is organised, other Ministers have to be consulted. Sometimes this is done bilaterally, but it is often convenient, and occasionally necessary, in a presentational sense at any rate, to get a formal 'chop' on a major matter. On the other hand, the Cabinet defence sub-committee has occasionally been used to deal with specially sensitive items and, in effect, to by-pass the full Cabinet.

Below this level, there is a set of specialist Cabinet Office committees, staffed by representatives of the relevant departments, which prepares the business for the more senior bodies. Additionally, there is an important specialist group, the Chiefs of Staff Committee, which is the formal method of preparing military advice to the Cabinet. The Chief of the Defence Staff, who is the chairman, now has extensive powers and responsibilities which amount, in many ways, to the role of a commander-in-chief, but the heads of the individual Services still retain the right to make any particular representations they want to insist upon. The Chiefs of Staff Committee meets regularly, and is illustrative of the extensive formal and informal links that exist between the FCO and the MOD, in so far as there is regular FCO representation and involvement, in the same way that in NATO, for example, the British delegation is made up of FCO and MOD components.

The Domestic Context

The work of the two departments is closely linked, in fundamentals as well as in day-to-day matters, and they are highly likely to see each other as allies in inter-departmental matters when perhaps overseas policy has to be balanced against domestic policy, and when the Treasury is looking for budgetary sacrifices. But there are major differences too, not least in costs. Foreign policy, except when it needs to be operationalised in programmes of aid and support, is not a very expensive activity, although some supporting activities like the BBC overseas services and information services, including the British Council, attract attention from time to time. Defence, on the other hand, is a hungry user of resources, manpower as well as money and high-technology industrial resources. The philosophy of deterrence, which puts a premium upon high states of readiness at the expense of less immediately available reserves of men and equipment, requires a cost to be paid for this readiness that is reflected in a defence budget which in recent years has risen from 4.5 to 5.25 per cent of the gross domestic product. This development happened in response to requests from the US and NATO to improve defence co-operation, but the growth occurred at a time of deep industrial recession, and very much complicated the Thatcher government's attempts to restrict the growth of government expenditure.

In some ways, therefore, the MOD can be seen, from a political point of view, to be a rather awkward department to handle. It spends a great deal of money; the military are a well-organised and efficient pressure group as well as being proficient at their jobs; and defence is a policy area where it is sometimes dangerous to be too radical or to take many risks (like, perhaps, signalling an intention to withdraw HMS *Endurance* from the South Atlantic, in order to save £3m). There is a continuing need, however, to try and make sure that the outfit is being run as efficiently as possible; which, in political terms, means stopping budgetary 'creep' and, possibly, saving money by reorganisations and redeployments.

The FCO, on the other hand, presents other difficulties. Although it is broadly true that for much of the post-war period there was a general consensus between the two main parties about Britain's foreign policy needs, there were always differences about emphasis and interpretation. During the 1970s more substantial differences emerged, as, for example, over continuing membership of the European Community and, later, over the place of nuclear weapons in British defence. One of the implications of this move away from consensus was that a new government was now much more likely to come into office with policy preferences that had been worked out in opposition, and might well have been formulated in a particular way that emphasised how different they were from the last administration's activities. They might, perhaps, not be all that different in fundamentals; but in this new atmosphere, both style and substance have to be represented as – at least – more appropriate. To give one early example to illustrate how style changes, Heath's government, between 1970 and 1974, when joining the Community was a major preoccupation, de-emphasised the connections and the 'special relationship' with the United States. By 1976, under Callaghan, and again after 1980, when President Reagan and Thatcher were in office together, the special relationship reverted to its accustomed place and, on occasions, assumed a high rhetorical visibility. These differences undoubtedly reflected the purely individual relationships between personalities; but they were also meant to reflect substance too. Thatcher not only found Reagan personally congenial, but British policy found some, at least, of the United States' policies more congenial. And the British government found American support, most notably over the Falklands, worthy of acknowledgement in this way.

For the FCO, this new move away from consensus creates problems. If there are differences between the political parties, the new objectives of a new government may take quite a long time to turn into visible change, even after the new government takes stock of its position, and argues its case through the layers of official opinion and precedent. If it is a question of recognising a controversial regime, or stepping up overseas aid, policy might be implemented quite quickly; but if it means building up new and profitable trade relations (for example with the Soviet Union and Eastern Europe, in the mid-1970s), there is quite a long time to wait, after the fanfare of signing ceremonies has faded, before results accrue. If there are differences within the party that forms the new government, however, then the position can become quite confused because nobody will be quite sure what the new policy is to be. If it is to be a compromise that sounds good but does not make much difference, it seems that it is difficult for the FCO to avoid collecting the criticism that will arise. It is not unfair to say that this is a difficulty which the Labour Party has more often had to deal with; the presence of a lively and traditionally strong strain of pacifist thought in the party has often created problems in agreeing a party-wide 'platform' on security policy goals. But differences affect the Conservative Party, too.

Soon after the Conservatives were returned to office in May 1979, it became difficult to hide the fact that the FCO's policy towards the settlement of the Zimbabwe question, and the policy preferred by Thatcher and a segment of party opinion to take a more direct initiative towards the refurbished Smith–Muzorewa regime in Rhodesia, were effectively irreconcilable. The FCO pattern was a 'continuous' policy, deriving from Commonwealth and Anglo-American discussions and proposals that went back several years under previous governments, and it was only under pressure at the time of the Lusaka Commonwealth Conference that Thatcher's desire to institute a novel policy was turned aside. It may well be that her determination to end the chronic state of affairs in Rhodesia/Zimbabwe stimulated the quite remarkable progress that was achieved in the ensuing months; but the incident drew attention to differences of style, at least, and led to a more persistent criticism in the Conservative party of the emollient performance and ethos of the diplomatic 'establishment' that resurfaced again over the Falklands campaign (*The Economist*, 27 November 1982).

The FCO is susceptible to criticism in this way since a large part of its function is to stay, as it were, in the middle of the road. This renders it vulnerable to the passers-by on the left and on the right. But the ability of any one state, except perhaps a super-power, to change the framework of international problems significantly or quickly is really quite limited, and it may well be necessary to manifest patience rather than initiative. This is where the FCO, as the permanent and professional group concerned with foreign policy in action, may fall out with a new government that wants to change the sign-posts along the road. To take the Zimbabwe issue as a case in point, although the eventual settlement was reached in a surprisingly quick way, during and after the Lancaster House Conference, and although that progress may have been stimulated in part by the new urgency of a new British government, the problem had existed, in its essentials, for nearly twenty years. An increasingly fierce and destructive war was debilitating the 'front-line' states as well as Rhodesia itself. But in 1979 it was generally perceived – by more of the participants than ever before – that the combination of circumstances at that time provided what might be a final chance for a negotiated solution. Even so, the final shape of the outcome, with Robert Mugabe's sweeping electoral victory, was largely unforeseen in Britain (even by the FCO). It was disappointing to the Prime Minister's faction that had urged a quick deal with Smith and Muzorewa: disappointing enough to foster resentment about the 'soft' FCO, and their negotiating champion, Lord Carrington, and to lead, in the aftermath of the Falklands campaign, to a resurgence of the arguments about whether the Prime Minister should not have a separate department of advisers and staff ministers, large enough to superintend rather than just to collate the co-ordination of affairs.

The normal pattern of diplomacy is, however, as much reactive as innovative; however well the national interests may be described in terms of general principle, it is unusual to find governments being able to move directly and consistently in pursuit of stated objectives. Their aims have to be set against the aims of other participants, who may have local advantages and more intense concerns, as British governments in the 1960s and 1970s found with Iceland, over a series of fishery disputes. The state may simply have to bide its time, as the British found out in pursuing the goal of membership of the European Community between 1961 and 1972. In the international arena, even more perhaps than in domestic affairs, the framework within which history and the pursuit of advantage have defined or limited the state's freedom of manoeuvre makes little provision for electoral timetables.

Some issues, too, fall across the boundaries which, either administratively or on precedent, we call foreign or domestic. Our relationship with our European neighbours is now, of course, one such issue; but the particular relationship with Eire, and the way in which the relationship is moderated by the fact, the existence – and the turbulence – of the province of Northern Ireland, is perhaps the classic example. To deal with the affairs of the province as if it was only of concern at Westminster is not acceptable (and not realistic to the FCO and the other ministries involved in Whitehall); but to try and engage the constructive participation of other actors, whether they be in Dublin, or Belfast, or even in Washington DC, is extremely difficult, without throwing up how clearly the differences may be between political and administrative perspectives (see Chapter 10).

In this sense, there is a distinct difference – certainly at the theoretical level, and sometimes in practice – between the interests of the state and the interests of any one particular administration. Reconciling such differences is the political prerogative of government; but 'carrying the can' seems to be an increasing burden for the FCO, at least in the sense that now the Conservative Party in office grumbles openly about the service it gets, almost as much as the Labour party used to.

Parties and Policy

There is another implication, too. What has been discussed so far is about what happens when a new government comes into office. But the decline of consensus also affects the relationship between government and opposition in Parliament, and therefore the extent to which differences in policies are openly debated. If a particular issue becomes openly controversial, then the government of the day, and the FCO as the executive agency through which policy is carried forward, may be disadvantaged as they seek to persuade other states or interests that British policy is fixed, and reliable. Other Foreign Offices read *The Times* too; and the other party in a negotiation may simply decide to wait and see what the next General Election will do to British policy. This is not really a novel problem, and is certainly not only a British problem; the democratic process, which provides for the peaceful transfer of power through an open electoral contest, necessarily provides for the open debate of differences. What is unfamiliar in Britain is the wider range of difference which now exists on foreign policy and defence issues between not merely two contending parties, but four. But this has occurred after a period of some thirty years in which one could reasonably say there had been a novel range of agreement and consensus.

Defence Differences

If we take one defence issue as an example, there is now very extensive divergence between the Thatcher government and the opposition parties about nuclear weapons policy. The parliamentary differences are extended by the presence of extra-parliamentary debate, too, which has been developed and focused, by groups which, although they represent a wide range of perspectives and concerns, we can call – in shorthand terms – 'The Peace Movement'.

The debate is concerned with a number of specific issues, but they have a common theme. The theme is the extent to which Britain should remain a nuclear-weapons state herself, and even includes the extent to which Western security should depend on nuclear weapons at all. Thus, some of the argument is about what British policy should be in the future, and some is more concerned with how Britain should influence her allies. One of the major issues is whether Britain should maintain her own nuclear deterrent. The existing force of four Polaris submarines dates from 1969, and in 1980 the new government announced plans to replace it in due course with another American submarine-based missile system, the Trident, under the same arrangements through which Polaris has been obtained. The plans have subsequently been altered to take account of the American decision to deploy a more advanced variant. It will be a lengthy and costly programme, even on the official estimates of some eighteen years and some £9,300 million pounds. Arguments about whether the policy should be implemented have developed on two levels. The first criticism is that the cost may grow (and may even be uncontrollable from an administrative point of view, since a large part of the expenditure will be incurred in the United States, and will therefore depend upon the pound/dollar exchange rate); and that it will be so large a programme that it might divert resources from other, conventional equipment, plans which are themselves large and ambitious. The second criticism is of a broader and more essentially political sort: that the United Kingdom should not try to remain a nuclear power and should certainly not go for a system that will militarily be more significant, and larger, than Polaris. The government response is that the estimates already include a large contingency reserve, that the programme is manageable; and that a British deterrent is still, nationally and collectively, an important security asset. Trident has, nevertheless, been caught up in the more general agitation about the high levels of nuclear armaments internationally, and is a prime symbol in the argument about the need for specific British initiatives towards nuclear disarmament.

The movements that include this advocacy derived a new wave of popular support in 1979–80, in reaction to the NATO plans for a 'twin-track' Cruise and Pershing missile policy that would, on the one hand, provide some counterweight to the growing Soviet superiority in nuclear weapons affecting Europe, by installing new missiles in a number of European countries and, on the other hand, would also induce the Soviet Union to limit its own programme and agree to some mutual limitations. It was a very complicated plan, that had important secondary implications in relation to reassuring European governments about the United States' long-term intentions. Whatever its intrinsic merit, it stimulated an unusually widespread range of concern, that led to large demonstrations in a number of countries. To a certain extent, governments were caught napping; few of them had made any recent attempts to lay before informed public opinion their own views about the intricacies, the risks and the benefits of what had by now become a traditional policy of deterrence, both nuclear and conventional. It was the first time for twenty years that nuclear weapons had become so wide and intense a source of concern; and governments were less ready with explanations about what their policies meant than their critics were with arguments about what would happen if those policies failed.

In the United Kingdom, the Peace Movement recaptured Labour Party support and gained sympathisers in the Alliance, with more extensive consequences for party policy objectives than was the case in 1959–60 (Capitanchik, 1977, 1983); and there is a wider basis of encouragement for them in similar Western European and American movements, and support for some of their morally derived concerns from senior churchmen. Politically, 'nuclear disarmament' is not a straightforward inter-party issue and it is a debate that is very difficult to focus, unlike arguments about education or agriculture where outcomes may be specific and measurable. All political arguments, of course, have symbolic as well as specific elements, but this particular debate touches on issues that all the protagonists accept are awesome, and fearful. Deterrence depends for its effect in part upon the same fear of the consequences of nuclear war that the disarmers share. The debate has both domestic and international implications (*The Times*, 29 November 1982) and featured in the 1983 election campaign. Although in form it is related to the relatively recent emergence of the Soviet Union as a military power of much increased potency, it also has some relationship to the apprehension that arises from United States policies that are perceived to be 'hard' and, therefore, more risky. In that sense, in the same way that there is an historical link between the CND of the 1960s and its rather different modern counterparts, there is a connection between the days of John Foster Dulles and 'brinkmanship' and the hard rhetoric of the early period of the Reagan administration.

The Thatcher government has been firm in defence of its policies, and has stressed their essential continuity both in national and alliance perspectives. They have been more concerned than earlier governments were to explain the nature of their thinking, both in debates in Parliament and in government publications. The opposition parties have a range of alternative policies, some of them still a matter of argument within the parties; but they share a degree of acceptance of the disarmers' case that reliance on nuclear weapons is not only dangerous but wrong. They differ on how far the reduction of such a reliance should be balanced by spending on additional conventional armaments, or by more actively pursued arms-control negotiations; it is not clear exactly what their policies in office would be, or how quickly they might move to radical alternatives. A crucial element seems to be the different interpretations which the protagonists all adopt about the nature and the immediacy of the security threat posed by the Soviet Union and its allies; it is not the only fundamental point of difference, but it is one from which spring a lot of the secondary arguments about how extensive defence preparations need, in any case, to be. And it is the issue which would probably decide what policies the opposition parties would actually follow, if they achieved power. Here, of course, one of the most interesting quandaries, for government and for opposition parties, is whether the new, and rather more modish, style of Gorbachev is anything more substantial than window-dressing. If a new Soviet government became more efficient, that would in itself be a new problem; but a new Soviet commitment to arms control would – or might – be worth taking very seriously.

It would be wrong to say that it is the existence, or growth, of extra-parliamentary agitation that has attracted the involvement of the political parties; there is, as has already been noted, a long and honourable pacifist tradition within the Labour Party, and a pacifist–internationalist element within the Liberal Party. But the proof which the Peace Movement offered, especially in the 1979–83 period, that there was a wide public concern that could be activated, was not lost upon the opposition parties, who were encouraged by it to try and define their own positions more radically than they might otherwise have been. Defence issues, and nuclear weapons, have not traditionally been seen by political party managers as easy vote-getting topics; but, in 1983, they probably did affect the way in which people voted (see Chapter 1). The opinion polls certainly showed that the issues were a matter of concern (*The Times*, 14 June 1985). But that may have had something more to do with the state of the major parties at that particular time – the Labour Party was in disarray, and the Alliance was not altogether convincing in its harmony, for reasons quite unconnected with defence policies. It is also true that, at that particular time, the focus and publicity which the Greenham

Peace Camps undoubtedly provided in relation to the installation of Cruise missiles also served to detract somewhat from the more general arguments advanced, for example, by E. P. Thompson, and Joan Ruddock of CND. The difficulty for the Peace Movement, as they sought to influence both party managers and the general public, was whether to concentrate upon a single issue or to argue on a wide front; and this, too, recalled the earlier days, and the earlier problems, of CND. The concern to which their arguments gave rise, however, does seem to have shifted public opinion quite significantly – at least to the extent that arms control and disarmament figure more prominently on any political agenda.

Foreign Policy Differences

Another example, on the foreign policy side, that illustrates party political differences is British involvement with the European Community. The United Kingdom joined the European Community in January 1973, but almost immediately the question of continued membership was tested, in a constitutionally innovative way, by a referendum in 1975. The new Labour government, which came to office in 1974, was divided, on Community membership as well as on other constitutional issues relating to devolution. A referendum was the device employed to resolve this division. It was, however, a rather unsettling experience which may have created more doubts than it resolved. For, with the advantages of hindsight, we can now see that the great economic problems of the 1970s, stretching into the 1980s, made any expectations of continuous progress that the enlarged Community had, and which it had itemised in a series of goals to be reached by 1980, too optimistic. The British experience of the Community, and its own role and attitude within the Community, have led to strains and disappointments that, even in 1975, would have seemed overdrawn. Consequently the question of continuing membership is a bone that elements in the opposition parties still chew on, and unrequited patriots everywhere still find tasty. Because the Community disappoints British interests from time to time, or does something which seems either silly or unfamiliar, it is a bone that can be tossed into the air, and it will always raise a few growls.

However, the serious undercurrent is that continued membership of the Community, and the increasing familiarity and expansion of common practices, effectively constrain the opportunity unilaterally to change British practices and structures in any radical way. In that sense, a measure of sovereignty really has rather drifted away from the United Kingdom; and the theorists, mostly on the left of the political spectrum, but with a few on the right, who postulate that a radical alteration is necessary to overcome British decline, are the serious protagonists of withdrawal. They are, in the proper sense of the term, extremists, because they advocate a development that would not only call for a major discontinuity in policy, but very extensive consequential structural change. Some of their arguments tend to play down the fact that the act of joining the Community was also a distinctive – some would still say extreme – commitment to change: change which had the objective of moving, with our European neighbours, collectively towards enhanced prosperity.

One change that happened quickly was to expose us to a new sort of politics. Politics in the European Community is not an exercise in traditional diplomacy; it is not 'foreign policy' in the conventional sense, and the FCO is the conductor of the orchestra rather than the composer of the symphony. The departments of Agriculture and Trade, and the Scottish Office, for example, play important solo parts over which the FCO may have relatively little influence. Some of the issues, like agricultural support, are intrinsically complicated and very large sums of money can be involved by way of transfer and adjustment. The raw politics of these matters affects, too, the way in which they are perceived and managed. Agriculture and food prices touch domestic concerns of great sensitivity: plans for regional aid, or for the introduction of new emission controls on motor-cars, affect employment prospects; and even within Britain a large number of governmental and non-governmental interests can be involved. The inevitable tendency is therefore that persistent disagreements in the Community have to be dealt with at a more senior political level, even if the issues are detailed and technical in form. Policy may finally have to be agreed at foreign-minister or head-of-government level. The string of Community summit meetings that have been dominated by specific issues have tended to emphasise disagreements within the structure. However unfair this might be, in relation to all the other things that the Community agrees about, it is in fact symptomatic of the difficulties that exist in not merely maintaining momentum in such a large and complex community but also in making notional agreements work out in practice.

After 1979, the Thatcher government brought a more direct, and bruising, style to Community politics. The willingness to make British viewpoints explicit and dramatic owed its origin in part to a sense that the need both for reform (particularly in agricultural support methods) and for short-term improvements was urgent. By the turn of the decade, the benefits of membership in day-to-day terms were seen by a range of opinion in the country, especially by segments of the Labour Party, as being so marginal as to justify a renewed agitation to withdraw from the Community. The generally worse economic environment, in which the Community's more general objectives were being slowed down, tended to highlight the significance of short-term gains and losses. The low level of active public interest in the Community was shown by the low turnouts in Britain for the European parliamentary elections in 1979, and again in 1984.

The tone of government involvement at Brussels and the lack of enthusiasm by the voters for the European parliament tend to confirm the suspicions of those other members of the Community who wonder whether the British have really made up their minds about the Community yet. In some ways, the suspicion may be justified; the popular sense of commitment and involvement is not high. But sectional support, in industry and agriculture, can be shown to be both high and committed; and one could argue that the brusqueness of some of the Thatcher government's negotiations demonstrated commitment at least as much as dissatisfaction. Beyond the tangled web of Community Agricultural Policy (CAP) affairs, the picture is often much clearer – but because the negotiations are less dramatic they do not always get the same attention. There has been a steady improvement in the way in which the

Community countries consult together about their common interests, and represent these interests to the outside world. The mechanism for European Political Consultation (EPC) that has now been formalised has produced some joint foreign policy agreements and declarations, even in its early days, that are quite significant advances – on policy relating to the Middle East, Afghanistan and the Falklands, for example. But some issues are still too difficult or divisive to agree upon: the American desire to impose sanctions on Poland and the Soviet Union in 1980–1 is an example here. A consistent European foreign policy is still a long way off, and the enlargement of the EC will almost certainly complicate the process. Foreign policy is so often tied so closely to security that to make foreign policy without being able to tie it in with defence policy is, philosophically as well as administratively, difficult; and attempts to mobilise European high-technology industries into forms of co-operation that would enable the EC to compete and co-operate with American high-technology initiatives highlight the problems.

Here the differences that exist between the EC and NATO – in membership, scope and perspectives – are major stumbling blocks. In theoretical and aspirational terms, the 'European ideal' which was so important in creating the EC provides for the Community to be developed so that it could eventually produce European union. The ideal is still there, and the aspirations still exist, and if some form of structure could be developed that transcended the limitations of the nation-state, then unitary action in all sorts of areas could follow. However, it is also possible to see the Community stopping short of that ultimate fusion, and still to be a foundation for better prosperity and co-operation. That seems to be the view of successive British (and French) governments: co-operation and involvement are indisputably necessary, union is a disputable and is certainly a very long-term goal. If there is a sense in which Britain has not yet joined Europe, it is at this level of ultimate aspiration.

The Special Relationship with the United States

If there is a single issue which symbolises why the United Kingdom is so cautious, it is the 'special relationship' with the United States. De Gaulle saw this link as being so fundamental that it was in his mind a major reason for blocking British membership of what was then 'the Common Market'. But although the world, and the United Kingdom, have moved on since those days, relations with the United States still form one of the staple items on the British foreign policy agenda, and are of great importance in security matters.

The United States connection seems so natural to most British people that it is easily misunderstood and mythologised, and Edward Heath's attempts to play down 'the special relationship' were both reasonable and disconcerting, in the light of our new European status. The condescension and mismanagement of Kissinger's 'Year of Europe' in 1973, and the paralysis induced by Watergate and Nixon's resignation in 1974, were disturbing; and the earnestness and confusion of the Carter period went even further to increase British bewilderment. At the official level, relations remained close; American support and assistance were instrumental in producing a new framework for resolving the Zimbabwe question and for engaging South African influence constructively, even when American policies towards Angola and Namibia were less coherent and less marked by consultation.

Britain had a higher expectation than some of the other allies of the capacity of American leadership (as well as a greater expectation of being listened to in Washington), and found it sometimes more difficult to participate in Joint European initiatives that were distinct from American viewpoints, as over Afghanistan and, to a lesser extent, over the Middle East after 1979. The renewed prominence that Callaghan and then Thatcher gave to a continuing 'special relationship' accordingly emphasised two parallel strands of political significance: first, the extent, the importance and the familiarity of Anglo-US friendship which, it was inferred, had not been displaced by EC membership. In so far as the 'special relationship' was based upon defence and foreign policy, it did not conflict with the limited spread of joint European activities. But, second, Britain was now in a better position to represent both British and, to a certain extent, common Anglo-European concerns to the United States; and that was both a good thing in itself and an enhancement of Britain's utility as a valued partner. The intricacy and some of the hazards of Anglo-American relationships were brought out in the 1982 Falklands campaign, when US concern, on grounds of principle as well as sentiment, to support a loyal ally, were intermingled with a national desire not to damage hemispheric relationships within Latin America. This tension illustrates for the United States the complexity which a leading role in international systems can have, when all relationships can be seen, at some times and over specific issues, as special in some way or other. It is fair to say, in this regard, that the British emphasis on its own special relationship still, from time to time, shows some wistfulness that it is now no longer unique.

In matters of defence co-operation, however, it has some claims to be considered unique. The sales of defence equipment, most notably Polaris and Trident, and a very wide range of co-operation at the political as well as at military levels, are tokens of a very long-standing and, it seems, mutually beneficial habit. Disagreements still occur, and consultation sometimes breaks down, but the fundamental ties persist.

The vivacity of this unusually sustained habit is partly dependent on a British willingness to accommodate to American priorities; it is a price to pay that cannot realistically be avoided, but is now sometimes more difficult to maintain since the United States has developed a direct set of bilateral links with the Soviet Union. In the period of emerging *détente*, between 1958 and 1965, the United Kingdom could sometimes act as a goad or an intermediary to both super-powers; now the super-power dialogue is perhaps the most significant 'special relationship' of them all. Britain may still be able to urge its own viewpoint about how necessary such dialogue is (and this certainly occurred in the matter of arms-control negotiations in the late 1970s and early 1980s), but her direct utility is now much lessened.

It may still be of some significance within the alliance

partnership however, in dealing with major American initiatives. President Reagan's attempts to review, and perhaps fundamentally alter, the structure of the military competition between East and West, which have been popularised as his 'Star Wars' concept, raise fundamental as well as structural issues of great potential significance. It is not merely a question of how to develop technically, and how to conceptualise doctrinally, the implications of a reliance on strategic nuclear defences. It also calls for an evaluation of what this would mean for conventional defences, and for alternative, non-ballistic, ways of threatening or avoiding conflict. The first British critique of the Strategic Defence Initiative (SDI), made in a speech by the Foreign Secretary (Sir Geoffrey Howe, Royal United Services Institute, London, 15 March 1985), followed a speech by the Prime Minister to the United States Congress a month earlier. It was a sympathetic but substantial criticism, that outlined both political and technical issues that would have to be resolved before so large and broad an initiative could credibly develop. It illustrates, perhaps better than any other recent example, what the responsibilities, as well as the benefits, are of loyal partnership.

PART TWO

Current Issues

10

Options for Northern Ireland

KEVIN BOYLE AND TOM HADDEN

Is Northern Ireland an integral part of the United Kingdom? Can it be effectively governed by the same democratic system as the rest of the United Kingdom? The answer to both these questions is 'probably not'. Northern Ireland is not an integral part of the United Kingdom in that successive British governments have declared themselves to be willing to cede it to the Irish Republic if a majority of the people in Northern Ireland so wish, and in that the Labour Party and many others have committed themselves to the active pursuit of the reunification of Ireland. Attempts to govern Northern Ireland by the same democratic system as the rest of the United Kingdom have patently failed, and there is a large measure of consensus that some other form of democracy will have to be established before effective powers can be devolved back from London to Belfast. In the meantime, pending the achievement either of Irish unification or of some other democratic system of government, Northern Ireland is governed by a more or less colonial system of 'direct rule', and the civil strife and terrorism which has been going on there since 1969 seems likely to continue indefinitely.

For all these reasons Northern Ireland is clearly an odd, though an interesting, corner of the United Kingdom. As a result, British politicians and students have tended to regard it as marginal to the mainstream of developments in British politics. That is understandable in the sense that Northern Ireland raises different issues from those which dominate British politics. But the issues which it does raise are fundamental to the nature of democracy and to the concept of state sovereignty and deserve more careful study than they are usually given. For it is precisely the failure of British politicians to get to grips with the essentials of the Northern Ireland problem, both in factual and in theoretical terms, that makes it so intractable.

The Genesis of the Problem

Northern Ireland was created in 1920 by a pragmatic British decision to partition Ireland. It comprises about one-fifth of the total land area of Ireland and six of its thirty-two counties. It is populated by just under one million Protestants or unionists who regard themselves as British and wish to remain part of the United Kingdom, and just over half-a-million Catholics or nationalists who regard themselves as Irish and aspire to the unification of Ireland. The remaining twenty-six counties form the Republic of Ireland, and are populated by some three-and-a-half-million Catholics and a mere 150,000 Protestants. It is easy to calculate that a very large majority of the people of Ireland as a whole would favour Irish unity. But within Northern Ireland itself an only slightly smaller majority favour the maintenance of the status quo. Northern Ireland thus provides a classic example of what is sometimes called a 'double minority', in that the majority in Northern Ireland is a minority within Ireland, while the minority in Northern Ireland forms part of that wider majority.

The purpose of the British government in adopting partition was to grant a measure of self-government to both the nationalist and the unionist majorities, each of which had shown itself to be ready and willing to fight to attain its objectives. The northern unionists had been mobilised in 1912 in Carson's Ulster Volunteer Force to oppose the grant of home rule on an all-Ireland basis. Nationalists had shown their commitment to independence first in the Easter Rising of 1916

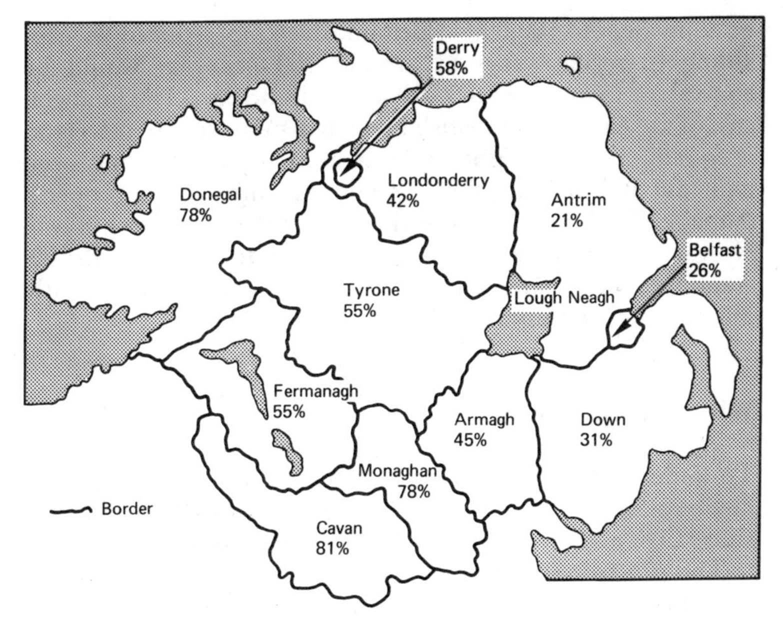

FIGURE 10.1 *The Catholic population of the nine counties of Ulster in 1901*

Source: Boyle and Hadden (1985).

and then in the overwhelming victory of Sinn Fein in the general election of 1918, and British security forces had been unable to contain the ensuing guerrilla campaign by the Irish Republican Army (IRA) in 1919 and 1920. The boundary of Northern Ireland was drawn on an *ad hoc* basis to include those counties in which there was a clear majority of Protestants or rough equality between Protestants and Catholics; in the remaining twenty-six counties the proportion of Protestants ranged from a maximum of some 20 per cent in the other Ulster counties to fewer than 10 per cent in the south and west (see Figue 10.1).

From the British point of view the partition settlement was relatively successful as a means of resolving the Irish question which had bedevilled British politics for fifty years. For the next fifty years the Northern Ireland problem lay dormant. In 1969 it re-emerged. Since then, some 2,500 people have been killed in Northern Ireland, a figure which in proportional terms would be equivalent to 100,000 in Britain. Even the 'reduced' level of killing in recent years at around 100 per year would be equivalent to more than 3,000 a year in Britain. The financial cost of the continuing British involvement in Northern Ireland is equally daunting. Since 1969 more than £1 billion has been paid out in compensation for personal injury and property damage. The annual cost of the overall British subvention to Northern Ireland, both for security and to maintain British standards in social security and other spheres, is currently more than £1.5 billion.

Why Partition Failed

To understand why the partition settlement did not produce lasting stability, it is necessary to consider the policies pursued since 1920 not only in Belfast but also in Dublin and London.

The most direct responsibility for the failure of Northern Ireland must be borne by the Unionist Party government in Belfast. In the fifty years following partition, the Unionist Party who were permanently in power made no attempt at all to integrate members of the minority community in the processes of government, or to solicit their support for the new state. Instead, they sought to rely on their built-in majority to create an impregnable political and economic regime. The development of opposition within the unionist community was discouraged by a continual emphasis on the threat from nationalists, and by the removal of the provisions for proportional representation which had been built into the 1920 settlement. And in border areas where there was a majority of nationalists, local government boundaries were gerrymandered to ensure Unionist Party control. To counteract the threat to the unionists' voting majority from the higher Catholic birthrate, discriminatory employment practices both in the public and private sectors were encouraged or condoned. The unemployment level among Catholics in 1971 after fifty years of Unionist Party rule was more than double that among Protestants, and the level of Catholic emigration was consistently higher than among Protestants. Any assertion of Irish identity by Catholics was discouraged, notably by the prohibition on the use of Irish street names and by discriminatory rules on the display of the Irish flag.

This sectarian approach to government was mirrored by that of successive governments in the Republic. Though politicians in the Republic continued to assert the objective of Irish unity, they lost no opportunity to emphasise the Catholic, Gaelic and non-British nature of their state and in so doing to confirm the fears and prejudices of the northern unionists. All links with Britain were progressively dismantled, and the views of the Catholic Church on divorce, contraception and censorship were enforced by law. These policies and the regular assertion of the Republic's right to extend its jurisdiction over Northern Ireland without taking any account of the British identity and commitments of Northern Protestants, merely served to strengthen their determination to have nothing to do with a united Ireland. Traditional Irish nationalism and Ulster unionism were thus mutually self-supporting. Successive governments in the Republic were, in practice, much more concerned with the assertion of Irish independence than with the realities, as opposed to the rhetoric, of unification.

Successive British governments, for their part, contributed to the problem, first by effectively abandoning all responsibility for the government of Northern Ireland to the Unionist regime without making any serious attempt to protect the interests of the nationalist minority. When the breakdown of law and order in 1969 forced the government in London to play a more active role, it initially sought to resolve the problem by persuading the Unionist Party to accept a package of not very radical reforms. Unionist resistance to the reforms and escalating violence from the IRA, however, shifted the balance from political to military action and the British Army was forced into its historic role of repression of Irish nationalists, notably in the introduction of internment in August 1971 and in the shooting dead of thirteen demonstrators in Derry on 'Bloody Sunday' in January 1972. The political repercussions of these developments led inexorably to the suspension of the Northern Ireland Parliament and the imposition of direct rule from London in March 1972.

Direct Rule: Why It Has Not Helped

Since then, successive British governments have sought to maintain a fair and effective security policy, to eliminate all remaining forms of discrimination, and to re-establish a form of devolved government in which representatives of both communities could share. They have been singularly unsuccessful.

On the security front, successes in identifying and bringing to justice those involved in violence on either side have been offset by a continuing series of abuses (Boyle, Hadden and Hillyard, 1980). The system of internment which was maintained until 1975 was based on a practice of mass 'screening' in Catholic areas, which involved the regular arrest and interrogation of thousands of ordinary people in suspect areas and occasional early morning 'headcounts' designed to build up comprehensive intelligence records. This antagonised even those Catholics who opposed the IRA, and ensured a continuing flow of recruits despite the arrest and internment of hundreds of IRA members. When internment was finally abandoned it was replaced by a policy of criminalisation and Ulsterisation – treating terrorists as if they were ordinary criminals and relying on locally recruited security forces. The policy of criminalisation got off to a bad start. Convictions in the non-jury Diplock Courts, introduced on the recommendation of the Diplock Committee in 1973, were based almost exclusively on confessions obtained during prolonged police interrogation, and the numerous allegations of beatings and ill-treatment were eventually recognised to have been justified. The attempt to treat convicted IRA terrorists as ordinary prisoners, by requiring them to wear prison uniforms and to do prison work, led inexorably to the 'dirty protest' and to the hunger strikes by Republican prisoners in 1980 and 1981. Though the problems were eventually resolved by allowing all prisoners to wear their own clothes and by a measure of tacit segregation from ordinary prisoners, the insensitive handling of the crisis from London led to the worst polarisation between the two communities since 1969, and provided a major boost to communal support for the IRA and its political counterpart Provisional Sinn Fein. The more-recent attempts to defeat terrorism by relying on uncorroborated evidence from a series of 'supergrasses', and by an apparent 'shoot-to-kill' strategy on the part of undercover police and Army units, likewise achieved only short-term successes and caused further damage to communal confidence in the administration of justice.

The achievements of direct rule in economic and political matters have been equally disappointing. Despite the creation of a wide range of official agencies to combat discrimination, notably the Fair Employment Agency, the returns of the 1981 census and other studies show that after ten years of even-handed rule from Westminster the unemployment rate among Catholics was still more than double that among Protestants. Successive attempts by the Northern Ireland Office to create a 'power-sharing' system of devolved government have been thwarted by distrust and disagreement between local politicians on either side. The 'power-sharing' government in Belfast established in January 1974 by William Whitelaw, the first Secretary of State for Northern Ireland, was brought down four months later by the Ulster Workers Council strike over the issue of a Council of Ireland. Similar attempts by his successors have been equally unsuccessful. Merlyn Rees's attempt to achieve consensus through the Northern Ireland Convention in 1975 failed, as did a series of inter-party talks arranged by Humphrey Atkins in 1980 and 1981. Jim Prior's attempt in 1982 to initiate a process of 'rolling devolution' by re-establishing the Northern Ireland Assembly has likewise been frustrated by the abstention of all nationalist members, both from the Social Democratic and Labour Party (SDLP), established in 1970, and from Sinn Fein, the political counterpart of the IRA. Unionists consistently refuse to share power with 'republicans' and nationalists consistently refuse to participate in a system in which there is not a meaningful Irish dimension.

Any hope that the gap between these two incompatible positions can be bridged by cross-communal centre parties is unlikely to be fulfilled. The figures in Table 10.1 show that both in first-past-the-post Westminster elections and in proportional representation elections within Northern Ireland unionist parties have repeatedly won around 60 per cent of the

TABLE 10.1 *Communal voting strengths in Northern Ireland 1973–83 (per cent)*

	Assembly Election 1973	Convention Election 1975	General Election 1979	Assembly Election 1982	General Election 1984
Democratic Unionists	11	15	10	23	20
Official Unionists	29	26	37	30	34
Other Unionists	22	22	12	6	3
Unionist bloc	(62)	(63)	(59)	(59)	(57)
Alliance Party	9	10	12	9	8
Labour/Workers	3	1	1	3	4
Centre bloc	(12)	(11)	(13)	(12)	(12)
SDLP	22	24	20	19	18
Republican/Sinn Fein	3	2	8	10	13
Nationalist bloc	(25)	(26)	(28)	(29)	(31)

Source: Boyle and Hadden (1985).

votes, nationalist parties some 25–30 per cent, and centre parties only about 12 per cent. (It should be noted that the remarkable increase in Sinn Fein votes within the nationalist community is partly attributable to the fact that until 1979 many committed republicans preferred to abstain from voting altogether.) It would be wrong to conclude from these figures that the adoption of PR voting makes no difference. It has encouraged the development of a wider range of political parties and ensured that seats in the Northern Ireland Assembly are more fairly distributed between the parties than at Westminster. But it has not had a major impact on the underlying communal division.

All these factors – continuing abuses in the operation of emergency powers, the continuing difference in levels of unemployment and thus of deprivation in the two communities, and the failure of the British government to make any progress in securing an effective role for the minority community in the government of Northern Ireland – have contributed to a deepening sense of alienation among Northern Ireland's Catholics, and to a deepening sense of despair in Britain and the Republic at the prospects for resolving the Northern Ireland problem within a purely British context. Many people in Britain have turned to more radical solutions. In 1981 the Labour Party committed itself to a policy of the re-unification of Ireland by consent. And there is increasing popular support for an immediate or phased withdrawal by Britain on a unilateral basis.

The New Ireland Forum and the Republican Ideal

Similar concerns in the Republic over the continuing impasse in Northern Ireland and over the increasing popular support for Provisional Sinn Fein led to the establishment of the New Ireland Forum in 1983. The Forum consisted of the leaders and delegations from each of the three main parties in the Republic – Fianna Fail, Fine Gael and the Irish Labour Party – and from the SDLP in Northern Ireland. Its task was to find a way in which 'lasting peace and stability could be achieved in a New Ireland through the democratic process'. Its report in May

1984 concluded that 'a united Ireland in the form of a sovereign independent Irish state to be achieved peacefully and by consent' was 'the best and most durable basis for peace and stability' (New Ireland Forum, 1984). The preferred option was a unitary state which 'would embrace the whole island of Ireland governed as a single unit under one government and one parliament elected by all the people of the island' (New Ireland Forum, 1984, para. 6.1). But two other options, a federal Ireland and joint authority over Northern Ireland by Britain and the Republic, were also discussed, and the participants declared themselves to be willing to consider other solutions which would meet the 'realities and requirements' identified in the report, notably the need to accommodate both traditions in Northern Ireland.

The choice of a united Ireland as the Forum's first preference was dictated by the need to secure the agreement to the Forum Report of Charles Haughey, the leader of the traditionally republican Fianna Fail party. But the prospects of achieving a united Ireland, whether as a unitary or a federal state, peacefully and by consent, are, in practice, no better in the 1980s than in the 1920s. It is quite clear that a substantial majority of the people of Northern Ireland are implacably opposed to Irish unity. In the last formal vote on the matter in 1973 some 58 per cent of the voting population voted to stay in the United Kingdom; less than 1 per cent voted for a united Ireland and 41 per cent did not vote. Allowing for the deliberate abstention of most Catholics and for the usual 20 per cent to 25 per cent of non-voters, it is clear that virtually the whole of the Protestant community turned out to vote for the union. More recent opinion polls have confirmed not only the solid opposition among Protestants to Irish unity in any form, but also the willingness of many Catholics to remain in the United Kingdom provided their rights and their Irish identity are properly guaranteed: for example, in a MORI poll carried out in 1984, 86 per cent of Protestants said they were strongly opposed to granting any say to the Republic over constitutional changes in Northern Ireland, while 50 per cent of Catholics said they were prepared to accept a devolved system of government within the United Kingdom with special guarantees for Catholics. There is little prospect of any rapid change in these figures, despite the continuing higher birthrate among Catholics than among Protestants. Until recently the emigration rate among Catholics was double that among Protestants, and the differential in the birthrate is declining. The latest estimates of the total number of Catholics in Northern Ireland based on the 1981 census vary from 37.6 per cent to 39.1 per cent compared with a figure of some 33.5 per cent in 1926.

Nor can there be much doubt about the willingness of many Protestants to fight to maintain their position. There are currently some 7,500 in the Ulster Defence Regiment (UDR), which was formed with the express purpose of defending Northern Ireland, some 8,000 in the Royal Ulster Constabulary (RUC), and some 4,500 in the RUC Reserve. Fewer than 10 per cent of the RUC are Catholics and there are very few Catholic members of the UDR. Though both the RUC and the UDR are officially non-sectarian there can be little doubt that the majority of those in them would resist any attempt by Britain or the Republic to impose unity. Many more Protestants, who hold the vast majority of licensed firearms in the province, would be ready to join official or unofficial paramilitary bodies, as they did in the 1920s and in the early 1970s, if there were any serious threat to the union.

These awkward facts show that the proposals for unification by consent made by the New Ireland Forum and the British Labour Party are impractical as immediate options. They also demonstrate the legitimacy of Northern Ireland in two internationally recognised senses: first in that a substantial majority of its citizens have consistently expressed their support in free and fair elections for union with Britain; and second in that in the event of an armed conflict within Northern Ireland, it is almost certain that the unionist community would be able to maintain an effective system of government, and thus to claim *de facto* and ultimately *de jure* recognition under international law. And if it is argued that Northern Ireland is not a legitimate entity because a substantial minority of its population do not accept its legitimacy, it must be remembered that the same argument would apply to a united Ireland.

An Independent Northern Ireland

The realisation that the unification of Ireland cannot be achieved by consent in the foreseeable future, has led some to consider the possibility of an independent Northern Ireland. Those who favour this idea, like the New Ulster Political Research Group – an offshoot of the Loyalist paramilitary Ulster Defence Association (UDA) – typically argue that the people of Northern Ireland are and always have been distinct and different from those in the rest of Ireland, and that if their conflicting loyalties to Britain and the Republic could be removed they would be able to realise their common identity and interests. They also argue that the obvious problem over the viability of an independent Northern Ireland could be overcome by a combination of continuing aid from Britain, which might gradually be replaced by aid from the European Community, and a heavier reliance on direct public borrowing, as in the Republic. A number of reputable economists have stated that this is not entirely unrealistic, provided that independence brought with it peace and stability. The problem is that this cannot be predicted with any certainty.

Independence has never been the preferred solution of any substantial number in either community. It is favoured by Protestants only in the last resort in the face of a threatened or actual British withdrawal. There has been no support at all for the idea from Catholics. On the contrary it seems much more likely that the IRA and other committed nationalists would see a British withdrawal as the first step towards ultimate unification and thus as a reason for stepping up their campaign of violence. The resulting need for stringent security measures in Catholic areas by what would inevitably be a predominantly Protestant government would then destroy any change that the bulk of Catholics would support or even accept an independent state. Even if the initial support of a majority in both communities could be achieved, which is doubtful, an independent Northern Ireland would thus be likely to degenerate into civil war.

FIGURE 10.2 *The distribution of the two communities in Northern Ireland in 1981, showing the proportion of Catholics in each district council area*

Source: Boyle and Hadden (1985).

Repartition

Some of the difficulties with independence – or with continued union with Britain – might be avoided if the boundary of Northern Ireland could be re-drawn so as to create a more exclusively unionist population. This seems an attractive idea and would follow the logic of partition. But it is almost impossible to achieve in practice. As Figure 10.2 shows, the two communities in Northern Ireland are intermingled in such a way that no boundary can be drawn between them. There is a heavy predominance of Protestants in the thirteen districts around Belfast that constitute the unionist 'heartland'. There is a further band of districts in which there is a rough equality between the two communities. Only in the border districts of Derry, Strabane, Omagh, Fermanagh, Newry and Mourne and Down is there a clear majority of Catholics. The border could be adjusted to exclude these districts, thus transferring, on 1981 figures, some 300,000 Catholics and 175,000 Protestants to the Republic. But this would still leave some 250,000 Catholics, most of them in Greater Belfast, out of a total population of around one million in the new Northern Ireland. The problem could only be finally resolved by arranging a very substantial exchange of population from Belfast to cities and towns in the Republic and by persuading large numbers of farmers to exchange their land. It is hardly surprising in this light that neither the Irish nor the British government has shown much interest in repartition. Marginal transfers may be defensible in themselves but would make no major impact on the problem. More substantial transfers and exchanges would be very difficult to achieve and might result only in the creation of two sets of embittered refugees.

Recognising the Two Communities

The straightforward solutions to the Northern Ireland problem – a united Ireland or the maintenance of the status quo, whether by continuing direct rule or setting up a new devolved administration – are based on the assumption that one or other of the two communities can be persuaded or coerced into abandoning its historic identity and allegiance. But it is precisely because both communities have shown themselves to be so resilient in maintaining their separate identities that the problem is so intractable. Both in Britain and in Ireland increasing attention is now being paid to a framework within which the rights and interests of both communities can be preserved. This was probably the true message of the New Ireland Forum, however much it was obscured by the apparent emphasis on unification. An approach of this kind, however, creates new problems of its own. It cuts across established conceptions of national sovereignty and demands a much more precise formulation of the democratic rights of majorities and minorities than has ever been attempted within the British parliamentary tradition.

Joint Authority

The model for a cross-communal settlement proposed by the New Ireland Forum was a system of joint authority, under which the London and Dublin governments would have equal responsibility for all aspects of the government of Northern Ireland so as to accord equal validity to the two communities and reflect their divided allegiances (New Ireland Forum, 1984, para. 8.1). The precise mechanisms for the operation of joint authority, however, were not worked out in any detail. What appears to have been envisaged was an essentially colonial system under which ministers appointed by London and Dublin would operate a more or less permanent system of direct rule. No attempt was made to show how the ministers were to be made democratically accountable either in London or Dublin or within Northern Ireland, or how legislation for Northern Ireland would be enacted. Nor was there any discussion of financial arrangements, other than the statement that 'the overall level of public expenditure would be determined by the two governments'. (The economic study commissioned by the Forum suggests that tax in Northern Ireland would continue to be levied at British rates, and that the cost of subvention would continue to be borne almost exclusively by Britain under a formula tied to the respective gross national products of Britain and the Republic.) Nor was there any concern to show how such a system might be made acceptable to Northern unionists and so meet the Forum's own criterion for any new arrangements.

An attempt to develop a less-colonial form of joint authority has since been made by the Kilbrandon Committee, an unofficial all-party body set up in 1984 to give a considered British response to the Forum Report. The whole Committee accepted the need for the removal of the Republic's constitutional claim over Northern Ireland as a precondition to unionist consent to any new structures. A majority then favoured a system of 'co-operative devolution', involving the appointment of a five-member ministerial executive comprising one British minister, one Irish minister, and three Northern Ireland ministers elected by proportional representation, two of whom whould be expected to be unionist and one nationalist. This would enable the British and unionist members to prevail where they agreed, but would remove the blocking power of unionists on measures agreed by the other members. It was also suggested that legislation for Northern Ireland might be enacted by an inter-parliamentary body, with members from the House of Commons, the Irish parliament and perhaps also from a directly elected Northern Ireland Assembly. This form of joint authority would not involve such a radical diminution of British sovereignty as that proposed by the Forum. But it seems likely that any system which would give unilateral authority to the Republic over the government of Northern Ireland would be unacceptable to unionists.

A further possibility would be a system in which authority on specified matters, such as industrial development, tourism and the control of terrorism, policing and judicial matters would be shared on a reciprocal cross-border basis between Britain and the Republic or between Northern Ireland and the Republic. This would allow the development of all-Ireland institutions on certain matters without denying in any way the legitimacy of the current constitutional status of Northern Ireland.

Minority Rights and Majority Rule

Whatever arrangements are developed on an interstate level, new structures for democratic government within Northern Ireland would also be required. It has long been recognised that the ordinary majority-rule system cannot work in a divided communal society. The alternative proposed by successive British governments has been 'power-sharing', in which all major parties would have a proportion of seats in a 'power-sharing' executive. This sounds eminently fair, but is unlikely to prove workable. The essence of power-sharing is that representatives of all major parties should be expected to shelve their differences and agree on a common programme. What is more, they would be expected to agree all the time, since the usual principle of majority voting within the executive could not be applied and since there would be no other way of resolving disputes. Any refusal to agree, even by a small number of members, could thus bring down the government. Politicians in London and Dublin who blame their counterparts in Northern Ireland for refusing to accept such a system, might usefully consider what their reaction would be to suggestions that the principle of power-sharing be applied within their own jurisdictions. It would make more sense to identify those matters on which the minority has an obvious communal interest, such as education and the siting of industrial development, and to prescribe suitable weighted majorities for legislation and other governmental decisions in such spheres. That is the standard method of protecting minorities in many other countries, and, incidentally, in British company law.

More explicit legal protections for the individual and the communal rights of members of both communities would also be required. It is already widely accepted that nationalists in Northern Ireland should be entitled to claim and exercise the rights of Irish citizenship granted to them under the Irish Constitution without losing any rights within Northern Ireland. This principle could be applied more positively to permit dual membership of the British and Irish Parliaments and perhaps also the election of representatives from Northern Ireland to the Irish Senate. Rights for local majorities to determine the names of geographical places, as in the notorious Derry/Londonderry dispute, and for members of the minority to use the Irish language in official dealings and in broadcasting, might also be created. The purpose would be to recognise the Irish as well as the British identity in Northern Ireland.

The Anglo-Irish Agreement

The British and Irish governments are currently engaged in an attempt to implement a jointly guaranteed settlement along these lines. In the communique issued after the Chequers summit in November 1984 Mrs Thatcher and the Irish

Taoiseach, Dr Fitzgerald, stated explicitly that they were seeking a framework within which 'the identities of both the majority and the minority communities in Northern Ireland would be recognised and respected and reflected in the structures and processes of Northern Ireland in ways acceptable to both communities'. In November 1985 they signed a formal Anglo-Irish Agreement, which has subsequently been approved by the London and Dublin Parliaments and registered at the United Nations. The central items in the Agreement are a renewed affirmation by both governments that 'any change in the status of Northern Ireland would only come about with the consent of a majority of the people of Northern Ireland' and the setting up of an Intergovernmental Conference, within which British and Irish ministers are to confer on political, security, legal and economic matters relating to Northern Ireland and on relations between the two parts of Ireland, though both governments have expressly reserved ultimate responsibility for decisions within their respective jurisdictions. It has also been agreed that responsibility for some matters should be devolved to Northern Ireland if the co-operation of constitutional representatives of both traditions can be secured, and that the Intergovernmental Conference should cease to have responsibility for any such matters.

This agreement has been generally welcomed in Britain, Europe and America as a balanced and pragmatic means of recognising the identities and interests of both communities in Northern Ireland. It has been opposed only by traditional republicans and unionists. Charles Haughey, leader of Fianna Fail, and Sinn Fein have denounced the apparent abandonment of Irish unification. Unionists of all shades have expressed resolute opposition to any form of involvement by the Republic in the affairs of Northern Ireland and have committed themselves to making the agreement unworkable.

It is difficult to predict at the time of writing precisely how this new initiative will develop. The Agreement is drafted not in the precise language of constitutional lawyers but in the flexible language of politicians. The choice of this mode of expression in itself shows the extent to which nineteenth-century conceptions of state sovereignty and parliamentary democracy make it difficult to resolve the problems of a small and unimportant place like Northern Ireland. It remains to be seen whether the two governments will be able to break free of these conceptual shackles, and whether the commitment of ordinary Ulster unionists and Irish nationalists to such simple notions will be strong enough to prevent them from doing so. Whatever the outcome, the exercise will be of theoretical as well as practical importance. The idea that a simple majority can do anything it wishes is, after all, a very strange and unsatisfactory conception of democracy. So, too, is the idea of the sovereign nation state, in a world in which many decisions must be taken on a multinational basis. Some changes in our conception of sovereignty have been forced upon us by membership of the European Community. Less progress has been made in the development of our conception of democracy. There is a corresponding need for much clearer thinking on the allocation of government functions between interstate bodies, state government and local or provincial government, and on the precise meaning of representative democracy at each of these levels. If the need to work out new conceptions of sovereignty and democracy in order to reach a cross-communal settlement in Northern Ireland helps to reveal the limitations in established political terminology, it may turn out not to have been such a marginal issue after all.

11

The Role of Law

MICHAEL J. ELLIOTT

The Law and Politics

In 1977 John Griffith published a short work (Griffith, 1977) which argued that the higher judiciary had, and revealed in their judgements, a set of beliefs that amounted to a consistent system of politics. The extraordinary reaction to his book – which was hailed by some as an unparalleled exercise in unpopular honesty, and by others as mischievous rabble-rousing – might have led the unwary to conclude that, hitherto, the politics of the law had not greatly exercised the minds of lawyers, law students, or legal academics.

Such a view is simplistic. Many of the great constitutional controversies of the seventeenth century were prosecuted by lawyers, deliberately using the concepts of the law in their political struggles. In the eighteenth century, too, the central political issue of property rights was one where the use of the law was of crucial importance. Maitland and Dicey, two of the greatest legal scholars of the nineteenth century, reserved their most renowned work for an assessment of constitutional law (Maitland, 1888; Dicey, 1959). Neither would have assumed that the discourse of the law was independent from that of politics – Dicey, indeed, meddled directly in political matters when he signed the Ulster covenant.

The consistency with which political issues have been defined and discussed in the terms of the law – rights, obligations, and remedies – is only one aspect of the linkage between the legal and political systems. Others are more mechanistic. The personnel of the legal system – judges and lawyers – have always exhibited close contacts with the world of politics. Lawyers, for example, have always been over-represented among MPs: they made up nearly one-sixth of the 1985 House of Commons; 7 of the 23-member 1985 Cabinet had legal qualifications; and the Lord Chancellor is a practising politician, the nation's senior judge, and the head of the legal profession – his department administers a budget of £500m on public legal services. The legal profession is closely tied to the institutions of the state. Some of its income comes from the Exchequer, through payment for Legal Aid work. Other income, like that which solicitors receive for their work in house transfer, has, traditionally, been a direct consequence of a state-sanctioned monopoly. Alone among all occupations, lawyers' work always consists of either mediating (in state-owned locations) between state power and private activity (as in criminal matters), or between competing claims to private activity in the shadow of state-promulgated or -sanctioned rules (as in civil law matters). The courts hold themselves open to arguments about political issues. They may have always insisted that those issues should be discussed in the language of the law and by reference to the modes of reasoning of the courts, but with rare exceptions British courts have never elucidated a theory of 'non-justiciability' that renders some matters too politically controversial for legal discussion. If a matter can be framed in legal terms, British judges will hear it.

Yet for all the history of shared preoccupations and problems between law and politics, and for all the links between personalities in law and politics, it is a common assumption of the ideology of British liberal democracy to claim that law and the legal system is, or should be, in some sense autonomous from the muddied field of politics. Politics can never be 'above' the law; we have a government of laws and not men. This 'autonomy value' appears to assume that there is something unique about the law, and legal values, that deserves protection. The process of distributing legal rights and duties should not, on this view, be sullied with political values which, implicitly, are based on qualitatively less objective, even moral, bases than those which underpin the law.

If Griffith was not the first to puncture that illusion (the hostility that his book provoked is some evidence of the fervour with which the illusion is felt) he at least chose a particularly good moment to do it. The 1970s saw the emergence in Britain of a phenomenon that had been familiar in North America since the mid-1960s – a 'radical' law movement made up of lawyers who denied the claim that their work could be divorced from their politics. In Britain, radical law made itself felt not only in the academic world (where the 1970s saw an outpouring of Marxist analyses of law), but in the growth of pressure groups like the Legal Action Group and, most important, in new forms of legal practice. Many practising lawyers, whether they worked in the new, community-based 'law centres' or not, increasingly saw their work as a form of political struggle in a legal arena. The subjects in which they specialised – and in which many of them still specialise – included housing law and the rights of tenants, employment law and the rights of workers, sex and race discrimination cases, and the rights of claimants to welfare benefits. For them, Griffith's argument did no more than state the obvious.

Many of the cases that the radical lawyers brought, and still bring, perforce, were claims against the government (central or local) or its agencies. At the same time, some government policies brought the law into a new and sharper focus. In the Heath Conservative government, this was particularly true of labour relations. The Industrial Relations Act 1971 attempted to introduce comprehensive legal regulation into the British system of industrial relations, whose hallmark hitherto had been a conscious voluntarism and disavowal of the utility of legal forms. The Labour government of the late 1970s, in a different way, continued the process, particularly by making access to legal services, like legal aid or law centres, a feature of the mild social reformism which was one of its ideological predicates. (It was that government which set up a royal commission on legal services in 1977.) To these developments the Thatcher governments have added an economic policy which, stressing the virtues of competition and market forces, has for the first time subjected to governmental scrutiny some traditional working practices of the legal profession; most obviously, the division of the profession into a specialist class of advocates (barristers) and office-bound general practitioners (solicitors), and the solicitors' monopoly of some economically valuable process in the system of house transfer. Taken together, this combination of a legal profession some of whose members have been overtly politicised, the acceptance by politicians of the law as a fit mechanism for, and tool of, political power, and a strongly held economic policy by central government, has given new edges to the old interface between the law and politics. The rest of this chapter looks at those new developments through three case studies.

Industrial Relations

Trade unions are, in law, private bodies, whose existence originally owed nothing to the state. To this day, they are not subject to state regulation in the same way, and to the same extent, as private limited companies. Yet their actions have, throughout much of the twentieth century, exercised the mind of the body politic, and that has been particularly true in the last two decades. What lessons does this concern for the activities of trade unions hold for the relationship between politics and the law?

To answer that question, it is necessary to go back to 1906, when the Liberal government legislated to reverse the effects of the Taff Vale case. The judgment in that case had made trade unions liable to actions by employers damaged through strikes. The 1906 act was taken by many to have confirmed an abstentionist approach of British law to industrial relations – or 'collective *laissez-faire*', as it was called by Kahn-Freund, the high priest of the doctrine.

There were two fundamental tenets to this approach. First, collective agreements between employers and unions were assumed not to be legally binding. Second, trade unions could not be sued in tort for damages, and their officials could not be sued if their actions were connected with a trade dispute – the 'golden formula'. The general line was confirmed by the Donovan Commission on trade unions and employers associations in 1968.

Yet even though an abstention of the law was often said to be the hallmark of British industrial relations, it was never more than partial. The courts continued to involve themselves in industrial disputes, whittling down, as trade unions would have put it, the immunities established in 1906. So the change wrought in 1971, when the Conservative government tried to force British industrial relations into a legal collar much like that which was familiar in America, was less dramatic than is sometimes portrayed. The experiment was, in any case, short-lived. The Labour governments of the 1970s repealed the Conservative legislation, and did their best to re-establish the old immunities. 'Did their best' is the operative phrase, for some judgments in those years interpreted Labour's legislation in a progressively narrow way. The Conservative government in 1979 thus had a base on which to build when it looked, again, at industrial relations law.

The Conservative strategy since then has been quite different from that adopted in the Heath years. The basis of the immunities has been kept, but the area that they protect has been whittled down – most obviously by requiring strike ballots before industrial action can be immune from legal challenge. By the Employment Act 1982, trade unions themselves (and not just their officials) may be sued for damages, for the first time since 1906. Of equal significance, the almost-secret garden of trade-union internal organisation has been invaded. Law now regulates the elections of union leaders, and stipulates the conditions under which unions may keep 'political' funds, which traditionally have been a method of financing the Labour party. The forms of abstentionism may still be there – the golden formula, somewhat tarnished, no doubt, is still on the statute book – but the reality is that the policy of the government towards the unions is one that has been implemented through a subtly (and not so subtly) restraining legal framework.

The new interpenetration of law and the politics of industrial relations is evident not only from legislation. The methods adopted by the government to combat the miners' strike of 1984–5 demonstrate the same point. Those methods included the use – with, at the very least, Home Office collusion – of a 'National Reporting Centre' for the police, which enabled reinforcements to be drafted from all over the country to wherever there looked likely to be violence on the picket line.

The officers of the law, meanwhile – magistrates, police prosecutors and judges – were refining a set of techniques whose impact was certain to be discussed in political terms (see Chapter 6). These included the use of onerous bail conditions for strikers convicted on the picket line, the sequestration of funds belonging to the mineworkers' union, and the selection of archaic charges such as 'watching and besetting' (with which over 400 miners were charged during the strike) for the prosecution of pickets. Some lawyers involved in the defence of the miners go much further. Judges, they argue, were uniquely creative in the uses to which they put the law during the dispute. They invented new rights of action against the union, found new remedies, and transformed the rules of procedure of English courts so that a union which challenged the legitimacy of legal determination of its affairs was severely handicapped.

The consequence of these developments has been dramatic. On the one hand, it has led some members of trade unions to identify the law as an instrument of the class enemy with a fervour that has not been seen for years (those marchers in South Wales who demanded amnesties for two strikers convicted of the murder of a working miner, is the most clear testimony to that). On the other hand, paradoxically, the traditional union and Labour Party belief in the virtues of abstentionism has been weakened. The partnership document launched by the Trades Union Congress and the Labour Party in the summer of 1985 spoke of the possibility of giving unionists 'positive rights' – code for that legally guaranteed right to strike without penal consequences which forms the basis of much continental labour law. Thus, from the union's point of view, the law has become at one and the same time despised and considered indispensable. From the government's point of view, there are murmurings that there might be yet more legislation in the future – perhaps to limit the right to strike in certain 'essential' industries and services. Of the old idea that the law should have little place in the politics and practice of industrial relations, there are now few defenders.

Administrative Law

No development in British law since the war has received so much academic comment as the rise of administrative law – which, for present purposes, can be taken to mean legal actions against any agency of the state. Self-evidently, such actions bring the legal process into close contact with the realms of politics. For that and other reasons, there have always been voices urging the defence of the autonomy of the law meant that the rise of administrative law must be challenged.

Those voices have normally concentrated on two matters. First, they have argued that if the courts are to play on this field, they need, but do not possess, an expertise in administrative processes. It is said that the judicial review of administrative action demonstrates only that lawyers are not sensible to the nuances of politics or the needs of good administration – things with which good administrators are familiar (see Chapter 15, topic 2). A variant of this objection states that law always reduces cases to an approximation of individual claims of right. The administrative process, it is argued, deals in the collective, public interest. Individual rights and wrongs are only part of that picture. The effective implementation of programmes and policies is equally important.

The second reason for a wariness of administrative law is the claim that in the British system of government, administrative decisions have been legitimated by the ballot box. Those who wield administrative power are ultimately responsible, so the theory goes, to those who have been elected. Judges have no such claim to deference. Thus the judicial overturning of an administrative decision represents the replacement of a democratic act of government with one that is less so. This view is usually associated with the political left, who have often believed that the judges' role in administrative law has been limited to striking down interferences with private property rights in legislation passed by Labour governments. It is, in fact, one more widely shared in the political community than many commentators seem to realise. It is a Conservative government, after all, that has come closest to removing Britain from the jurisdiction of the European Court on Human Rights, whose judgments often trespass on the same territory as that traversed by domestic administrative law.

How have these worries about the growth of administrative law manifested themselves in recent developments? It is necessary, first, to go back to the period immediately after 1945, when the legislation – on town and country planning, or social security – that has been meat and drink of administrative lawyers, was relatively new. At that time, it seemed as if the courts had become as infected with the omniscience of administrators as any good Fabian. A case involving Wednesbury Corporation in 1948, the *locus classicus* of a judicial discussion of administrative discretion in those years, appeared to adopt a role of self-denial on the part of the judges that left them with little to do in constraining the actions of government. That attitude did not last. First, the judges themselves, with Lord Denning to the fore, progressively tightened the legal shackles on the procedural methods by which government decisions might be taken. This development, known to administrative lawyers as the rediscovery of the 'rules of natural justice' (which are rules that stipulate that both parties to a dispute must be heard, before an independent tribunal), had, as early as 1964, transformed the decision-making procedures within both central and local government. It is only in decisions involving national security that government can be sure that procedural fetters will not be placed on them by the law. Administrative lawyers have also been prepared to look at the merits of decisions, as well as the procedures by which they are taken.

Taken together, procedural and substantive checks on government have limited the freedom that administrators thought they possessed. For most of Britain's 'social' legislation depends on the exercise of discretionary powers by officials. The rebirth of administrative law provided a method of structuring that discretion in certain ways. That opportunity was seized in the 1960s and 1970s by those lawyers who had noted the interventionist role played by their American cousins. They started to use the courts as a way of guaranteeing those welfare benefits that, apparently, could not be delivered by the administrative process.

For a time, then, administrative law was associated with

radical politics. At least at the level of commentary (practice, as we shall see, is another matter) it is no longer. A recent article refers baldly to the 'ruse' of administrative law (Hutchinson, 1985), arguing that its impact on the administrative process has been marginal, its claims mere ballyhoo.

If there is one case that is used by radical commentators to support that position it is that which involved the Greater London Council (GLC) and the London Borough of Bromley in 1982. Following a promise in the manifesto in which they had successfully fought an election, the ruling Labour group on the GLC introduced a new policy of subsidy of fares on London Transport. This required the raising of extra funds through rates from the London boroughs. Bromley claimed that the policy breached a legislative requirement that London transport should be run economically and efficiently, and the House of Lords judicial committee agreed. The case turned in the end on the 'fiduciary duty' that the courts held that local authorities owed to their ratepayers. That duty, the argument seemed to suggest, deserved precedence over any obligations that might be owed to voters or the travelling public.

In one almost orgiastic moment, all the suspicions of administrative law were gathered together. There was a lack of expertise in the judges' woeful misunderstanding of the system of local government finance; there was a choice between economic values and those of public service; there was the clear preference for the property rights of ratepayers over the rights to cheap transport of users. Above all, perhaps, there was the disregard of a manifesto commitment; and the overturning of the policy held by a campaigning Labour authority.

Those who were most suspicious of the intrusion in law into the new local politics appeared to have many of their worries confirmed when, in the wake of Bromley, local authorities all over the country found themselves challenged by ratepayers claiming a fiduciary duty. The government was not slow to capitalise. In a little-analysed policy innovation, the Transport Act of 1982 attempted to control local authority revenue support for public transport not by financial limits, but by the deliberate use of the fact that ratepayers could, and would, sue an authority that appeared to be wasting their money.

Yet here, as in industrial relations, the case for a withdrawal of the law from the political scene was never convincingly made out. The 'fiduciary duty' that was used against those campaigning local authorities was as old as the hills; indeed, it derived from the experience of those very county boroughs to whose memory those who wished to appeal for freedom of action on the part of local authorities had, perforce, to appeal (Elliott, 1983). Moreover, whatever commentators might have said, practitioners did not assume that the lesson of Bromley was that the law could not play a useful role in defending radical policies. At one moment in 1985, those local authorities who were 'ratecapped' had initiated no less than eight cases against central government's interpretation of the relevant legislation. In the same year, the courts, at the motion of the GLC, overturned a decision by the transport secretary which had sought to render of no effect a lorry ban, and scrapped regulations that had established a draconian regime for 'bed and breakfast' payments to social security claimants.

This use of the law is hardly surprising, and it is testimony to the unreality of much modern administrative law scholarship that it is sometimes assumed to be so. For legal cases against the government – whatever the lack of expertise in the judiciary, whatever their preference for property rights over social policy – have one distinguishing feature. If successful, they award a legal right to the person challenging government action, and bring that action to a stop. True, that stop may be only temporary, but this determination of rights against the government represents a qualitatively different form of accountability from that which can ever be exercised by, for example, parliamentary debates, pressure-group activities, or the work of ombudsmen. In those circumstances, those who have the interests of recipients of welfare benefits at heart, as well as those in public authorities who wish to pursue policies antithetical to the government of the day, will continue to muddy law and politics by asking for the judicial review of administrative action.

The Legal Profession

In the case of both industrial relations and administrative law, the claim that law should remain at one remove from politics is made because it is felt that the particular political process concerned has certain requirements of expertise, and a certain ideological base. Similar claims are made in the third case, that of the legal profession. But whereas in industrial relations and administrative law it is political controversy that has forced law and politics together, in the case of the legal profession it is economic policy – albeit a policy closely identified with a particular political ideology.

The politics of professionalism has produced some excellent academic analysis in the past decade (see, for example, Johnson, 1977). Like architects (Dunleavy, 1981), lawyers are intimately connected with the activities of the state. This is not primarily because they sell their services to state agencies by contract (though some younger criminal-law barristers now derive almost all their earnings from the public funds of the legal aid scheme). The reason for the close links between lawyers and the state is a direct consequence of the nature of 'law' – Laws are norms that are dignified by the presence of a state-sanctioned 'signifier'.

It is thus not surprising that political decisions should have interfered in the private processes of the legal profession, in terms of who may practise law, how legal education is organised, and the internal rules of the profession (or, at least, the solicitors' branch of it). Yet when a politically motivated economic policy increased the scale of that intervention, the profession seemed stunned that it could happen at all.

Solicitors have traditionally obtained approximately 30 per cent of their annual income of about £3bn from their legally protected monopoly over the 'conveyancing' of houses. That monopoly was always likely to be challenged by the competitive, market-oriented economic policy of the Thatcher governments. In fact, possibly because of the gap between the rhetoric of that policy and its reality, nothing was done until, in 1983, Austin Mitchell, a Labour MP, introduced a private members bill to abolish the conveyancing monopoly. The government was then forced to take up Mr Mitchell's proposal, and from 1986 licensed conveyancers will be able to compete with solicitors for the housebuyers' business.

The knock-on effects could be dramatic. Faced with the

possible loss of a third of their income, solicitors have begun to press for the removal of other restrictive practices, in particular that which reserves rights of audience in the higher courts for members of the bar – who in turn have been threatened by the government's commitment to a national system of salaried prosecutors. These may be expected to take some of the junior bar's work in the criminal courts.

The oddity of these developments has been the way in which the legal profession seemed utterly unaware of the threat to its economic livelihood. So secure were the solicitors in their belief in the autonomy of their profession that they seemed almost incapable of heading off the threat that economic liberalism posed for them. The bar, too, appeared to believe that their traditional 'independence' from the state could guarantee them a waiver from the normal rules of politics. Thus neither branch of the profession has been able to mount a convincing defence of its traditional position, and the momentum is now with those who would introduce yet further changes to professional practices.

Conclusion

What lessons can be learned from these three case-studies of recent developments in the relationship between the law and politics? The first is that the autonomy principle is – as it ever has been – more honoured in the breach than the observance. When both trade unionists and radical local authorities are prepared to use the law to prosecute their political aims, it becomes clear that the rhetoric of legal abstentionism from politics has little grounding in real-life practice. But, as a second conclusion, it would be most surprising if that rhetoric disappeared. Judges, and the legal system more generally, are a convenient scapegoat when radical policies go wrong. The antipathy of sections of the labour movement to the law is unlikely to disappear, even as that same law is used to further the interests of workers.

The third lesson is that much more attention needs to be paid not to the effect of politics on the law but to the effect of political economy. It is not fanciful to suppose that the results of the abolition of the conveyancing monopoly will be a radical restructuring of the legal profession, in ways that cannot yet be clearly outlined. But one likely consequence is that as the income of solicitors from conveyancing starts to shrink, so their ability to cross-subsidise less profitable work, especially legal-aid work, will diminish. Thus there will be a reduction of legal services to the public, which will at some point become a political issue. The state may then step in once more, perhaps by means of a system of state-salaried lawyers, to fill the gap. The professional autonomy of lawyers will be exposed as something that cannot be sustained, at its historical level, in modern conditions of economics and politics.

The most fundamental lesson that flows from the weakness of the autonomy principle is that it makes possible some creative links between law and politics, which hitherto have been considered almost impermissible. One of the reasons often given for the absence in Britain of a bill of rights, for example – or even a 'repatriated' European Convention on Human Rights – has been that judges lack the expertise to handle political decisions. Their reputation would be tarnished, it is said, if they were seen to meddle directly in politics. By the same token, politicians' freedom of action would be needlessly – possibly dangerously – curtailed. But that argument only holds water if the autonomy value is sustained as a matter of practice. In recent years, it has not been; it is unlikely to be in the future. A little honesty about the constant and deep interpenetration of law and politics would be no bad thing.

12

Industrial Relations

MICHAEL MORAN

Power and Industrial Relations

Industrial relations have always been political because they have always concerned power – the allocation of power in the workplace. In Britain over the last two decades, however, they have become political in more obvious ways: the state has been increasingly involved in the regulation of the workplace and this involvement has prompted intense debate about the distribution of power at work. For most of the 1970s this debate was dominated by arguments about trade-union influence. Some observers thought unions overwhelmingly dominant, combining as they did historically established influence in the Labour Party, a special role as one of the partners of government and a unique ability to disrupt economic and social life through industrial action (Middlemas, 1979; Finer, 1973; Brittan, 1975). By contrast, others were sceptical of the unions' ability to control the Labour Party, denied that they occupied any privileged position as a pressure group, and argued that industrial action provided limited and negative sanctions compared with the economic power exercised by business (Coates, 1980; Marsh and Locksley, 1983).

This debate developed in an era when unions operated under full employment and under a succession of governments anxious for their co-operation. The atmosphere of the 1980s has been very different. Economic crisis and hostile government policy have subjected unions to immense pressure. Most observers now believe that union power has declined in recent years; debate has switched to arguments about the causes and extent of decline.

Casual observation might suggest that any debate is unnecessary. Mass unemployment, falling union membership and a series of restrictive laws passed by the Conservative government suggest a decisive change in the balance of power, to the unions' disadvantage. This suggestion seems to be confirmed by a series of defeats inflicted in recent years on public sector unions, notably on steel-workers and miners fighting closures in their industries. But concentrating on these spectacular conflicts misleadingly implies that the struggle for power in industrial relations is a war decided by a few pitched battles. On the contrary, it is more like trench warfare. A wide range of combatants are engaged in a variety of actions over a long front. Huge battles – like the miners' strike of 1984–5 – inflict serious casualties and force the losers to retreat on one part of the front. These highly public conflicts are surrounded, however, by a host of lesser engagements whose cumulative outcome is immensely significant. The struggle for power in industrial relations certainly involves the great conflicts between government and unions which occupy the front pages of newspapers; but much of the contest is hidden away in a multitude of small engagements at work. It is in the workplace, therefore, that an examination of the pattern of power in the 1980s ought to begin.

Changing Patterns of Work

The most dramatic change in the economy in recent years has been the rise of mass unemployment. In March 1975, there were under 750,000 registered unemployed; by June 1979, just after Mrs Thatcher became Prime Minister, the figure was 1.175 million; by September 1985 it exceeded 3.1 million.

The most obvious measure of the impact of mass unemployment on unions can be seen in figures for membership, their lifeblood. In 1979, total membership of unions in the Trades Union Congress reached a historic peak of just over 12.1 million; by the beginning of 1985 the figure was 9.8 million. There has also been a related decline in the 'closed shop' (a system under which union membership is an obligatory condition of employment). Between 1978 and 1982 – during which time the full force of mass unemployment was released – numbers covered by closed-shop agreements fell by over an eighth, to 4.7 million (Dunn and Gennard, 1984, pp. 146–7). This suggestion of unions in retreat is heightened by evidence about strikes, the most formidable weapon in their armoury: in 1984, for instance, the number of stoppages in the United Kingdom was less than half that recorded in 1978.

These suggestions of dramatic change must, however, be treated with caution. The national figures conceal the fact that union decline in the workplace has been far from uniform. The widest variations are revealed by the figures for membership. In six years after 1979, for instance, Britain's biggest union, the Transport and General Workers, lost over a quarter of its members, while in a single year (1982–3) the small agricultural workers' union lost a fifth. By contrast, the local government officers' (the fourth largest TUC affiliate) and unions organising banking employees and workers in the health sector, all recorded modest rises.

Membership loss has thus been very unevenly distributed. Indeed, the very language of 'loss' is misleading because it implies, wrongly, that workers have been deserting unions. This did indeed happen in the great depression of the 1920s and 1930s, but in the present recession most of the decline has resulted from union members lapsing on losing their jobs; few in work have left. The simplest sign of this fact is provided by figures for *density* – the proportion of the working population in unions. In 1979, union density reached a historic peak of 55 per cent; in the mid-1980s it had fallen to just under 50 per cent. By comparison, density at the end of the 1960s – thought by many to mark a post-war peak of union power and militancy – was only 44 per cent (Brown and Sisson, 1984). A similar tale lies behind the decline of the closed shop, almost all of which is due to loss of jobs – and therefore numbers – rather than to the ending of closed-shop agreements (Dunn and Gennard, 1984).

The strike record is also ambiguous. Measures of industrial conflict are notoriously unreliable and difficult to interpret. To set against the falling total of strikes we have to recall a succession of bitter disputes in declining industries like steel, railways and coal-mining, as workers tried to resist the consequences of the recession. We should also observe the spread of the strike habit to previously non-militant workers, like civil servants and teachers, both of whom have been in prolonged disputes in the 1980s. The contradictory character of the evidence can be illustrated by observing that whereas the number of strikes in 1984 was less than half that of 1978, the number of days lost in strikes was – because of the miners' great dispute – nearly three times the 1978 level.

Union membership, the closed shop and industrial militancy are therefore not quite so much in retreat as a first inspection of national figures might suggest. But this limited consolation is all that is available to unions; in other respects gross national figures understate the extent to which changes in labour markets have weakened their bargaining power in the workplace.

Two aspects of change merit emphasis. First, the four-fold rise in the numbers unemployed since the mid-1970s is only one measure of how the changing labour market has weakened the bargaining position of workers. An even more dramatic sign is provided by figures for the long-term unemployed: in 1974 just over 30 per cent of the jobless had been out of work for six months or more; a decade later the figure was just over 60 per cent.

The second change concerns the structure of job losses and job creation. In the last decade jobs have been both lost and created on a massive scale; but whereas losses have been concentrated in areas of traditional union strength, new jobs have been concentrated in sectors where unions are weak. Two contrasting illustrations make the point. Unions have traditionally been best organised and most militant among men doing full-time manual work; weakest among women working part-time in 'service' jobs like catering, cleaning and routine clerical work. The biggest loss of jobs in recent years has been among manual occupations, especially in 'heavy' industries dominated by men; the biggest expansion has been in part-time jobs for women. Since 1979 the numbers of women in part-time work have risen by nearly 10 per cent; the numbers of men in full-time work have fallen by the same amount. Even more ominous for the unions is the fact that these trends, though accelerated by the recession, are part of a long-term shift against full-time male employment in manufacturing in favour of part-time jobs for women in 'service' work: between 1971 and 1976, for instance, fully 95 per cent of new part-time jobs were in service work usually done by women, like catering and cleaning (Dix and Perry, 1984). Even in the unlikely event of an end to recession, the structure of employment will continue to work against trade unions.

The Employers' Offensive

The long-term expansion of part-time work, especially for women, is linked to more general changes in the labour market and in the pattern of bargaining power. Managers are mounting a sustained offensive to recapture control over the pace and pattern of work. Three signs of this offensive are notable.

First, labour forces are being segmented. A key division increasingly separates a 'core' of full-time, well-organised, well-paid and comparatively secure workers, mostly men; and a 'reserve army' of poorly paid, ill-organised part-timers, disproportionately women. (Over 80 per cent of part-time workers are female.) This segmentation gives managers flexibility in the disposition of labour, because a 'reserve army' of poorly organised part-timers allows ample scope to contract or expand the labour force according to market conditions.

A second sign of the drive for control can be seen in the continuing decline of industry-wide collective bargaining. Once the norm in the private sector, by the early 1980s it covered only about a quarter of workers (Sisson and Brown, 1983). Pay and conditions are now characteristically settled at firm (or even plant) level; in this way employers are fixing conditions according to their own distinct market situation, rather than as the result of national power contests between trade unions and representatives of whole industries.

This drive for greater control by individual companies over pay and conditions is connected to a third set of changes, best summed up as the more rational and scientific organisation of work in the interests of maximising the productivity of labour. The signs of change include the wider application of work study techniques to settle work rates and pay levels; persistent pressure to eliminate specialisations based on traditional craft skills and to replace them with fewer and simpler job categories; and the increasing incorporation of shop stewards, traditionally a focus of resistance to managerial authority, into bargaining hierarchies within firms (Sisson and Brown, 1983; Purcell and Sisson, 1983). The drive for nationalisation also extends to the creation of 'single union' agreements, an arrangement where one union is given a monopoly of bargaining rights, in place of the complex and competitive pattern of multi-union bargaining customary in British industrial relations. A study of the north-east of England – an area noted for its traditions of militantly competitive unions – showed the existence in that area alone of 239 single-union agreements (Bright, Embridge and Rees, 1983).

This drive for greater managerial control over the organisation of work has culminated in a number of highly publicised agreements involving new enterprises in high-technology industries. Most have been entered into between a single union – the 'electricians and plumbers' (EETPU) – and Japanese electronics firms establishing new assembly plants in Britain. These agreements are notable for the way they break with established bargaining patterns. They characteristically offer an exchange: in return for sole bargaining rights the union binds itself to a 'no strike' clause, often involving 'pendulum' arbitration, an arrangement under which arbitrators, forbidden to split the difference between contesting parties, are forced instead to choose between the two opposed positions. But while single-union bargaining, no-strike agreements and pendulum arbitration clauses are the most commonly noticed features of these arrangements, they are only part of an attempt at a wider reform of the character of workplace relations. Other changes include the incorporation of shop stewards into 'company councils', the abolition of traditional skill differences, the systematic application of work study techniques and the conscious fostering of a company 'culture' to transcend the traditional class and occupational cultures of workers.

The creation of a 'reserve army' of weakly unionised part-time labour; the rigorous application of work study techniques; the elimination of specialist craft skills; the introduction of single-union agreements and no-strike clauses: all these would lead us to expect a significant shift in workplace power. That a drive to shift the balance of power exists is certain; where we examine the evidence, as we do next, the success of that drive is less sure.

The Balance of Power at Work

Power in the workplace has an elusive equality because it is dispersed through tens of thousands of firms and plants. We cannot observe all workplace relations, so we are driven back to summary measures of the outcome of bargaining. Two of the most revealing are pay and productivity: pay, because it indicates what workers have been able to squeeze from employers; productivity, because it may indicate the quality and quantity of effort which employers have squeezed from workers.

The national figures for pay increases in the 1980s suggest no dramatic collapse of union bargaining power: between 1979 and the start of 1984, for instance, real earnings (pay increases in excess of price increases) rose by 17 per cent. But this simple measure conceals great variations between industries and categories of workers. One group – the police – used their political alliances and their strategic role in the maintenance of public order to push real earnings up by nearly a third; other well-organised groups, for instance in construction, barely managed to beat inflation. The average earnings of non-unionised manual workers actually grew more rapidly than did those of union members (Metcalf and Nickell, 1985).

The evidence of pay suggests two conclusions. First, workers performing strategic social functions can still exercise leverage, but the strategic groups are no longer those in the industrial economy, like miners or car workers; they are groups (like the police and water workers) concerned with the maintenance of order. Second, a significant amount of leverage has been transferred from collective bargaining power to influence conferred by particular job shortages signalled by markets.

The record on productivity is more ambiguous still. It was widely observed that the depth of recession in 1982–3 coincided with a marked rise in the productivity of both labour and equipment, especially in manufacturing industry. It was plainly possible that these figures reflected a breakthrough by management in the reorganisation and control of work practices. But other forces were also at work: closing unprofitable and unproductive plant, for instance, raised average productivity without necessarily transforming work practices. The figures for different sectors indeed show substantial variations: in the chemical industry, for example, output per head (a standard measure of productivity) rose nearly 25 per cent in the five years to the end of 1983; but in mechanical engineering, once the heart of manufacturing industry, output per head actually fell.

The impact of technical change and new work practices on the traditional skills and independence of workers is also uncertain. Argument exists about how far changes involve deskilling (the transformation of work into routine and repetitive tasks) or reskilling (equipping workers with new capacities and oportunities to perform tasks cutting across old craft boundaries). Case studies of the process suggest that the 'deskilling/reskilling' issue is peripheral. The most important function of the introduction of new technology is to help managers push forward the frontier of control over workers; but the success of this push is highly variable because, while 'deskilling' is a common motive, practical implementation on the shopfloor allows workers scope to evade and manipulate the new regime (Wilkinson, 1983; Rubery, *et al*., 1983).

The limited significance of the highly publicised 'single-union, no-strike, pendulum arbitration' packages should, finally, be noted. They typically involve foreign firms establishing new plant, on fresh sites, employing new workers, in economically buoyant high-technology industries. The agreements are a model which the most economically advanced employers would like to adopt; but their widespread use runs counter to a native culture of industrial relations characterised by deep suspicions between managers and workers, fragmented and competitive unions and strong class and occupational loyalties. By mid-1985 the much publicised EETPU deals covered only 10,000 workers.

The politics of industrial relations in the workplace have changed radically in the 1980s. The great agents of change have been mass unemployment, falling union membership and structural changes in the job market. The result has been an employers' offensive on established work practices and bargaining arrangements. The 'trench warfare' of the workplace has nevertheless not always gone the employers' way: the frontier of managerial control has been pushed forward, but on an erratic and limited scale. Few of these matters are discussed in everyday accounts of the politics of industrial relations. For most, the phrase refers to the unions' relations with the government of the day. This is our next concern.

The Government's Offensive

The 1980s have seen a sustained government offensive against unions; but it is worth noting in passing that this offensive began in 1976, when the Labour government abandoned full employment as an object of economic policy. The distinguishing characteristic of the two Thatcher administrations has been that the offensive has been more determined, wide-ranging and sustained. Five features are worth emphasising.

First, the Thatcher administrations have been more willing than their predecessors to sacrifice jobs in order to drive down union bargaining power: that is the lesson of the doubling of unemployment since the start of the decade.

Second, where government has been an employer it has systematically tried to give a lead in the 'managerial offensive' described earlier in this chapter. The search for manpower cuts in the civil service, the insistence on linking pay bargaining to new work arrangements in schools, the pressure on the boards of the public corporations of rail, steel and coal to cut inefficient plant and labour practices: all these illustrate the government's desire to lead by example in the offensive against the unions.

The most radical illustration of this process is the 'privatisation' programme, part of whose purpose is to strengthen the hands of managers by exposing workers to wider competitive pressures (see Chapter 5).

Third, some efforts are being made to dislodge unions from the institutional positions which they occupied in the heyday of their power: for instance, the abolition in 1982 of sixteen statutory Industrial Training Boards, and their replacement by one hundred non-statutory organisations, involves supplanting a system where the TUC had a central role by one where the representatives of the Confederation of British Industry are dominant. The massive expansion of youth training in the 1980s has been dominated by a public body, the Manpower Services Commission, with the trade unions divided over how far they should co-operate with the youth training schemes.

Fourth, a wide range of individually small changes in the social security system have been designed to weaken the resistance of the unemployed to taking low-paid work, and to reduce the supply of public money to strikers.

These four prongs of the Government offensive are all important, but it is the fifth, legislation, which has attracted most attention.

Unions and the Law

When the first Thatcher administration came to power in 1979 it looked back on an unhappy history of Conservative legislation on industrial relations. Mr Heath's Industrial Relations Act (1971) proved worse than useless: it failed, but provided a precedent for legislation passed under the Labour governments, 1974–9, expanding the legal rights of unions. The present government has made three main legislative attempts to correct this unhappy history (at the time of writing a fourth is contemplated).

The first attempt, the Employment Act 1980, made available public money to finance secret ballots on strike calls or union elections; gave protection against dismissal to workers who refused to join a union operating a closed shop; made picketing other than at the point of industrial action liable for civil damages; and severely restricted the grounds on which 'secondary' industrial action (blacking, sympathy strikes) was immune from actions for damages in the courts. These provisions were reinforced by two 'codes of conduct': on picketing, recommending that numbers on a picket line be limited to six; and on the closed shop, recommending a pre-existing union membership of 80 per cent of a workforce as a condition for establishing a closed shop.

The second piece of legislation, the Employment Act 1982, was notable for three key features: it gave additional protection to workers threatened with dismissal for refusing to join a closed shop; it provided that (with effect from November 1984) existing closed-shop agreements could be renounced by employers unless at least 80 per cent of affected workers supported the arrangement in a secret ballot; and it laid unions open to claims for damages in the civil courts if they incited industrial action beyond the range of a lawful trade dispute as defined in the 1980 legislation (thus effectively removing their legal protection if they called 'secondary' disputes).

Finally, the Employment Act 1984 required unions to hold secret ballots on a wide range of matters: on national executive committees; on strikes, as a condition of retaining immunity from civil actions for breaches of contract; and on the setting-up and retention of political funds.

Compared with previous Conservative attempts at legislation these laws have enjoyed a fair success – indeed, they could hardly have been less effective than the fiasco of the 1971 Industrial Relations Act. Ministers have learned from the earlier experience. Shifting the onus of enforcement on to employers, suing in civil courts, avoids the direct confrontation between government and union which was so damaging to the 1971 legislation. Relying on suits for damages and seizures of union funds to enforce the laws, has removed the symbolically inflammatory sight of trade unionists being jailed for defiance of the courts.

The Trades Union Congress, which has tried to organise a boycott of almost all the new laws, has found it most difficult to maintain a united boycott at the points where unions have faced penal financial sanctions in the courts. The most illuminating instance came in the Stockport Messenger dispute when – on the issue of unlawful secondary picketing – a small employer spectacularly defeated one of the best-organised printing unions: pursued civil actions through the courts; broke the boycott of the law by instituting contempt proceedings, thus causing union assets to be frozen; and by testing the resilience of the union boycott exposed bitterness and divisions inside the TUC (Gennard, 1984). Fines for contempt of court, and threatened actions for damages, began to break resistance to holding secret ballots before strikes, while the lure of over one million pounds in public funds has induced the second biggest TUC union, the engineers, to apply for public funds to conduct its elections. By December 1985, the TUC seeing its boycott campaign in disarray, backed down.

It is tempting to conclude from these experiences that unions are reeling under the impact of the new laws. The government's record of success is nevertheless uneven. The court injunction has certainly now become one of the battery of weapons used by employers, but the biggest firms remain cautious in turning to the courts, preferring to let smaller, more-aggressive enterprises test the unions' resolve (Marsh and King, 1985). This caution may disappear with time (British Rail and British Telecom, for instance, have announced that they will no longer enforce a closed shop in the absence of a supporting ballot). Yet the greatest uncertainty about the long-term prospects of the legislation lies not in the immediate caution of employers but in one of the key assumptions underlying the new laws: that there indeed exist 'moderate' majorities in unions whose views will be revealed by such democratic procedures as secret ballots. The assumption is apparently supported by opinion polls among the wider population which show majorities against practices like the closed shop and union political funds. It is much less certain that 'moderate' views will guide union members in secret ballots. Of course, ballots on strikes often produce rejections of strike calls; but where majorities support strikes they give added legitimacy to industrial action and make difficult the task of weakening and dividing the strikers. Nor is there any evidence of a systematic relationship between internal democracy and moderation. The most democratic of the large unions, the engineers, has in the last quarter-century gone from moderation to militancy and back to moderation without any significant change in its internal constitution. Two particular examples from the present legislation make the point in more particular ways.

The first concerns the closed shop. The TUC boycott of the legislation means that no industry-wide ballots have yet taken place, but of eighty local ballots monitored by the government's own Advisory Service to the end of 1984, only eleven failed to produce the necessary majority for its continuation. The decline of the closed shop in the 1980s is almost entirely due to the recession and owes little to legislation (Advisory Conciliation and Arbitration Service, 1985, p. 10).

The limited success of the ballots on political funds is even more obvious. The requirement to ballot members as a condition of maintaining political funds is an obvious attack on the Labour Party's main source of income. The government have plainly calculated that ballots will force many unions to disband their political funds, since opinion polls have repeatedly shown a majority of trade unionists opposed to levies and to the unions' special link with Labour. The early returns cast doubt on this calculation. A well-organised campaign in favour of keeping the funds, together with careful staging of the order in which unions ballot, has so far produced clear majorities favouring political funds in all cases. Even if some unions are forced to wind up their funds, the overall result is still likely to strengthen traditional practices. 'Yes' votes will give added legitimacy to the link with Labour; will give union executives a mandate to generate more income by raising the individual political levy above its present trivial level; and will put moral pressure on the minority who presently 'contract out' of payment to cease doing so.

The laws passed by the Thatcher government have been in many respects remarkably successful. The obligations to hold ballots are now widely obeyed; the offer of money to conduct postal ballots threw the TUC into tactical confusion; the rights conferred on individual workers (notably in respect of the closed shop and secret ballots) will almost certainly endure if a Labour government is ever returned. Doubts about the efficacy of the reforms turn not on their tactical success but on their strategic effectiveness. Their long-term purpose is to give a voice to the 'moderate' majorities believed by the Conservatives to exist in the ranks of members. On the evidence of balloting so far, this moderation is much less thoroughgoing and persistent than Conservatives believe.

The Miners and the Conservatives

The long-term impact of Conservative legislation is uncertain. Yet the government has enjoyed one clear victory over the unions. In 1984–5, using the National Coal Board as an auxiliary, it crushed the National Union of Mineworkers at the end of a strike lasting over a year. The victory was momentous not only because it supported the 'managerial offensive' elsewhere in the economy, but also because it defeated the very organisation whose ability in the 1970s to impose social and economic sanctions had seemed to typify union power.

How far this outcome was the result of a coherent strategy is unclear. Those who believe it was can point to the *Ridley Report* of 1978, an internal Conservative Party document outlining plans for defeating the miners (*The Economist*, 27 May 1978). Those who doubt the existence of a coherent strategy can point to the *Carrington Report*, a Conservative Party document warning of the dangers of opposing groups of key workers, and can point likewise to the Conservatives' retreat from challenging the miners in 1981 (Taylor, 1985). Whatever the particular calculations of ministers, it is plain that three wider changes have tilted the balance of power against formerly strategic groups like miners.

First, there have been advances in policing. In the early 1970s the police were baffled by flying pickets and by mass picketing (Taylor, 1984) (see Chapter 6). By the mid-1980s, after a decade of crowd disturbances, they were mobile, well equipped and experienced in crowd control. Second, throughout the economy, enterprises have over the last decade responded to the threat of strangulation at the hands of one occupational group by diversifying their sources of raw materials and services; it is now much more difficult than hitherto for a particular group of workers to cripple a single enterprise, let alone a whole economy. Finally, the recession has produced a significant cultural change. The miners' success in the two great disputes of the 1970s rested heavily on persuading other groups of workers not to cross official picket lines. One of the distinguishing features of the 1984–5 dispute, and of other great disputes of the decade, is that this taboo has declined in force. In the 1980s, workers certainly cannot be trusted consistently to display moderation in their views and actions; but neither can they be relied on to show solidarity with each other. The refusal of the Nottinghamshire miners to support the 1984–5 coal strike, and the subsequent willingness of 20,000 mineworkers to support a breakaway from the National Union of Mineworkers, shows how far the culture of solidarity has been eroded even in the most traditional sections of the working class.

Unions on the Defensive?

The traditional debate about trade-union power in market economies has turned on judgements about how far unions are essentially offensive or defensive organisations. But, in truth, unions are not 'essentially' either of these. Like any other organisation a union's behaviour is contingent on the circumstances in which it operates. In the trench warfare of industrial relations different circumstances will dictate very different strategies. In the 1980s, circumstances have dictated defence or occasional retreat; but only in the case of the miners in 1985 could this be said to have turned into a rout. Half-a-decade of mass unemployment, a managerial offensive in the workplace, and a political offensive by government, have not dislodged unions from positions established in the economy over the preceding half-century.

But if union power remains significant, the culture in which it is exercised has changed profoundly. The Conservatives' laws requiring secret ballots have proved successful because they reflect a rising individualism and a declining sense of class solidarity. Future governments, including Labour governments, will have to take account of this change. The decline of class solidarity will not necessarily bring a decline in militancy. The Conservatives have adopted a highly dangerous course in giving a greater voice to individual trade unionists. The new laws weaken the power of trade-union bureaucracies – full-time officials who, for the most part, have been patriotic, cautious and well integrated into the dominant political culture. The rise of the rank-and-file at the expense of officials poses great problems for the national leadership of the Labour Party which, of course, is closely allied to the trade-union movement. But Labour's problems will be in some degree shared by any party of government. Trade-union bureaucracies are a known quantity: cautious, fairly cohesive and well integrated into the prevailing pattern of politics. The rank-and-file are something else: fragmented, unpredictable, a fertile breeding ground for all kinds of novel ideas. Conservatives may yet rue the day they undermined the trade-union officials.

13

Race and Gender

CHRISTOPHER T. HUSBANDS

The British political system does not easily accommodate the aspirations of groups of the electorate that see themselves as having particular collective interests. This has been a feature of the past as well as of the present. Such ethnic and religious groups as the Irish and the Jews – although they have eventually been more or less fully politically assimilated – had initially to press their claims for the representation of their interests by means of some form of group-based political action.

The contemporary period sees a number of groups in a similar situation. Some of these are marginal, their possible effectiveness reduced by their small size, and by covert or overt hostility among the wider public to the realisation of their specific claims. On the other hand, during the last decade, activism by vociferous minorities among two particular groups, blacks[1] and women, *has* succeeded in pushing their concerns closer to the mainstage of British politics. This chapter examines the effects of blacks and women upon the British political system and, in particular, the steps that black and female activists are taking to insert their group's specific concerns on to the national political agenda.

[1] Throughout this chapter the word 'black' is used to refer to both Asian and Afro-Caribbean Britons.

The Political Circumstances of Black Britons

The concept of a 'black political agenda' – meaning the existence of a set of issues that concern black people exclusively – has to be used with some slight circumspection when describing black politics in Britain. There is certainly evidence to support an argument that, at least in many respects, the political concerns of black people as a whole do not greatly differ from those of whites; the high level of unemployment, for example, was the major preoccupation of both groups during the 1983 general-election campaign. On the other hand, there are *some* noteworthy differences between the two groups; black people as a whole are rather less likely to be concerned about issues with an international or foreign-policy dimension and, especially if Afro-Caribbean, they are far more likely than whites to think of the police and policing as a political issue. Many Asians face problems related to immigration and nationality. Racial attacks are also an ever-present source of concern to many black people, especially many inner-city residents; those affected are particularly likely to come from certain Asian groups, but Afro-Caribbeans also frequently have to contend with these highly distressing experiences. Of course, for different groups of black political activists the police and policing, immigration and nationality, and racial attacks are all major issues around which their activities are variously focused.

It would be a mistake to regard Britain's black population as homogeneous in organisation or interests. There are the obvious differences between Afro-Caribbeans and Asians, as well as within these respective groups. However, one can exaggerate such differences, which may have been less significant in politics than in certain other areas of black social organisation.

Initiatives of the Major Parties on Black Issues

The Conservative Party has long had an image of being 'hard' on race. This was actively promoted in the late 1960s (Layton-Henry, 1984, pp. 75–80) and again in the late 1970s. On the other hand, there has also been an uneasy feeling in the party that it should attempt some overture to black Britons; the result has been an ambivalent policy described by one author as seeking to 'hold on to the support of the racist right while trying to increase support among blacks' (FitzGerald, 1984, p. 23). The Conservative Party's earlier major post-war experience of appealing to an immigrant group, to the Poles by means of its Anglo–Polish Conservative Society, was reasonably harmonious because of their considerable ideological congruity with Conservatism. This Society was then the model, perhaps rather inappropriately, for the foundation by the Conservatives in 1976 of an Anglo Asian Conservative Society and an Anglo West-Indian Conservative Society. The former has undoubtedly been the more successful, although members are increasingly wanting to move from it into the party proper. In mid-1985 there were thirty-two branches in England. It is reasonably represented in all locations with major Asian populations. The Conservative Party concedes that its initiative to the Afro-Caribbean population has been less successful, the Anglo-West-Indian Conservative Society having only ten branches in mid-1985; these were almost exclusively in London and comprised members from such occupations as the law and business. There are unlikely to be any attempts, formal or informal, to persuade a local association to adopt a black parliamentary candidate in a winnable seat. However, it is seen as 'only a matter of time' before there is an Asian Conservative MP, even though little prospect seems to be held out for an Afro-Caribbean Conservative MP in the foreseeable future.

The Labour Party's dilemma on this subject has been a similar one, although it has recently assumed a more acute form as black activists have become more vociferous within the party. In the 1960s and 1970s it attempted to resolve its ambivalence on race and immigration by a dual approach emphasising tight immigration control and a commitment to full equality for those blacks already in the country (Husbands, 1983a). This stance, it was hoped, would prevent erosion of Labour's white racist vote, while simultaneously attracting black voters to the party. After the loss of the 1979 general election, the emphasis did shift and in 1980 the party's National Executive Committee (NEC) issued an advice note, *Labour and the Black Electorate*, which acknowledged that the party had to do more if it was to retain its strong support among black voters. There followed a questionnaire to constituency parties and meetings with representatives of black groups; the response to the questionnaire and the limited evidence of activity to attract black people's support were considered disappointing. On the other hand, Labour's 1983 election manifesto did mention a number of policies that were intended to appeal to black voters.

Since the 1983 general election the issue of 'black sections' has dominated Labour Party debate on this subject, although even among black activists there is no unanimity about how desirable these might be. They have been criticised by some as attempts by middle-class black people to create an avenue for their own political ambitions and as correspondingly irrelevant to the needs of working-class blacks. The 'black sections' issue first emerged in a major way at the 1983 party conference. At the 1984 party conference in Blackpool there were various motions on black sections, which Jo Richardson (for the NEC) sought to have remitted. When the movers refused this, conference voted by 4,018,000 to 2,019,000 to accept a resolution opposing black sections; two resolutions supporting black sections were defeated overwhelmingly. A working party that was examining black representation in the party, chaired by Richardson, had issued a consultative document to local parties, and its final report, completed in early June 1985, contained a majority recommendation for the formation of black sections with delegates at local, regional and national level and some representation on the NEC. A minority report, however, wanted merely a Labour Black Rights Campaign open to all party members. Neil Kinnock and most of the party's parliamentary establishment had early on made very public their opposition to black sections. Such evidence as there is suggests that most grass-roots Labour activists are also opposed. The 1985 party conference endorsed by 5,337,000 votes to 1,181,000 the NEC's policy statement calling for a Black and Asian Advisory Committee. Two composited pro-black-sections resolutions were heavily defeated, even the more favoured one falling by a margin of over four to one. Even so, the black-sections campaign achieved a notable success in December 1985 when Diane Abbott, one of its most prominent activists, was chosen as the prospective Labour parliamentary candidate for the safe seat of Hackney North and Stoke Newington, deposing the sitting MP, Ernie Roberts.

The Alliance parties perhaps have less public salience on race issues than do the other two major parties. The Liberal Party has in fact opposed all immigration legislation passed since the Commonwealth Immigrants Act 1962, although this fact does not feature in the public's image of the party (which is in general seen as having a strong commitment to restriction). However, the Liberal Party does have an active Community Relations Panel, which is responsible for contributing policy initiatives and for publicising the party's concern about ethnic issues; it has also been given the responsibility for conducting an enquiry into ethnic-minority involvement in the party. Liberal Party policy calls for the repeal of the British Nationality Act 1981 and its replacement with a more generous alternative. The Social Democratic Party has a Social Democratic Campaign for Racial Justice, established by a number of activists. Its role is that of watchdog over party policy and it is a source of information on race and ethnic issues, although it has deliberately sought to avoid a direct concern with immigration and nationality questions. The SDP has tended to implement its domestic policy initiatives on race through its general policy on urban issues. The party calls for 'positive action for members of ethnic minorities to remove the effects of past discrimination and inequality' as well as for a policy on equal opportunity employment. However, its policy on immigration asserts the need for controls, although calling for them to be operated in a non-discriminatory manner. It is undoubtedly intended to be seen as restrictionist, more so than that of the Liberal Party.

Blacks' Participation in Politics

Access by black people to elected political office has increased, but only to any marked extent within the past four or five years and very noticeably only at the local level. In the post-war period there have, of course, been no MPs defining themselves as from ethnic minorities, although there have been some black candidates. Both Labour and Liberal Parties put forward a handful of such candidates in general elections from 1959 to October 1974, the best-known being David Pitt (since 1975 Baron Pitt of Hampstead) in 1970, whose Labour candidature suffered a quite disproportionate counter-swing. In the 1979 election the Conservatives had two black candidates (both Asian) for the first time in recent history; Labour had one and the Liberal Party two. In the 1983 general election there were eighteen in all: the Conservatives had four Asians; Labour had four Asians, one Afro-Caribbean and one Afro-English; the Liberals had three Asians and one Afro-Caribbean; and the SDP had four Asians. However, only one of these (Labour's Paul Boateng in Hertfordshire West) ran in a situation that was even theoretically winnable. By December 1985 a number of black prospective candidates had been adopted for Labour in safe or winnable marginal seats and so there are likely to be a small number of black MPs after the next general election.

On the local level, blacks have been slightly more successful, though still under-represented in comparison with local demographic characteristics. Of the 1,914 councillors in the thirty-two London boroughs who were elected at the May 1982 elections, eighty-two (4 per cent) were from ethnic minorities, seventy-nine being Asian or Afro-Caribbean (FitzGerald, 1984, p. 97). Thirty-five of the former and thirty-four of the latter represented Labour, but the Conservatives had five Asian and two Afro-Caribbean councillors, the Liberals had one of each, and there was a lone Asian independent (from Spitalfields ward in Tower Hamlets). Since May 1982 there has been a further increase in the number of black borough councillors in Greater London.

Blacks' Electoral Behaviour

Black people are less likely to be registered to vote, despite their eligibility. In a study done in 1981, 6.5 per cent of all eligible people in Great Britain were unregistered, but the figure was 31 per cent among New Commonwealth citizens and 19 per cent among those born in the New Commonwealth whose citizenship was not established (Todd and Butcher, 1982, pp. 13, 22). Some Labour-controlled local authorities have recently undertaken registration drives that report spectacular results (for example, an additional 10,000 electors, many black, in Haringey).

Given that they are registered, Asian voters are frequently more likely to vote than whites in the same constituency, while registered Afro-Caribbeans are least likely to vote. Except in exceptional situations (for example, among Asians in Rochdale), both Asian and Afro-Caribbean voters have tended to be heavily Labour, although there has been some dispute about precisely how monolithically Labour they are. The 1984 CRE report on ethnic minority voting produced the following figures for vote intention among all sampled Afro-Carribeans naming a party: Conservative 8 per cent; Labour 86 per cent; Alliance 5 per cent. For Asians the figures were: 6 per cent, 80 per cent, and 12 per cent (Anwar, 1984, p. 11). On the other hand, in the BBC/Gallup election study, whose sample of black voters was more nationally representative, the corresponding figures for all such voters were: 21 per cent; 64 per cent; and 15 per cent. The discrepancy between these data for the 1983 general election led to speculation about whether some black voters, particularly non-manually employed Asians, might have deserted Labour for the Alliance (Studlar, 1983; Layton-Henry, 1983). However, this seems to have happened, if at all, only on a small scale.

Reactions to Black Candidates and to Blacks' Locational Proximity

The politicians' received wisdom about the reaction among the electorate to black candidates has been severely influenced by the disproportionate swing of 10.2 per cent against David Pitt in the Wandsworth Clapham constituency in 1970. Although during the 1970s black candidates became a normal phenomenon in local elections in a number of cities, it was not until the 1983 general election that enough candidates were put forward by the major parties to permit something approaching a systematic test of the electorate's reaction to them in national contests. The four black Liberals suffered a net decrement of over 5 per cent and the four black Social Democrats one of over 3 per cent. However, the four black Conservative candidates and the six black Labour ones suffered only the slightest loss, an average of well under 1 per cent for each group. Thus, only the

Alliance's black candidates performed less well than the other characteristics of their contests would have predicted, doubtless a consequence of the less committed nature of much Alliance voting.[2]

The pattern of support for racist candidates is an appropriate index for the assessment of how locationally specific is any backlash by white voters. Such candidatures were collecting not insignificant support in high-immigrant constituencies as long ago as the late 1950s. There has been some dispute about the precise nature of any locational effect in accounting for support for racist candidates, particularly whether reactions occur at such micro-spatial levels as the neighbourhood, or even the individual street. The concentration of support for the National Front was much more obviously at the city-wide than at the neighbourhood level, being reasonably dispersed throughout those cities where it was present. Local levels of New Commonwealth settlement were, of themselves, a poor indicator of National Front support (Husbands, 1983b, p. 25), which was much more a consequence of particular processes at the city-wide level in cities with at least a critical minimum presence of black residents.

Reactions to Issues Associated with Blacks

The complement of reactions to black people that are a consequence of proximity to them is those that occur more generally, independent of whether such proximity exists. We have become accustomed to thinking that the consensus of opinion among white Britons towards black people has always been exclusionist, if we mean by this that 'we wish you had not come'. However, although this was the case by the early 1960s, it was probably not so in the immediate post-war period,

[2] These estimates have been derived from a variation of the model used by Dunleavy and Husbands (1985, p. 205), except that in the present case the respective 1979 percentage of party support has been included among the independent variables, the dummy variable for women candidates has been omitted and all relevant constituencies (including those with two women major-party candidacies) have been used in the calculations. The Liberal vote has been used as the 1979 vote in the SDP estimate. The differences account for the slight discrepancy between the results reported here and those in Dunleavy and Husbands (1985).

certainly not when the issue was stated with a caveat about labour shortages. Banton (1983) reviews a number of studies done in the late 1940s and early 1950s, including a nationally representative one conducted by the Government Social Survey in 1951. In response to the question 'Providing, of course, that there is plenty of work about, do you think that more coloured people should be encouraged to come and work over here?', responses were: 'yes' or 'yes, I think so', 38 per cent; 'yes, providing there is work, etc.', 8 per cent; 'no', 38 per cent; other answers, 10 per cent; and 'don't know', 6 per cent.

As we have seen already, by the 1964 general election race and immigration had become significant in a number of contests. It is generally agreed that there was no noticeable impact from these issues in 1966, perhaps because the Conservative leader Edward Heath discouraged party candidates from indulging in these sorts of tactics (Layton-Henry, 1984, p. 65). However, by 1970 public perceptions of both Labour and Conservative Parties had been altered and, in the restrictionist political climate of that election, the Tory party used the issue in various local contests, such as Eton and Slough, and among the electorate as a whole was able to gain from it sufficiently to win thirty to thirty-five additional seats. Although Enoch Powell's statements about the Conservative government (concerning the EEC rather than race) may have had some impact upon certain results in the 1974 elections, particularly the February one, it is generally agreed that race and immigration had little noticeable impact upon the overall outcome of these contests. Their direct effect upon the 1979 result is uncertain, although the Conservatives had made every effort to capture these issues and had been able to gain some temporary extra support in the opinion polls as a result of Mrs Thatcher's widely publicised speech about 'swamping' in late January 1978. In the 1983 general election, according to Crewe's analysis (1983), immigration had no detectable impact upon the outcome, being explicitly mentioned as an issue by only 1 per cent of voters.

Race and immigration exist as issues with a usually low profile, but high potential for electoral influence, even if such influence has usually been only short-term. Certain race- and immigration-related stories, when widely disseminated in the mass media, have given a substantial temporary boost to the percentage of the electorate that regards immigration as 'the most urgent issue facing the country at the present time'. However, the net effect of such incidents upon levels of major-party support in the polls has been short-term, seldom having the momentum to sustain itself for more than a month or two.

The Political Circumstances of Women in Britain

The story over the last half-century of women in British politics has been one of rather limited success, of occasional gains punctuated by further setbacks till the present situation where women, although reasonably well represented among party members and in local politics, are still a comparative rarity among national politicians. The increased feminist consciousness of the past fifteen years has perhaps changed attitudes about women being professional politicians but it has had little impact on the national political stage, despite a woman as Prime Minister since 1979.

Even so, several women-related issues have been placed on to the political agenda largely because of the efforts of feminist political activists. Foremost among these are childcare rights, parenthood leave and equal employment rights. The government is resisting efforts from other countries in the European Economic Community for the introduction of a statutory right to parenthood leave, whose provision in the United Kingdom compares very unfavourably with that in most other countries of western Europe. Equal rights in employment have been a significant political issue since at least the early 1970s, although deepening economic recession has meant a deterioration since then in some aspects of women's position in work. This is despite a number of recent individual successes using the Equal Pay Act, as strengthened from January 1984 at EEC insistence to allow cases to be based on the 'equal value' of work done rather than its being exactly the same job. Greater public discussion of some other women-related issues has come about partly because of feminist agitation. Violence against women is no longer quite the unspoken subject that it once was; to some extent as a result of publicity on this matter, there has been a marginal improvement in the way the police handle rape cases, although their traditional non-interventionist approach to domestic disputes seems largely unaltered. Moreover, nearly twenty years of sporadic feminist agitation against sexist advertising seem to have had little effect, except perhaps in the liberal quality press.

Initiatives of the Major Parties on Women's Issues

It has become axiomatic among politicians and writers on the Left that policies of the Conservative government have worsened the lot of many women, especially working women. Certainly, the Conservative Party has done little that suggests any explicit special appeal to women voters and has undertaken no particular initiatives to increase the number of Conservative women MPs. However, the three other parties have formulated practices or policies in both these areas.

The Labour Party's 1982 programme contained a section on women's policy. This was designed to appeal to working women in particular. Sections of the 1983 election manifesto promoted similar and related themes. However, in such party publications as *New Socialist* there have been contributions criticising the excessive male-orientation of some of Labour's economic and industrial policies. On the other hand, in mid-1985 Neil Kinnock made major pledges on women's issues, including the formal commitment to establish a minister for women's rights, and Larry Whitty (the party's General Secretary from June 1985) promised action to increase the role of women in the party's organisation. In October 1985 a working group chaired by Jo Richardson produced a party discussion paper for a fringe meeting of the party conference that sought to formalise Kinnock's earlier proposal for a Ministry for Women's Rights; the proposals were then agreed by the NEC at its November 1985 meeting.

The two Alliance parties have also sought to establish a significant profile on women's issues. Each has a constituent women's section, the Women's Liberal Federation and Women for Social Democracy; in 1982 they jointly produced *The*

Alliance Charter for Women, which calls for equality of treatment with men in a number of spheres and for a range of improvements in women's positions. In 1983 the Liberal Party established a working group on the status of women, which produced a report for a party Commission on the Status of Women at the 1984 Assembly. The Commission moved a series of statements of principle about the position of women in the wider society and in the Liberal Party itself; these were passed by the Assembly. Section B of its motion, containing specific proposals, was more controversial and was referred back to the working group for consideration in its second report. This was debated at the 1985 Liberal Assembly in Dundee, which – despite a number of critical speeches from the floor – unanimously passed a comprehensive motion that demanded the end of discrimination against women and urged the Liberal Party to be in the forefront of women's rights campaigning. Likewise, the SDP has published an extensive *Policy for Women*, originally the product of a Working Party on Women's Policy; its major concerns are for women in the wider society. The SDP claims a larger proportion of women members than other parties and it is proud of the high proportion of women in the top levels of its party organisation, although there are of course no SDP women MPs. However, the SDP is the only one of the major parties whose constitution provides that a minimum number of women be included in any parliamentary shortlist; it was the campaign within the party in its early days to have this provision included in the constitution (as well as four women and four men as the 'national elected members' on the party's National Committee) that generated Women for Social Democracy. More controversially, there are attempts within the SDP to persuade it to take policy positions in favour of controlled embryo research.

Women's Participation in Politics

In the 1983 general election the Conservatives had forty women among their 633 candidates, a mere 6 per cent. Among all male Conservative candidates 65 per cent secured election, whereas only thirteen (33 per cent) of the forty women did; thus, male candidates were almost twice as likely as female ones to be successful. For Labour the picture is both better and worse. Labour had seventy-seven women out of 633 candidates (12 per cent) but success rates were greatly different. Thirty-six per cent of Labour's male candidates were elected, compared with a mere ten (13 per cent) of women; male candidates were therefore almost three times more likely to be successful. Thirty-two (10 per cent) of Liberal candidates in 1983 were women, as were forty-three (14 per cent) of those for the SDP, the highest of the four parties. However, no women Alliance candidates secured election or came very close to doing so; Shirley Williams was defeated in Crosby.

There are a number of suggested reasons for the limited participation of women in electoral politics in general and for their minimal presence in Parliament in particular. First, there have been straightforward assertions that most women prefer things this way, although some feminist writers have used a similar argument, treating participation in mainstream politics with disdain and hence imputing rationality to a desire not to become involved. Rather differently, Randall (1982, pp. 12–34), while seeing biological sex differences as the root cause of male dominance, insists on the significance of culture in accounting for limited female participation, particularly aspects of female socialisation. On the other hand, Vallance (1984) has taken her to task for this emphasis, and the former author's explanation for this phenomenon is much more in terms of institutional barriers and discrimination that women face. Both these aspects seem important in accounting for the small number of women MPs. Even when adopted on to a party's panel of potential parliamentary candidates, women may be at a disadvantage compared with men. Because of domestic commitments many such women restrict their search for seats to a few constituencies that are readily accessible from where they live; men, on the other hand, are far more likely to be willing to seek a seat anywhere in the country.

At the local level, where activists and councillors are unremunerated (except for some reimbursement of expenses), participation by women is considerably more widespread, in some respects exceeding that of men. All political parties have substantial proportions of activists, often majorities, who are women. This fact has frequently been especially noted in the context of the Conservative Party, although this level of participation has not filtered through to the party's parliamentary candidate panel, which is still nine-tenths male (Vallance, 1984).

There has been some research upon women in local government, both as employees (for example, Webster, n.d.) and as elected councillors (for example, Bristow, 1980). Webster pointed out the extent to which local government services rely on the labour, often the poorly paid labour, of women. Bristow reported that, in 1977, 17 per cent of local councillors in England and Wales were women, being best represented in London. Outside London women were best represented in more affluent areas, a phenomenon that Bristow tends to explain by the availability of sufficiently sizeable pools of middle-class housewives. Although London may in some respects be exceptional, it is worthwhile examining the more recent success of women in entering London borough politics. Slightly more than a fifth of all councillors elected in the thirty-two Greater London boroughs in the 1982 council elections were women, a figure that varies little among the major parties. The Conservatives had 472 women among their 1,826 candidates, a percentage of 26 per cent. Among all male Conservative candidates 57 per cent secured election, while 45 per cent of the party's women did; thus, male candidates were just a little more likely (1 to 1.3) to be successful than were female ones. For Labour the picture is remarkably similar. Labour had 481 women out of 1,914 candidates (25 per cent). Forty-three per cent of Labour's male candidates were elected, compared with 36 per cent of its women; male candidates were therefore a bit more likely (a ratio of 1 to 1.2) to be successful. Four hundred and forty-four (24 per cent) of Alliance candidates in these elections were women. Seven per cent of male Alliance candidates were elected, compared with 5 per cent of female ones – a ratio of greater male success of 1 to 1.3. Thus, in Greater London women are local candidates far more frequently than is true in national politics, though still less than their proportion of the electorate would predict. Once chosen as candidates, the level of their disadvantage suffered relative to men is much less, although it still exists.

Women's Electoral Behaviour

The subject of female electoral behaviour in Britain is a complex one. Let us first dispose of the matter of electoral participation. Whatever evidence there may have been in the past that women were less likely than men to vote has long since been superseded. In fact, in the survey used by Dunleavy and Husbands (1985) women reported that they were slightly more likely than men to have voted in the 1983 election, although the difference (2 percentage-points) has no substantive significance. We move on now to consider actual voting, first considering simple comparisons between women and men in the aggregate, unembellished by variables other than gender. In 1979 the aggregate figures for each gender found by the British Election Study were (Särlvik and Crewe, 1983, p. 92):

	Women	Men
Conservative	48%	45%
Labour	38	38
Liberal	13	15

These authors say (1983, p. 91) that this 'difference was probably smaller than at any time since the war'. In 1983, according to the data of Dunleavy and Husbands (1985), the comparable percentages were:

	Women	Men
Conservative	45%	41%
Labour	25	30
Liberal/SDP	29	30

During the period when, considered merely at the two-variable level, there did exist a noticeable relationship between gender and voting (albeit never a large one compared with certain other variables), political scientists sought third variables that would explain or interpret this difference, thus showing the gender-voting relationship either to be spurious or the result of the impact of gender upon some intervening variable that itself was the real determinant of voting behaviour. Age has perhaps been the favourite third variable for attempting to show the spuriousness of the gender-voting relationship (for example, Hills, 1981), thus regarding gender as potentially an extraneous variable and thereby testing a model of conjoint influence.[3] Older voters tended to be more Conservative-inclined, and women, with their greater longevity, were on average older than men. Taylor (1978), in a test of Parkin's theory of working-class conservatism, also argued that the engagement in manual work outside the home was the intervening variable accounting for the gender-voting relationship among working-class voters; working-class women who worked outside the home had a profile of party support identical to that of similar men.

However, now that there is almost no significant two-variable relationship between gender and voting, the earlier strategy has been superseded by a search for suppressor variables, ones which, when controlled, reveal that there *is* a genuine relationship between gender and voting. Foremost among these suppressor variables is social class, assigned from the present or most recent previous occupation of the individual voter and not of the head of household. According to Dunleavy and Husbands (1985, p. 128), there is a clear interaction effect of gender and social class upon voting behaviour. Women manual workers were much more Labour- and less Alliance-oriented than were similar men in the 1983 election; in fact, as many as 52 per cent of women manual workers who voted supported Labour. On the other hand, women non-manual workers were more Conservative-, less Labour- and less Alliance-inclined than comparable men; as many as 53 per cent of these women voted Tory.

Reactions to Women Candidates

Dunleavy and Husbands (1985, p. 205) show that in 1983, controlling on other relevant variables, there was a net

[3] Rosenburg (1968) is perhaps the most accessible account of the various roles that third variables may play in understanding the two-variable relationship between a dependent and an independent variable. Of particular relevance in the present respect are his discussions of extraneous variables (pp. 27–40), intervening variables (pp. 54–66), suppressor variables (pp. 84–94), conditional relationships (pp. 105–58), and conjoint influence (pp. 159–96).

decrement suffered by the women candidates of all four major parties, although it was not large: 0.3 percentage-points for the SDP, 0.4 points for Labour, 0.8 for the Conservatives, and 1.6 for the Liberals. Thus, women candidates – fighting in most cases very barren political terrain – may have suffered a very slight degree of voter discrimination in 1983. In any case, decrements of these magnitudes, even in marginal seats, would affect the actual outcome in no more than one or two constituencies.

On the other hand, there does seem to be one way in which women candidates are discriminated against by voters. Once they have won a seat, women may find it more difficult than sitting male candidates to accumulate the electoral benefits of incumbency. Rasmussen (1981), examining earlier female candidatures, has noted that women, once elected, were more vulnerable to subsequent defeat than comparable men, almost as though a section of the electorate, while willing temporarily to suspend its displeasure, moves against those women who seek to persist with a parliamentary career. Dunleavy and Husbands (1985, p. 205) found that Labour women incumbents were disadvantaged relative to similar men on account of their gender; surprisingly, however, Conservative women incumbents were much less affected on this account.

Reactions to Issues Associated with Women

We have seen the electoral potential of race- and immigration-related issues. Not only have they affected electoral outcomes in a significant way but, because of the strength of feeling that they arouse in a large part of the electorate and/or among party activists, they are subjects upon which governments themselves have been forced to introduce their own legislation. On the other hand, issues concerning women have not usually had the same status. Either, like certain pieces of social security legislation or divorce reform, they tend to have been universally regarded as issues affecting women rather than as 'women's issues' and have been incorporated relatively inconspicuously into a government's general legislative programme: Or, like abortion or domestic violence, if they have been seen as 'women's issues', governments have been able to avoid becoming embroiled in the controversy likely to surround them because strong feelings upon either side have been held only by minorities of the electorate. Such legislation has usually been initiated by private members' bills. This is true of the issues of abortion reform and domestic violence. Even attempts to make current abortion legislation more restrictive were introduced as private members' measures, as was the bill from Enoch Powell to outlaw embryo research. Governments usually intervene on 'women's issues' only when forced to do so under pressure from some important group or groups. Thus, the British Medical Association urged the Department of Health and Social Security to appeal – in the event successfully – to the House of Lords against the 'Gillick ruling' prohibiting doctors from prescribing contraception to girls under sixteen without parental consent. However, as long as most 'women's issues' remain the focus of concern only for minorities of the electorate, their potential for electoral impact will remain far less than issues associated with race. Paradoxically, the fact that women are more than half the population but are not segregated from men contributes to the 'invisibility' of 'women's issues'. On the other hand, black people, although less than 5 per cent of the population, tend to be residentially segregated from whites and this works to increase the salience of 'black issues'.

14

Mass Media

KENNETH NEWTON

Most people receive most of their information about the political world through the press and broadcasting. The potential political influence of these media is therefore considerable. Some observers, and most apologists for the news media, do not think that any undue influence is, in fact, possessed by them; they think the press and broadcasters offer the public and the politicians a plurality of sources and views. Others believe the news media reflect and reinforce the dominant values of British society and serve to filter out any important dissident views. The argument between these two positions has become increasingly important in British politics as the political allegiance of so many voters has become volatile (see Chapter 2). There are also three different kinds of change occurring more or less simultaneously within the news gathering and disseminating world: first, there have been important changes in the pattern of ownership and control; second, there have been important changes in the techniques by which news is gathered and broadcast; third, there have been changes in the ways the parties and the government manipulate and are manipulated by the news media. In this chapter the evidence about each of these kinds of change – and they are interrelated – is examined in order to gain a purchase on the argument between the pluralist and the dominant-values views.

The Concentration of Ownership and Control

In the twentieth century news media have shown a persistent trend toward the consolidation of mass markets. The total number and type of news outlets was of course transformed by the advent of radio and television. In addition, however, there has been a proliferation of specialist publications designed for minority markets. But the production of the mass-audience media has tended towards concentration. While there were thirteen national dailies in 1922, by 1975 there were only ten. In 1985 there were nine, and one of these, *The Financial Times*, has a very small circulation. There was a similar decline in the number of Sunday, evening, provincial, and weekly papers. At the same time, the circulation of the surviving papers has increased over the long term. In 1937 just under ten million national dailies were sold; in 1975 the figure was just over fourteen million. By 1985 the circulation figure was just under fifteen million.

The decline in the number of titles and the increase in mass circulation has been combined with an increased concentration of ownership and control. Thus those newspapers which have survived are not merely few in number but their ownership is in the hands of an even smaller number of companies. In 1983 it was calculated that 85 per cent of daily and Sunday sales was in the hands of seven companies, three of which accounted for over half the total. Five companies produce half the total circulation of regional papers. Moreover there is a marked and apparently increasing tendency for the big newspaper companies to have their fingers in other media pies – especially commercial radio and television. Recent developments in cable and satellite broadcasting have exacerbated this tendency and there is, of course, a good deal of overlapping ownership between newspapers and magazines. The following thumbnail sketches give some indication of the interests of the largest of the media magnates and their organisations.

Lord Matthews, Trafalgar House Ran the Express Group (until it was sold to United Newspapers in 1985), which included the *Daily Express* and the *Sunday Express* in England and in Scotland, plus the *Star*, the *Standard*, and eleven local newspapers. The total circulation is over 7 million. Trafalgar House also has interests in TV-AM, Capital Radio, and in publishing companies abroad. It also has a controlling interest in Cunard Shipping and Cunard Hotels as well as large property and insurance holdings.

Rupert Murdoch, News International Murdoch publishes the *Sun*, *News of the World*, *The Times*, *Sunday Times*, and the various *Times* supplements. The total circulation is about 10 million. Murdoch is known to be actively involved in the editorial policies of his papers and has strong conservative opinions. His newspaper chain stretches over three continents and he also has financial interests in Satellite TV, William Collins (publishers), Reuters, Sky Channel (cable TV), and in transport, oil and gas production.

Robert Maxwell, BPCC Maxwell bought the *Mirror* in 1983 and he also publishes the *Sunday Mirror*, *Sunday People*, *Daily Record* and *Sporting Life*. The total circulation is over 10.5 million. Maxwell also has an interest in the publishing company of E. J. Arnold, *Observer* Magazines, Central TV and Rediffusion Cablevision, as well as in a number of other companies at home and abroad.

Lord Rothermere, Associated Newspapers Publishes the *Daily Mail*, *Mail On Sunday*, and *Weekend*. The total circulation is about 5 million. He also has an interest in Northcliffe Newspapers and various local presses, and in broadcasting companies and various oil, transport and other businesses.

These bare outlines suggest multi-media empires stretching far beyond Fleet Street, and into television, both conventional and cable, as well as into a host of related activities – radio, entertainment, property, shipping and finance. Most of the large newspaper companies are involved in commercial television: Maxwell in central TV, SelecTV and Rediffusion Cablevision; Murdoch in satellite and cable TV; Reed International (from whom Maxwell bought the Mirror Group) in ATV; the *Guardian* in Anglia TV; and International Thomson Organisation in STV. It is not just that newspaper ownership is highly concentrated but that ownership and control of the media as a whole is increasingly concentrated.

It might be argued that recent technological developments in printing – notably the development of powerful computer-driven techniques of typesetting – will break up the Fleet Street oligopolies. Mr Eddie Shah, who was successful in introducing this new technology a few years ago, is now challenging Fleet Street for the production of the first daily paper in colour. Various other possibilities, including a new national daily and another London evening newspaper, are being explored, although much depends on the extent to which new technology can in fact be employed and existing outlets and channels of distribution utilised.

Eddie Shah's success in breaking union opposition to the new technology became a news story in its own right. In this it was similar to the extensive coverage now given to stories about restrictive practices and exorbitant wages in the industry. A few of the printers and typesetters left in Fleet Street do make high wages, but Fleet Street employs a relatively small proportion of workers in the industry. Part of the special nature of Fleet Street working conditions and wages derive from its peculiarly perishable product. Change may not have been as rapid as some would wish, but there has been reform and rationalisation there. Although it is extremely difficult to obtain reliable figures, the one thorough and reliable study of the numbers employed in the national newspaper industry showed a decline from 41,590 in 1970 to 37,367 in 1975, a fall of 10.2 per cent, in under five years (Royal Commission on the Press, Cmnd 6680, 1976, p. 232.

It is important also to note that the pioneering role of small companies in developing new technologies does not necessarily spell the death of the giant conglomerates. Experience with other technological developments (for example, with micro-computers) suggests that the small companies kill each other off in the early stages of experiment as they compete to produce better equipment more cheaply. Once there are few major innovations left to develop, the big companies then swing into mass production. After a generation in which small and innovative concerns have battled over the computer and word-processor market, a few multinationals have entered the market and are beginning to consolidate their control. The newspaper and publishing industries are no strangers to technological revolutions but this has not prevented the emergence of a few dominant corporations. It is likely that when the current round of technological fighting is over, these same corporations will be in the best position to exploit its fruits.

Quite probably a similar pattern will emerge in relation to the free sheet, which has constituted a major change in the newspaper industry. The free sheet was hardly known in most parts of the country fifteen years ago. Now the number and circulation of such sheets is increasing rapidly and at the expense of traditional local newspapers. In part its emergence has been made possible by new technology; but it has also been encouraged by the huge advertising expenditure in the local and regional press, which is a third larger than the national press and expected to grow rapidly. At the moment, many free sheets are run by small independent companies (some of them by the same companies that produce more orthodox local newspapers). However, when these companies have tested the new market – with all the risks that involves – it seems likely that the larger companies will muscle in on this area. Reed International, one of the big six publishing companies, launched the first free daily in Birmingham in 1984. It has already achieved a circulation of over a third of a million. Neither new technology nor free sheets are likely to disturb the trend towards further concentration and control of the industry.

Newspaper Partisanship and Media Bias

Bias and partisanship are not of course consistent with the pluralist theory which claims that a large number and diverse range of sources of news and comment exist. Indeed the political bias of the media seems to have become increasingly pronounced in the last twenty years. There are good reasons for this. First the media now play a more important part in politics, especially elections, than before. The old style of electoral campaign at the hustings has been replaced by carefully planned and expensive electoral strategies. The Conservative Party has taken this process furthest by holding election briefings in the staged atmosphere of its advertising agency, and the closest that Mrs Thatcher came to an encounter with the general public during the 1983 election was on a television programme in which a member of the audience disturbed her public composure by persistent questions about the sinking of the Argentine battleship the *General Belgrano*.

The media has also increased its coverage of political news in the past ten to fifteen years and large sections of the national press have developed an increasingly partisan tone. Before 1955 there was little television coverage of politics, and indeed the '14 day rule' actually obliged the BBC to avoid broadcasting on current legislation and matters due for parliamentary debate during the following two weeks. By 1970, TV coverage of the news was quite extensive and, with the growth of TV ownership and extra channels, there were may more people watching it (Tunstall, 1983, pp. 9–13). Both radio and TV lost some of the distinctive features which they had acquired under Lord Reith; their public broadcasting mission was complemented by a strong emphasis on entertainment.

The arrival of television as a news and current affairs medium had a major impact on the daily press. Television is, in theory, bound by law to provide a balanced account of events and may not exhibit partisan bias. This has encouraged Fleet Street to take a different role, involving strong partisan attitudes. The tabloids have used every opportunity to press their own party political views, and the quality newspapers have increasingly augmented front-page news with centre-page opinion pieces, leader columns and in-depth comment. The tabloids and the quality press in their different ways have moved away from simply presenting the news, towards moulding, slanting and interpreting it. The papers also now give more space to elections than they did fifteen years ago.

There is the further point that the media generally and the dailies particularly may now have greater opportunity to influence the political attitudes and voting intentions of the public. For much of the post-war period, the party allegiance of most voters was settled and therefore not easily influenced by the media. But with a volatile and de-aligned electorate the media's potential for influence is likely to be appreciably greater. In other words, if the electorate does not have firm and decided views of its own, then the way in which news is reported may help the public to make up its mind.

Fleet Street Partisanship

Some recent events indicate that the media are not as balanced as pluralist theory suggests they both are and should be. The sale of Reuters, the international press agency which provides much of our overseas news, was one example of a change which had implications for political impartiality. In Fleet Street, both press and journalist unions have expressed concern about biased reporting – a bias which was reinforced by the sale of the *Sun* and the *Mirror* to press magnates with explicit political views which they wished to communicate through their newspapers. Similarly, concern has been expressed about the independence of *The Times* and *The Observer*, which although always active in politics have now adopted a more controversial line.

In fact, the party politics of Fleet Street has become more monolithic as its ownership has become concentrated (see Figure 14.1). In the 1945 election, four national dailies – representing a readership of almost 4.5 million – supported Labour. Those newspapers accounted for almost 35 per cent of sales. The Conservative Party was then supported by four papers, but these had over half the national circulation. Since 1945 the number and circulation of Labour papers has declined, while Conservative support and circulation has increased (see Seymour-Ure, 1974, pp. 166–7). In every election since the Labour Party was formed, there has been an imbalance between its voting strength and the proportion of the newspaper market which papers supporting Labour could claim. In the period since 1970, the imbalance of partisan support has been overwhelming. Indeed, in 1983 only the *Mirror* (with 22 per cent of newspaper sales) recommended voting Labour, as compared with six strongly Conservative papers which together accounted for 75 per cent of national circulation.

In 1977 the Royal Commission on the Press stated: 'There is

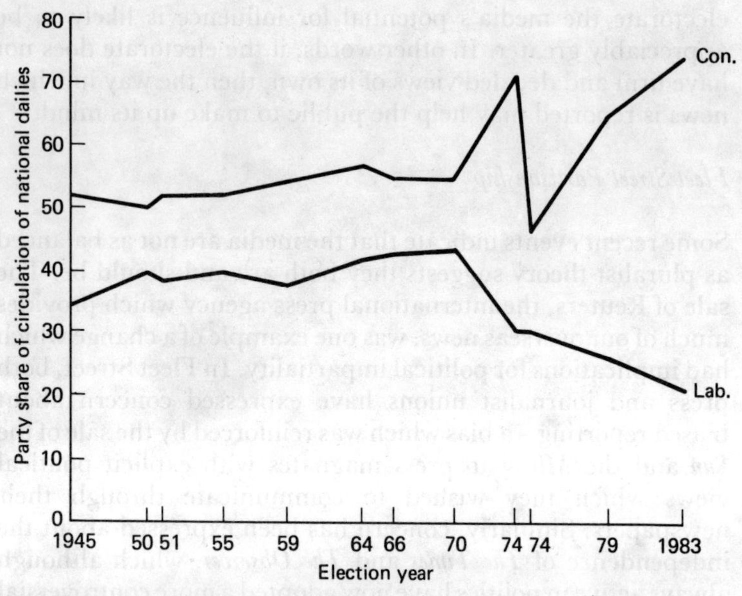

FIGURE 14.1 *Conservative and Labour Party newspapers: share of the circulation of national dailies, 1945–83*

Source: Tunstall, 1983, p. 12.

no doubt that over most of this century the Labour movement has had less newspaper support than its right-wing opponents and that its beliefs and activities have been unfavourably reported by the majority of the Press' (Royal Commission on the Press, Cmnd 6810, 1977, p. 99). Since then, Labour's situation has worsened. In an era in which effective political communication with voters seems to be vital to the success of the political parties, the Labour Party seems to have been all but excommunicated by Fleet Street.

Television Independence

The question of the neutrality of television reporting has been raised frequently over the last decade. Several events have caused concern about the extent of its independence. Controversy flared up when the radical campaigner against nuclear weapons, Professor E. P. Thompson, had an invitation to lecture withdrawn. Other figures, such as Edward Heath, seem to have suffered this fate, also on political grounds. Matters touching official secrets and the security services always have to be handled carefully, although there was some surprise in 1985 when it was revealed that a number of BBC posts were subject to vetting by a member of MI5, who was actually installed within the premises for the purpose. (This concern has since been alleviated by the Director-General's public determination to reduce to a minimum the number of posts which should be subject to such vetting.) There has also been severe criticism of the way material relating to such sensitive areas as the police, the Falklands War, Welsh Nationalism, official secrets, national security, the Middle East and Zimbabwe have been covered, and in relation to Northern Ireland there was fierce controversy over the refusal to screen a documentary – 'Real Lives' – in which two terrorists were the principal subjects. That programme was withdrawn after a letter from the then Home Secretary, Leon Brittan, and an unusual expression of opinion by the Governors which itself precipitated an internal constitutional conflict in the BBC.

The row over the 'Real Lives' programme came at a particularly unfortunate time for the BBC, since it was in the midst of a controversy over the level of the licence fee – a controversy which had led an unsympathetic Conservative government to appoint the Peacock Committee to enquire whether alternative sources of revenue, such as advertising, might be used for the BBC.

Bad News or Not-So-Bad News?

No single event can be taken as conclusive evidence for anything, but a succession of such events may indicate that television news is not quite as independent as it should be. Other research takes a different line in arguing that television news and current-affairs programmes reveal a systematic and persistent bias in their treatment of political events. It is, of course, extremely difficult to maintain perfect impartiality in these matters, since even the words we use are loaded. Thus, to speak of the 'miners' strike' is different from speaking of the 'coal dispute', just as 'Mr MacGregor declined to comment' is different from 'Arthur Scargill refused to answer'.

Evidence to support the claim that TV news is biased in a conservative direction has been collected by the Glasgow University Media Group in its four volumes entitled *Bad News*, *More Bad News*, *Really Bad News*, and *War and Peace News*. In the first and probably the best of their books, the Glasgow research team reported the results of their coverage of both BBC and ITV. The study – which was based on a few months news coverage in 1975 – concluded that the news reflected an elite consensus on the causes of economic and industrial problems. Thus the news generally assumed a direct link between inflation and wage rises, rarely mentioning other potential causes of inflation. The news also displayed an obsession with strikes, although as the study pointed out, the number of days lost through such causes as illness or injury is much greater than the number lost through industrial disputes. Stoppages caused by strikes at British Leyland were invariably reported, but those caused by management failure were not. In 102 news bulletins covering a strike of dustcart drivers in Glasgow, the research group found that the reasons for the strike were rarely mentioned, and when they were, it was in a confusing and vague manner. Moreover, the number of strikers was wrongly reported on occasion; and not a single striker was interviewed during the three months of the dispute, although other people were. When the dispute finished, the return to work was inevitably presented as a defeat for the workers.
Bad News concluded that:

> Our analysis goes beyond saying merely that the television news 'favour' certain individuals and institutions by giving them more time and status. Such criticisms are crude. The nature of our analysis is deeper than this: in the end it relates to . . . the laying of blame for society's industrial and economic problems at the door of the workforce. This is done in the face of contradictory evidence which, when it appears, is either ignored, smothered or at worst is treated as if it supports the inferential frameworks utilised in the producers of the news. (pp. 267–8)

This view is certainly not held universally and it has been strongly attacked by some critics. For example, after re-examining the scripts of ITN's news coverage for January to April 1975, Martin Harrison found two major flaws in the *Bad News* study (Harrison, 1985). First, Harrison claimed that the research team had ignored evidence that contradicted its own interpretations. For example, the news did report employers' actions such as lockouts, withholding pay and refusing overtime; and it did tell its viewers who was on strike and why. It also told viewers whether the strike was official or not. Second, *Bad News* persistently interpreted words and images as prejudicial to union interests, whereas they might have been interpreted by viewers in different ways, some of which might have been favourable to the unions.

Harrison concluded that *Bad News* was a 'classic example of how not to conduct scholarly research'. Time and time again, he suggested, the *Bad News* critique was wrong, exaggerated or ignored material that did not fit the authors' 'unvarying inferential frame' – which appeared 'to work to the principle that if something might be conceivably taken as unfavourable to the unions, then that is the only possible interpretation' (Harrison, 1984, p. 10; see also, Hetherington, 1985).

Meanwhile, the last but by no means the conclusive word should go to the Annan Committee, if only because it was both thorough and disinterested:

> that the coverage of industrial affairs [is] in some respects inadequate and unsatisfactory is not in doubt. Difficult as the reporting of industrial stories may be, the broadcasters have not fully thought it through. They too often forget that to present management at their desks, apparently the calm and collected representation of order, and to represent the shop stewards and picket lines stopping production, apparently the agents of disruption, gives a false picture of what strikes are all about. (Report of the Committee on the Future of Broadcasting 1977, Cmnd 6753, p. 272)

Perceptions of Bias

The last argument of the pluralist view of the media, is that the bias displayed by any given news source is likely to be spotted by its audience, which can therefore make allowances for it. Thus, bias matters little, because it is both perceived and discounted. There is some truth in this, especially in relation to the partisan leanings of the daily press. Recent surveys show that almost half the *Daily Express* and *Daily Mail* readers acknowledge these papers' Conservative bias (Kellner and Worcester, in Worcester and Harrop 1982 (eds), pp. 56–67); moreover, 92 per cent of *Daily Telegraph* readers perceive that paper's Conservative support and not one is under the illusion that it is a Labour paper; 75 per cent of *Mirror* readers are aware that the paper is Labour and only 8 per cent believes it to be Conservative. Nearly a quarter of the paper-reading public believes the press is generally biased towards the Conservatives.

On the other hand, large minorities get things wrong, either because they think the press is generally neutral or because they misperceive the nature of its party bias. Well over a third of readers think the daily press is generally neutral, and another third does not know. A fifth of all *Mirror* readers think the daily press has a Labour bias, and one in ten of *Sun* and *Daily Mail* readers are under the same illusion. When we turn to views about particular papers, a large minority also get them wrong: 22 per cent of *Daily Express* readers and 34 per cent of *Daily Mail* readers believe them to be unbiased. In 1979, when the *Sun* was solidly Conservative, a third of its readers thought it was Labour. Half a million (8 per cent) of *Mirror* readers thought that paper Conservative. And 3–4 per cent of *Daily Mail* and *Daily Express* readers think their papers are Labour. In sum, a large minority think the daily press is neutral; and smaller, but significant, minorities misperceive the actual party allegiance of the papers they read.

When we turn to television coverage of election news, there is agreement that both BBC and ITV are unbiased. Seventy per cent holds this view, and 20 per cent does not know. Only 13 per cent can see a bias in BBC news coverage (9 per cent believe it to be pro-Conservative) and only 8 per cent in ITV news (4 per cent towards the Conservatives). In this respect the majority are probably right, since the great bulk of election coverage maintains a party political neutrality as the law requires. However, there is a big difference between party political bias and general political leanings – it is perfectly possible to be conservative without being Conservative. While party preference is easy to spot, especially in election-day headlines and leader columns, the general tenor of the media is more difficult to discern. The main criticisms of TV news, and certainly the criticism from *Bad News*, is that the news conceals a consistent conservative bias beneath its party political neutrality. If this is the case then it seems that few viewers perceive it. The great majority believe that they are getting a fair and impartial account of the news without party political or any other kinds of bias.

Ultimately it seems that most readers recognise the bias of the paper they read, even though a large minority believes the press to be generally neutral and a small minority fails completely to recognise the party political bias of their own paper. The great majority believe TV news to be unbiased in its coverage of elections. To this extent, a large minority of the population fail to recognise the political bias of the media and are therefore unable to make any allowance for it. The independent impact of the media on these people is unlikely to be large. On the other hand, elections are often won by small majorities, and it is increasingly difficult to believe that the conservatism of television and Conservative monopoly of Fleet Street are electorally unimportant.

The Future

It is difficult to assess future trends in the British media, if only because the pace of technological change is so difficult to gauge. For example, the local radio stations, of which so much was expected when they were established a short while ago, now seem to be in some difficulty. However, the last century of media history is certainly consistent with the prediction that the capital cost of entering the media business will continue to rise, and that it will increasingly be controlled by a small number of multi-media giants. These large concerns are likely to consolidate their hold over the mass circulation papers and over broadcasting; and they are likely to extend their interests in cable television and other new forms of television-related enterprise. Meanwhile, both the technology and the content of satellite TV is likely to become increasingly dominated by the Americans, whose influence on the British media is already pervasive. All this suggests a further decline in the pluralism of news sources and perhaps a further increase in the conservatism of media politics.

One topical possibility concerns the only remaining ripe but unpicked plum in the world of media commercialism – the BBC. The Peacock Committee, due to report in the summer of 1986, is currently considering different ways of financing the BBC. Advertising and sponsorship may appear as alternatives or additions to the licence fee. The chairman of the committee, Professor Alan Peacock, is a distinguished economist who is well known for his belief in the free market and who was an early advocate of a school voucher system.

Though not as large and powerful as in the days of its broadcasting monopoly, the BBC is still an attractive source of profit and prestige for investors and sponsors. However, we should not overlook the experience of the United States, where news consultants have developed a form of 'tabloid television' in which 'shock–horror' and cosy human-interest stories are mixed together in 90-second doses, rounded off with a comforting small child or animal story. However, talk about the commercialisation of the BBC in one form or another has so blossomed in recent years that it is not likely to disappear. And whether or not any radical change occurs immediately, it seems likely that this aspect of the media at least will remain on the nation's political agenda.

PART THREE

The Political Science of British Politics

15

Topics in British Politics

PATRICK DUNLEAVY

Previous chapters have been primarily concerned with describing and trying to make sense of 'real life' developments in British politics. The focus in Part Three of this book is on mapping the debates among academic authors about political institutions and processes. Political science is an inherently multi-theoretical discipline. On questions of interpretation and evidence, there are a number of rival approaches. Some controversies concern only one aspect of the political process. Others form building blocks in much broader patterns of political argument.

This chapter examines eleven specific topics of debate. Two of these concern the constitutional basis of British politics, and the relationship between law and politics. The next four examine political input processes, voting at elections, party competition, interest groups, and the mass media. Finally, five topics explore particular institutions of government, covering Parliament, the core executive, the civil service, quasi-governmental agencies, and local politics. For each topic, alternative views or models (these terms are here used interchangeably) are presented in summary form, followed by a brief discussion of how academic debates in the area have evolved in recent years. Chapter 16 draws on these topic approaches in discussing how three overall perspectives or theories of the state have been applied to British politics.

Topic 1: The Constitution in Britain

Key questions: What is the basis of the constitution? And how important is it in structuring British political life?

Institutional–Legal View

The constitution in Britain is uncodified, but a clear body of rules about the powers of government institutions and their relations with citizens can be distinguished, which derive from four sources:

(1) Parliamentary statute regulates the holding of elections, the powers of the Lords and the Commons, and many other issues of constitutional significance. Statute law overrides all other constitutional sources, and the current Parliament cannot bind its successors: existing legislation on constitutional as on all other issues can be amended or abolished without any special procedures by a future House of Commons/House of Lords majority. Taken together, these twin principles constitute 'Parliamentary sovereignty'.

(2) Crown prerogative confers significant executive powers on British governments. It consists simply of those privileges of the mediaeval absolute monarchy which have not been assumed by Parliament in the intervening centuries, such as the ability to declare war or peace, hire civil servants, and reorganise government departments. Crown prerogatives are now exercised by government ministers and not the monarch personally – with the possible exception of the power to dissolve Parliament or to nominate a Premier when no party has an overall majority in the House of Commons. Crown prerogative powers are residual: they can be reduced by new legislation, but not extended.

(3) Common law is the cumulative mass of court decisions where judges, and to a lesser degree juries, interpret Parliamentary legislation, regulate the boundaries between statute and Crown prerogative powers, and make fresh law where existing provisions are ambiguous or missing entirely. Common law is heavily influenced by judicial values of 'reasonableness' in the use of governmental powers and the principles of 'natural justice'. Common-law decisions can be

overridden by new Parliamentary statutes, since the courts can only interpret legislation, and not pass judgment as to its 'constitutionality'. Even the central principles of common law can be negated by Parliament, for example, by passing retrospective legislation.

(4) Conventions are general principles or maxims governing constitutional issues which are widely recognised by 'authoritative' practitioners, constitutional lawyers and the public. Their precise status is a little obscure. Their value is in systematising behaviour and debarring one-off or lightly undertaken violations of common expectations. Some authors view them as 'trip wires' which alert politicians to the possible long-term damage or even illegality which may be implied in a course of action. However, they cease to apply with any force if they are frequently breached or no longer perceived as important. Conventions can be overruled by statute and common law, but many aspects of Crown prerogative are regulated by them – for example, the convention that the monarch does not exercise most prerogatives personally.

The interaction between these four elements makes the constitution a dominant influence on British political life, structuring even institutions outside the formally constituted government. For example, conventions about the office of Prime Minister greatly influence the powers of party leaders, and the common-law requirement that ministers make new regulations in a 'reasonable' manner influences consultations with interest groups.

Behavioural View

Reliance on an uncodified constitution, plus the historical evolution of Parliamentary sovereignty, means that a great concentration of power accrues to a government with control over Parliament. Since Britain is a unitary state, majority governments can remodel or abolish sub-national governments, make wholesale changes in the law, and closely control non-governmental institutions such as businesses, the professions, the trades unions and the mass media. The fact that these powers have been used only in legitimate ways owes very little to any formal checks and balances in the constitution.

Instead it demonstrates the absolute importance of political culture – citizens' attitudes and politicians' values – in determining political development. Britain is the limiting case of an unbalanced constitution, with little effective separation of powers, which none the less remains a stable democracy because its citizens and institutions adhere to democratic values, and heavily punish any government which oversteps the limits of acceptable behaviour. The constitution at any time is a crystallisation of the outcomes of elections, party competition and representative government. It is chiefly the effect, not the cause, of democratic behaviour.

Ideological Front View

The constitution is a 'dignified' mask, a set of fictions which legitimate the operations of a capitalist ruling class. Provisions for periodic elections are 'rituals' which give ordinary citizens the illusion of control over their own destinies. In practice the preponderance of non-elected institutions in British government severely circumscribes the ability of any government to carry out radical changes. The set of arrangements signified by the otherwise meaningless label 'the British constitution' is not inherently democratic. The Lords, civil service, judiciary, security services and armed forces all constitute independent power centres. 'Parliamentary sovereignty' functions chiefly to stigmatise any effective strategy taking an 'extra-Parliamentary' route. The practical impotence of the House of Commons is disguised by elaborate procedural charades and an ideology of 'Parliamentarianism' – both of which divert left parties into symbolic struggles and ineffective reformism.

* * * * * *

Britain's constitutional arrangements are among the most idiosyncratic of any Western democracy, although much British debate sees them as some kind of democratic norm. The institutional–legal view is the orthodoxy of textbooks of constitutional law, of which de Smith (1975) is the best example. The age of serious debate about the sources of the constitution has passed, although echoes still occasionally reverberate about Dicey's view of the importance of judge-made common law, or Leo Amery's traditional assertion of the primacy of the Crown (Dicey, 1959; Amery, 1953). But, for the most part, controversies within this tradition revolve around details or fairly fanciful conjectures – as when Lord Denning claimed that judges could in some unexplained way 'strike out' or refuse to implement legislation which radically redefined Parliament as a single chamber legislature.

Far more so than in the USA, liberal writers have played down the significance of constitutional arrangements. Early pluralist thinkers stressed citizens' democratic values as the foundation of the British system. Ivor Jennings' 1933 book, *The Law and the Constitution*, is a devastating critique of Dicey and exaggerated constitutionalism. Post-war political science moved towards behavioural studies, neglecting constitutional issues as incapable of systematic analysis. However, the reappearance of a multi-party system in the 1980s has reawakened interest in these dormant issues (see, for example, Norton, 1982). The distinctive importance of constitutional issues was also denied in Harold Laski's inter-war analysis of the constitution as an instrument of class domination. Modern Marxist work shows little development of Laski's ideas (see Miliband, 1983).

Topic 2: The Law and Politics

Key question: How does the legal system and law enforcement structure civil liberties, public order and political life?

Institutional–Pluralist View

The British legal system has three key features:

(1) the primacy of the rule of law, which requires that laws are implemented even-handedly and uniformly, especially as between state agencies and ordinary citizens;
(2) the independence of the judiciary from control by government; and

(3) a developed system of 'precedent' and several layers of appeal courts to sift and check judicial decision-making.

Further safeguards of impartial law enforcement are the separation of policing and prosecution functions from direct control by government – particularly the professionalism of police forces. Regional police authorities ensure a measure of citizen input into local-level policing. The use of juries in criminal law cases and some civil actions also introduces a key direct input by ordinary citizens. In any democracy, the legal system must strike a delicate balance between maintaining public order, and safeguarding the most extensive feasible civil liberties. The British system is a stable and popular, if still flawed, compromise, which has adjusted successfully to changing social conditions – such as increased armed crime, political protest, and industrial militancy, in a more ethnically diverse community, characterised by greater social malaise.

Neo-Conservative View

The legal system in Britain is the most important of a range of institutions which protect essential principles of society from being eroded, even by a House of Commons majority. It is natural (that is, both inevitable and perfectly defensible) that the judiciary and other law enforcement agencies should take a substantive position in two respects. First, judges assert the primacy of protecting citizens from arbitrary government, implying for example a strong judicial commitment to uphold private property rights even if an egalitarian Commons majority should legislate for their removal. Second, the courts and law-enforcement agencies cannot be 'neutral' where the security of the state is involved, but are bound to take a firm stand against determined groups bent on subversion, embracing even 'extraordinary measures' to defeat them. Because maintaining a working centre of state authority is a primary good, no coherent sense can be given to the concept of abstract 'natural rights' of the individual. However, citizens do have legitimate expectations that their government will preserve the maximum feasible extent of social life free from state interference, and that any interventions are made deliberately and following customary procedures.

Radical View

The British legal system and law enforcement agencies do not protect citizens in conflict with state agencies or capitalist values, for three reasons.

(1) Many key safeguards against unreasonable or illegal behaviour by state agencies or corporations require civil law actions, which are (a) initiated by private plaintiffs rather than by the police or Crown prosecutors, (b) decided by judges sitting without juries, (giving a key role to judicial values and precedent), and (c) enforced by court orders and further civil actions rather than being automatically policed by law-enforcement agencies. Because legal fees are so high, and legal aid is restricted to the very poor, access to civil liberties not guaranteed under the criminal law is effectively confined to the very wealthy and those with access to large corporate funding (such as private businesses and perhaps trade unions). Most civil actions are initiated by companies or rich individuals, so that the development of legal knowledge, case law, and the values of judges and barristers are all systematically biased towards capitalist interests. By contrast, civil law relevant to defending ordinary working people's rights can be virtually inoperable, with prohibitive legal costs making them purely 'paper' rights and entitlements. Where trade unions or left councils seek to defy biased civil law rulings, the law of contempt allows judges to levy open-ended fines or even to imprison dissentors. In addition, the police and government agencies frequently help to enforce judicial decisions in contempt cases.

(2) In criminal law cases, judges influence juries by interpreting the law, have discretionary sentencing powers, and use their speeches to assert particular social values. Judicial values are overwhelmingly formed by civil actions, and hence defer to a particular view of the interests of the wealthy and corporations. Because a private income is needed to become a barrister, virtually all judges are drawn from privileged social backgrounds, educated at public schools, etc. Even well-

meaning judges have very restricted and atypical social experience upon which to draw, and they routinely display extreme social conservatism. In lower courts, magistrates are drawn from a rather broader social base, but their limited powers and amateurism do nothing to reduce the class-biased character of court decision-making or sentencing. In civil liberties cases, especially those about the misuse of police or prosecution powers, judges and magistrates almost uniformly view criticism as likely to undermine the effectiveness of policing. Judicial decision-making consistently construes the authority of the state as a higher good which must be protected from challenge.

(3) Police forces in Britain are almost uncontrolled by anyone outside the force. Semi-elected regional police authorities have very limited powers over Chief Constables, who control all operational decisions. In London the Home Secretary notionally controls the 42,000 staff and £800 million budget of the Metropolitan Police. National co-ordination of policing occurs but is controlled by no one. There is no independent investigation of complaints against the police. Police value systems are inherently authoritarian and conservative. Conservative governments have privileged police pay, taken 'firm' stands on law-and-order issues, and involved the police actively in winning their political and industrial battles. Political policing and the unregulated surveillance of potential radicals are major infringements of civil liberties.

* * * * * *

Controversy over the law and politics has recently been increased by a rash of civil law actions in central–local relations (Elliott, 1983) and the fall-out from the Conservatives' trade union legislation (see Chapters 11 and 12). The legal/institutional view sees these developments as maintaining an impartial and independent legal system, which none the less responds directly to clear Parliamentary decisions. Diverse organisations use the legal system to pursue 'political' objectives, a testimony to its even-handedness as between social interests. The substantive commitments of the common law, judges and the police are to principles enjoying near-universal support – such as the rule of law, preserving public order, preventing crime, ensuring a fair trial, assessing evidence fairly, resolving ambiguities in the law in line with existing social expectations. Such commitments are important to almost anyone, whatever their political beliefs. Although the laws that Parliament requires the courts and the police to enforce may be partisan in intention or effect, the fundamental commitments of the legal and law-enforcement systems are not disputed in mainland Britain.

The neo-conservative position renounces this consensual approach by associating legal principles explicitly with a far-right, inegalitarian defence of established privilege. Roger Scruton's *The Meaning of Conservatism* (1980) emphasises that the English legal system is a key institutional check on the 'disease' and 'contagion' of democracy. Claims for the essential neutrality of the legal system are pious cant, essential for maintaining legitimacy but empirically insupportable. Since a pragmatic 'nation state' conservatism is the 'natural' ideology of any rational person sensitive to the extreme value of an established social order, judges or courts cannot fail to act in a manner biased towards maintaining existing social arrangements against the careless, ungrounded reformism of liberals and the left. The similarity between Scruton's analysis and that of modern Marxist writers (for example, Miliband, 1983), is worth noting.

The radical model of law and politics has rarely been set out systematically. Many socialists and Marxists make anecdotal criticisms of the legal system, based on cases where judicial decisions or police action have harmed trade unions, Labour councils or other left causes. Premature appeals are made to the inherently conservative or capitalist values of lawyers, judges or the police to explain these outcomes. Little attention is given to cases resolved in favour of the left. John Griffith's *The Politics of the Judiciary* (1977) is a good example of the genre. A more plausible case would rely less on the biased social backgrounds of judges, and more on the cumulative impact of the civil/criminal-law split, the high costs of access to the courts, and hence a strong differential development of legal knowledge and expertise. Attention would focus on the systematically class-biased development of legal and police professionalism, rather than on individual-level explanations of particular cases.

Topic 3: Electoral Behaviour

Key question: Why do people vote the way they do?

Party Identification View

Voting involves clarifying preferences, taking moral positions, and making empirical judgments. People must know both what they want, and what can feasibly be achieved – that is, how different public policy programmes ` will map out. Unsurprisingly voters find deciding between party programmes on their merits a daunting task. So they short-circuit a closely rationalised decision by following cues encountered in everyday social relations with workmates, neighbours, family or friends. Occupational class is still a key indicator of differences in these social networks. Voters are often socialised into political viewpoints by their upbringing and life experiences, rather than making very explicit partisan choices: for example, developing long-term 'identifications' with one party by the time they are in their thirties. Socially mobile people (whose family background and current occupation diverge), or who live in mixed-class households, tend to have weaker allegiances or to switch parties more often. Since the early 1970s, people have been realigning their views, producing an appearance of 'class dealignment' and greater electoral volatility. But once allowance is made for the transition from a two-party to a three-party system, differences between occupational classes in their voting behaviour remain as strong as ever.

Issue Voting Model

Voters in political markets behave as rational actors, rather like consumers in economic markets, by choosing the party programme closest to their own views on important issues. They maximise the benefits they receive from party programmes, after allowing for any costs. Key issues normally include inflation, unemployment and income growth, since people are basically self-interested. Some voters with clear-cut economic interests are consistently better off under one party than with its rivals, and develop 'brand loyalty'. But this attachment is always reconsidered if a better alternative appears or the 'brand' gives declining satisfaction. Voters discount unfeasible promises, however, and treat manifesto pledges sceptically. Instead of assessing conflicting promises about the future, they can decide to reward or punish the incumbent government for its performance, basing their judgments mainly on short-term economic variables. So a prosperous election year can make up for harder times in the past.

Radical Model

Aggregate shifts in alignments result from two influences.

(1) The first is voters' positions in a complex structure of inequalities and conflicts of interests. Social class cleavages remain the most important lines of division, but their impact has increasingly been cross-cut by public/private sector conflicts triggered by the post-war expansion of the welfare state. State employees have grown to 30 per cent of the workforce; the state-dependent population has mushroomed with increased unemployment and more pensioners. Some types of consumption (housing, transport and health care) are marked by strong public/private sector differences in government subsidies and quality of service. Labour has been increasingly identified with public sector interests, while the Conservatives' 'anti-statism' is a key feature of Thatcherism's success. Since public sector groups are generally in a minority (except on the NHS and state schooling), the growth of sectorally based alignments has heretofore benefited the Conservatives and tended to residualise Labour support.

(2) Dominant ideological messages, generated particularly by the mass media (but also by political parties, business and other social institutions), selectively provide voters with information about their interests *vis-à-vis* other social groups. For example, home owners receive more state subsidy than

council tenants (via mortgage interest relief), yet dominant ideological messages portray home owners as on the 'private' side of public/private sector conflicts. People tend to vote instrumentally in line with the collective interests of their social location as these are represented to them. Some groups whose well-defined interests are incongruent with dominant ideological messages can screen out mass-media influences, basing their alignments directly on their life experiences. However, increased volatility and third-party voting reflect the effects of state intervention in producing more complex patterns of social cleavages, where more voters are cross-pressured by influences from their social locations, and hence more open to dominant ideological messages.

* * * * * *

In the 1950s and 1960s, electoral studies was dominated by the party identification approach, the classic British statement being Butler and Stokes (1969). This view successfully explained electoral performance in a stable two-party system, with a dominant middle-class/working-class cleavage. From 1970 onwards, however, the approach was undermined by evidence of a weakening association between occupational class and voting (class dealignment), and between voting for a party and support for its major issue commitments (partisan dealignment). Increased third-party voting, greater electoral volatility, and the 'haemorrhaging' of Labour support in 1979 and 1983 all seemed to indicate an extensive reshaping of electoral behaviour. The conventional wisdom by the late 1970s became the issue voting model. Increasing political information, better-educated voters, and the growth of issues which cross-cut party ideologies were seen as creating increased space for voting on individual salient issues. *The Decade of Dealignment*, analysed by Sarlvik and Crewe (1983), has not completely invalidated the party identification model. But longer-term identifications have been hollowed out as influences on voting behaviour, and have little application for a large section of the electorate. A sceptical note about the extent and meaning at least of 'class dealignment' is struck by Heath, Jowell & Curtice in *How Britain Votes* (1985). Their implicit claim is that alignments could settle down into a stable three-party pattern, marked by the same predictable occupational-class influences as the older two-party regime. By contrast, the Crewe and Sarlvik model has been touted in the media as forecasting a more extended process of transition to a new two-party system, in which the Labour Party plays a decreasingly important role.

The radical model originated in the late 1970s as a critique of party identification accounts, which tend to assimilate all aspects of social differentiation into a single occupational-class cleavage. Its fullest statement is Dunleavy and Husbands's *British Democracy at the Crossroads* (1985). The approach denies that issue attitudes are key causal factors in voting change, arguing that the same social and ideological shifts which restructure alignments also promote a rethinking of attitudes, without their being any causal connection between attitudes and voting. In addition, many voters adjust their issue positions to fit in with their decision to vote for a particular party. Voters are heavily influenced by their exposure to mass-media messages. Only a small part of what is normally seen as 'public opinion' is unambiguously the product of voters' autonomous choices.

Topic 4: Party Competition

Key question: Does competition between political parties provide an effective mechanism by which citizens' votes can control government policies?

Responsible Party Model

The key threat to effective democracy is that of tacit collusion between party leaderships to keep issues off voters' agenda. Vigorous competition between clearly differentiated parties is the best way of guaranteeing citizens real choice. A two-party system normally delivers a secure majority to one party, which is unambiguously responsible to voters for government performance. Citizens can help set the political agenda through the mechanisms of internal party democracy – mass membership organisations, open membership policies, and direct or indirect pyramiding of elections from grass-roots branches used to select leaders and/or policies. Activists and leaders have strong but shared ideological convictions, so that tensions between them are usually manageable. Parties do not win elections by simply accommodating themselves to the current preferences of a majority of voters. Rather, they must persuade unattached voters to see the issues their way, while retaining the support of their existing 'identifiers'. An effective party campaign and leadership team can exercise some 'opinion leadership', convincing voters to support them despite some specific disagreements with their policies.

Economic Model

Party leaders are rational actors who maximise votes in the electoral 'market', rather as businessmen maximise profits in economic markets. Since voters act rationally (as in the issue voting model above), the winning party is that which successfully tailors its position to match majority preferences. Two-party competition produces convergence by both parties on the median voter's position, the person who has as many people to the left of her views as there are further right. In a three-party system, the middle party is squeezed by its more polarised rivals tempering their positions. Convergence on the median voter is ethically desirable, since whichever party wins the elections it will implement policies as close as possible to majority voter preferences.

Adversary Politics Model

Issue voting dominates electoral behaviour and party leaders maximise their electoral support. But party competition none the less can produce over-polarisation of the major parties.

(1) If parties are internally democratic, support from grass-roots activists (or MPs, or financial backers) is a prerequisite for becoming party leader. Activists want to see their party campaigning for an ideological standpoint close to their views, rather than pragmatically courting electoral popularity. Since leaders must retain party office to have any chance of entering

government they are constrained to appease activist opinion, and may be locked into commitments which are unpopular with voters. Party leaders converge on their median activist's position. Since left- and right-wing parties draw supporters from at best half the ideological spectrum of voters, neither of their median activist positions can approximate that of the median voter. However, centrist parties have no such problems, since they recruit activists whose views cover a range around the median voter position.

(2) The plurality-rule ('first past the post') elections used on the British mainland insulate the Labour and Conservative parties from the effects of advocating unpopular policies. Any third party must surmount a high threshold level of support if it is to win Parliamentary seats even roughly proportional to its votes. Consequently, voters may be continually forced to choose between equally 'extreme' major party manifestos, prompting increasing cynicism about electoral politics. The transition to a more evenly balanced three-party system, under way since 1974, has progressively weakened the protection offered by the electoral system, however.

Radical Model

Parties do not compete just in terms of what they say; if they did, their influence would be small when set against other opinion-forming institutions, such as the mass media, business, or the labour movement. Party leaders have little room for manoeuvre in presenting a case to voters. They must maintain continuity with their party's history, record in government, and established voter expectations. Consequently party leaders put most effort into trying to reshape voter preferences, to bring them into line with party policies. For the incumbent party of government a key mechanism for reshaping preferences is the use of state power to reward people in social positions supporting the party, and penalise positions or areas of the country associated with the opposition. Control of the state machine also allows party leaders to shape political agendas, create internal crises or external threats which rally support behind the government, or manipulate the economy along a path favourable for re-election. Major opposition parties may counter by using party power – for example, pledging to repeal legislation, or utilising control over sub-national government to frustrate implementation of domestic policies.

* * * * * *

Political scientists have generally agreed with Schumpeter (1944) that the dynamic of partisan change in government is a key influence in keeping rival political elites responsive to a mass electorate. Party competition keeps public policy broadly in line with citizens' views and is a key source of change in governmental objectives. But the mechanisms by which competition is achieved, and the fit between policy outcomes and voter preferences, remain controversial.

The post-war orthodoxy was the responsible party model, which provides a strong defence of the two-party system, and is closely linked to party identification accounts of voting behaviour. The distinctiveness of rival manifestoes was modified by the use of mass membership organisations to select policies and leaders, even if both major parties were pushed by constitutional and electoral factors into similar modes of operating, with the leadership exercising most policy control (McKenzie, 1963). Declining party memberships by the 1970s, plus moves to greater internal party democracy, threatened to make the approach anachronistic – as did the setting-up of the SDP as an explicitly cadre party. Yet the stability of the Conservative leadership under Thatcher and the shift back under Kinnock to a traditional style of Labour leadership have both revived the model's plausibility.

Trends towards reduced two-party differentiation up to 1970 have also been reversed by increasingly ideological party politics in the late 1970s and the 1980s, squeezing the economic model of party competition into a minor position. The model was formulated by Antony Downs's *An Economic Theory of Democracy* (1957), and its heyday was the period of 'Butskellite' consensus on economic growth, decolonisation, and extensions to the welfare state. However, the model still remains influential as an account of how government parties manipulate the economy. It predicts that incumbents use the first part of their term of office to deflate the economy, so as to create space for a reflation in the run-up to the next election.

The adversary politics argument is a British variant of the economic model. It became a fashionable label for the newly widening gap between the Conservatives and Labour in the late 1970s. Critics on the right, such as S. E. Finer (1975, 1980), used the account to pinpoint the inadequacies of social democratic pluralism. More recently Gamble and Walkland (1984) set out two versions of the thesis, one similar to the Alliance rhetoric about superseding the era of Conservative/Labour competition by a three-party system, and the other close to Marxist accounts. The approach predicts that left-wing governments will begin their term of office by reflating the economy, while right-wing governments launch themselves on a course of public expenditure reductions and deflation. Both governments initially spent much time on reversing each others' legislation, before 'U-turning' in mid-term towards more consensual policies designed to secure their re-election.

The radical approach emphasises the calculations of political advantage underlying key policies of the Thatcher governments – such as the privatisation of state companies, the contracting-out of public service functions to private firms, the sale of council houses, increases in council-house rents, the phasing-out of state earnings-related pensions, abolition of the metropolitan county councils, the imposition of balloting for unions to retain political funds, etc. Each of these changes reduces the size of the public sector groups or attacks institutions associated with support for Labour. Together with the news management during the 1982 Falklands war (which produced a 17 per cent growth in Conservative opinion-poll support), these measures herald a new level of partisanship in state policy-making. A countervailing use of similar tactics by Labour councils resisting rate-capping and abolition of the metro counties merely strengthens the argument. For example, between 1983 and 1986 the Labour GLC spent £10 million on a media campaign against abolition legislation, with favourable political consequences for Labour in London. Dunleavy and Husbands, *British Democracy at the Crossroads* (1985) sets out the radical model and compares it with alternative approaches.

Topic 5: The Interest Group Process

Key question: Does the interest group process provide equal access to policy-making for groups in proportion to their numbers?

Group Politics Model

Citizens join interest groups for an enormous variety of reasons – to protect common interests, act on moral commitments, or assert particular identities. Since participation is voluntary, and groups handle issues close to people's own experiences, members have more control over leaders than with political parties. Groups usually have representative structures to attract members and to legitimate their claims to mirror members' views. British political culture is supportive of interest group activity. Politicians, the mass media and the public recognise the value of group activity and the need to consult diverse interests in policy-making. Other things being equal, larger groups are more influential than smaller groups; groups with a high participation from potential members are more influential than less well supported groups; and groups whose members publicly demonstrate their intense preferences are more influential than those where people are apathetic about group campaigns. Governments gauge all these variables from group activity, since numerous, well-mobilised or intense groups could represent an electoral danger. Group leaders try to secure low-cost ways of making their views known to government at early stages of policy formulation, rather than campaigning against publicly announced policies. Minority groups try to win over sections of the majority to their viewpoint, hence the group process is more effective where there are cross-cutting conflicts of interest. The universe of interest groups is unstable, with new groups constantly being formed, winning or losing influence, and shifting group alliances across issues and over time. No group occupies a dominant winning position, and strongly organised groups tend to stimulate 'countervailing powers' to organise and offset their influence.

Logic of Collective Action Model

People join interest groups only to secure 'public goods' – that is, benefits which require collective action if they are to be achieved, but which once provided are freely available, whether campaigned for or not. For example, if a factory wage increase is gained by union action, the benefits accrue to all the workers, not just union members. People are rational actors, who support a group only if the benefits they receive are greater than the costs of joining. The key problem for all groups is to prevent 'free-riding' – where individuals let others bear the costs of group action, reaping public-good benefits at no cost to themselves. Group leaderships combat free-riding by developing private benefits available only to group members, or sanctions against non-joiners. Groups where this is not feasible remain latent or weakly developed. Groups are most easily organised where the potential membership is small, non-joiners are visible, and group effectiveness is clearly reduced by one individual's non-participation. Large anonymous groups are hardest to organise since non-joiners are invisible and group effectiveness is negligibly impaired by one person's free-riding. Collective organisation in large groups is a 'by-product' of the ability to offer private membership benefits, so there is little relation between groups' effectiveness and the intensity of their members' views, a fact of which governments are well aware. The universe of groups is very stable over time, with persistent inequalities of group influence.

Log-Rolling Model

Groups are chiefly organised not around genuine public goods, but in order to secure sectional private goods from the government budget. By building alliances with other minorities a group can become part of a winning coalition. Each group agrees to vote for an overall bundle of measures, including its own key preference, so long as other parties to the agreement vote the whole deal through as well. Log-rolling means that group demands that would be rejected if voted on singly can none the less pass as a package. All log-rolling is strategic voting which 'exploits' people not included in the winning coalition, so it may substantially reduce the social welfare. Alliance-building is crucial to group success, and entrepreneurial group leaders are the conduit via which private benefits are siphoned off the government budget into meeting members' private interests.

Corporatist Model

Major interest groups command resources essential for economic activity. Their influence is based less on electoral clout than on their organisational power and ability to tap strong solidaristic loyalties in the population. Trade unions and business associations are particularly organised into single hierarchical peak associations. Together with some professions these producer groups have a special bargaining position with government. Their close and permanent 'insider' relations with state agencies create a 'corporate bias' in policy-making which relegates other interests to a secondary position. Corporate group leaderships both represent their members to government and control their members on behalf of government. Group leaderships are often extensively integrated into government functions, for example via quasi-governmental agencies. They are also insulated from much direct accountability to their grass-roots members. Corporatist interest bargaining is decisive on economic issues. A more open and free-wheeling pluralist group process still handles secondary issues.

* * * * * *

There have been two periods of intense debate about interest groups in Britain. A wave of pluralist writing in the 1950s accumulated some case-studies of group behaviour, some basic generalisations about group strategies, and a claim that the 'group process' dominates modern British politics. However, the group politics approach defines virtually all actors – such as professions, the mass media, individual companies and government agencies – as 'groups' (Richardson and Jordan, 1979), rendering this claim almost tautologically true.

A second wave of debate in the 1970s focused on

corporatism, and derived from Western European arguments that negotiation between government, businesses and trade unions was increasingly indispensable in managing advanced industrial economies – for example, an effective incomes policy requires consent from unions and employers. Schmitter and Lehmbruch (1979) saw strong *Trends Towards Corporatist Intermediation*, and Britain under the Wilson and Callaghan Labour governments was no exception. The 1975–8 'social contract' gave trade-union and business leaders a voice in many domestic policy issues, in exchange for delivering their members' compliance with government wage norms. Longer-term trends towards corporatism in British government were detected (Middlemas, 1979), but their significance was disputed. Authors on the left saw corporatism as the latest strategy by which capital/state interests could co-opt labour-movement leaders into sustaining capitalism. Liberals argued that peak associations (such as the TUC and CBI) remain weak organisations, linked by little more than tripartite meetings, and always liable to collapse back into a more-conventional group process.

Any corporatist account must explain the reorientation of policy under Conservative governments since 1979, which has distanced ministers from any direct interest bargaining, allowed corporatist institutions to fall into disuse, faced down union opposition, and been indifferent to business pleas for a mitigation of monetarist policies. As corporatist theory has gone out of fashion only neo-pluralists such as Galbraith have argued that the 1979–86 experience changes nothing. Government–union–business co-operation is still essential for economic prosperity in advanced industrial economies, but the possibility always exists that an ideologically motivated 'rogue' government might try for a period to dispense with corporatist arrangements. Yet by renouncing incomes policies the Thatcher governments have been forced to rely on monetary instruments to combat inflation, with predictable consequences in terms of mushrooming unemployment, high real interest rates, and stagnating industrial production.

The roots of the Conservatives' anti-corporatism lie partly in the public-choice literature. In *The Logic of Collective Action* (1968), Mancur Olson demolished pluralists' optimistic view of the interest group process. Although enormously influential in theoretical terms this approach remained conspicuously unapplied in British empirical studies, because little systematic research into group activity has been carried out. In the USA most of the development from Olson's work was by New Right economists, who adumbrated in technical terms a long tradition of liberal hostility to sectional interests. Their log-rolling models describe not just group activity, but also how political parties compose manifestos, and how legislatures and cabinets make public policy. Olson himself has shifted substantially towards a New Right position. His 1982 volume *The Rise and Decline of Nations* argues that where a political system has been stable for long periods (as in Britain), multiple interest groups develop veto powers over social change which effectively choke off economic growth. By contrast, countries where established veto groups are destroyed by revolutions or wartime defeats are much more successful in achieving growth. Some contemporary Conservatives echo this analysis by calling for a radical assault on 'vested interests' that inhibit economic dynamism.

Topic 6: The Mass Media

Key question: Do the mass media communicate adequate, unbiased political information for citizens to make autonomous judgments on party alignment and policy questons?

Public Opinion Control Model

The British mass media are organised as a system of countervailing powers. Newspapers are partisan and privately owned, operating in a free market environment. Concern to expand sales ensures that papers express diverse political positions, so there is normally a rough partisan balance in press readerships. Since Fleet Street is financially fragile, consumers can effectively control newspapers' propagandising by switching papers. In addition, most voters rely on television news produced under a 'public service' ethic of partisan balance, enforced by state regulation through two quasi-government agencies, the BBC and IBA. Television and radio news is consequently better trusted by citizens, who discount much of what they read in newspapers. Private ownership of newspapers provides an important guarantee against state control of media content. The competition between newspapers and television and radio ensures a vigorous searching-out of news stories, while their contrasting approaches to analysis maintain a creative tension which continuously defines the limits of plausibility in media views.

Market Model

As in any industry, the only way in which citizen-consumers can effectively control what is supplied to them is via open competition in a market. This condition is partially met in the newspaper market, despite the oligopolistic power of the national press corporations and the restrictive practices of the Fleet Street unions. But free competition in broadcasting has been eliminated: because of unfounded fears about the propaganda potential first of radio and then of television; by historical technological limits on frequencies; and by politicians' anxiety to preserve a 'nanny state' role. Because broadcast political news has been artificially homogenised by state regulation, channel-switching by consumers is ineffective, and only the availability of newspapers keeps the system from severe distortion. The result is a fake, non-interactive system of news production unrelated to genuine popular views, creating a media system skewed towards fashions among a media elite and deference to government sensitivities.

Dominant Values Model

The mass media in Britain are effectively unified by a common journalistic value system and the hegemonic influence of private capital. The existence of a 'free press' confers no effective control on citizens, because of the stranglehold of a few national media corporations, directly controlled by press barons. In addition, the buying-power of a papers' readership is crucial for advertising, while the numbers of copies sold is a secondary determinant of a title's viability. Hence there are multiple 'quality press' titles but less choice for ordinary working people. Private proprietors routinely interfere in news production, so that journalistic professionalism is undermined and journalistic values are structured to the right. Newspapers remain a critically important influence on political news reportage and on their readers' political alignment. Because press and broadcast journalists share the same 'news values', are frequently linked by personnel interchanges, and feed off each other's stories, supposedly neutral broadcast news is pulled into the same right-wing slant. It displays the same deference to establishment institutions (such as the Royal Family, the armed forces, police and established churches); the same idolisation of wealth and corporate power; and a strikingly similar paranoia about trade unions, strikes and left-wing 'subversion'. Direct interference by ministers to control broadcast news and comment is relatively rare, since journalists and news managers routinely anticipate adverse reactions and tailor their output accordingly. But overt censorship is practised on some key topics where there is an elite consensus – for example, the dangers of a nuclear arms race, the impossibility of a settlement in Northern Ireland, and the operations of British security services.

* * * * * *

Pluralist writers conventionally rely on 'the liberal theory of the press' for their optimism that the mass media are responsive to citizen views, a position qualified only slightly before mass access to television by a defence of the BBC as a unique institution. Around 1960, TV was seen as a potentially decisive political medium. When a first wave of media research found no evidence of dramatic impacts on political behaviour, the conventional wisdom came to emphasise the offsetting influences of a partisan press and bi-partisan TV news. However, some pluralist writers in Britain and the USA have recently become more concerned about the extent to which the mass media have become 'para-Parliaments', taking on the functions of representative institutions such as the House of Commons, or displacing political parties as key determinants of who becomes a major political leader (Polsby, 1980). The market model of the mass media used by New Right authors also challenges the conventional wisdom, by arguing for the deregulation of television news, along with all other aspects of broadcasting. The pluralisation of media systems in the late 1980s, especially the growth of cable TV and satellite broadcasting, all pose difficult problems in maintaining non-partisanship in broadcast news coverage.

The dominant values model was reinvigorated in the late 1970s by a shift of emphasis in media studies away from the political effects of coverage, and towards the conditions of its production. A succession of *Bad News* studies (Glasgow University Media Group, 1976, 1978) claimed to have demonstrated systematic anti-union and anti-left bias in television news output, thereby generalising an already plausible left critique of Fleet Street to cover the entire media system. More recently the credentials of the original research have come under pluralist criticism (Harrison, 1985), while some interest among radical authors has shifted back towards the anti-Labour consequences of mass-media outputs.

Topic 7: Parliament

Key question: How important is Parliament as an institution influencing policy-making?

Legal–Institutional Model

The Commons retains the potential to act in the manner suggested by the doctrine of Parliamentary sovereignty and put into practice during the 1832–68 'golden age' – when Parliament could change governments between elections, force the removal of individual unpopular ministers, decide each piece of government legislation on its merits, and initiate some significant legislation itself. These powers have rarely been activated in this century because there has been continuous majority support for the government within the Commons. But conceivably the transition to a new three-party system marked by coalition or minority governments will demonstrate that these capabilities have not lapsed.

Westminster Model

Parliament has not played a significant policy-making role for over a century, despite three minority governments in the 1920s and late 1970s. Instead it provides a forum for debate between the government and its critics, publicising the executive's legislative proposals and executive actions, and allowing the party battle to be kept in focus for the mass media and the public at large. Strong partisanship, the whipping system, and the Commons' standing orders mean that governments with a majority can control debate and maintain a strict legislative timetable, while excluding private members' bills with financial implications. Floor debates and questions to ministers are key Commons activities, but committee scrutiny of legislation is disorganised and relatively ineffective. Even a minority government remains the prime initiator of legislation, although its success rate will be less complete. Strong party loyalties means that there is little sense of the corporate identity of Parliament. For example, opportunities for scrutinising government finances in detail have been remorselessly converted into yet more slots for mainstream party controversy. These long-run trends reducing the Commons' capacity for any involvement in detailed policy-making cannot be significantly changed by a transition to a multi-party legislature.

Transformative Model

Parliament is emerging again as a policy-making body in its own right, in three main ways. First, the 'Parliamentary rule' (where a government defeat on a major vote automatically precipitates a general election) has been relaxed. Majority party MPs are more willing to defy blanket discipline by dissenting on particular bills, even when this causes government defeats in standing committees or floor debates. Second, MPs have used departmentally organised select committees since 1979 to extend their scrutiny over the government's executive actions. Although the committees' performance has been patchy, they are influential on detailed policy questions. Third, the range of 'conscience' issues decided by MPs voting independently of party discipline now includes many controversial issues – such as capital punishment, abortion law, divorce provisions, the regulation of pornography, embryo research, or surrogate motherhood – plus some more conventional questions such as house conveyancing, or Sunday shopping. In a multi-party Commons the executive will remain the prime initiator of legislation and there will be no general disappearance of party discipline among MPs. But there should be more intra-party debates and cross-party voting on contentious legislation, more forceful independent scrutiny of executive actions, and genuinely Parliamentary decision-making across a wider range of issues.

* * * * * *

Legislative studies has been a neglected branch of British political science for many years. However, pioneering work by Berrington, Finer, and most recently Norton (1985) has provided increasing evidence of the role played by backbench dissent in shaping policy, especially in weeding out the proposals which governments bring to the Commons at early stages in the decision process. The survival of many of the dissenting traits of the late 1970s into the period of large Conservative majorities in the early 1980s has considerably strengthened the plausibility of the transformative model, especially its scenario of a future 'hung' Parliament.

The Westminster model remains the orthodox interpretation, however, for as writers such as Griffith (1974) have exhaustively documented, any government with a Commons majority will see the vast bulk of its legislative proposals implemented with few amendments. There is a conflict of criteria for assessing Parliamentary influence here. In the Westminster model the Commons' marginal role is charted by the trickle of successful private members bills, compared with the steady torrent of public legislation. But for the transformative model there is no expectation that a powerful Parliament will necessarily be a pro-active one. The power of the legislature may be indicated as much through the vetoing of legislation as through its enactment.

The legal–institutional model remains an important set of fictions to which the courts refer in refusing to get involved in deciding matters of 'policy', rather than interpreting legislation or safeguarding procedural correctness. British public law has never successfully adapted to the realities of a party-dominated legislature, hence the counter-factual insistence that the long-dormant practical sovereignty of Parliament could be revived to match the legal primacy of statute. However, some New Right authors now prescriptively assert the importance of building up Parliamentary scrutiny again as a check on the 'Leviathan' state – for example, by enacting only 'sunset' legislation, with built-in deadlines for terminating new spending programmes.

Topic 8: The Core Executive

Key question: Which actors or institutions in Britain's core executive fix the direction and co-ordination of government policies?

Cabinet Government View

British government places a strong emphasis on collegial government by ministers. The Cabinet and its committee system remain key institutions for extending the scope of rationalistic policy debate at the centre of Whitehall. They constrain any exercise of monocratic power by the Premier, because Cabinets usually represent different party factions, departmental ministers control key resources of information and expertise, and the committee structure cannot be wholly rigged to favour the Prime Minister's priorities. The principle of 'collective responsibility' denotes the ability of a significant minority of ministers to debate key issues before they are promulgated as government policy, and is thus much more than a requirement of ministerial solidarity. Cabinet government pluralises the viewpoints considered in policy-making, multiplies access points to the core executive, encourages more considered and systematic decision-making than a monocratic premiership, and provides a key safeguard against the kind of 'groupthink' which produces policy fiascos.

Prime Ministerial Government Thesis

British premiers occupy a very strong position in the core executive, since they can remodel Cabinet procedures, committee structures, decision agendas and personnel. Modern Cabinet conventions have strengthened pressures for group solidarity, and external changes in the standing of party leaders against their colleagues have accentuated media and public concentration on the Prime Minister. The growth of the Cabinet Office and the premier's personal staffs has combated the lack of a departmental back-up. Except in periods of extraordinary personal weakness a PM can be sure of a secure majority for her or his policies. Prime Ministerial government may imply either the hegemonic position of the premier personally in key decision-making, or the centralisation of effective decision capabilities in a clique of ministers and advisers selected by and working to the Prime Minister.

Segmented Decision Model

Controversy focused on the relative power of premiers and Cabinet is misleading, since the two are characteristically not in conflict. As in other liberal democracies, the British core executive shows a segmentation of policy decisions into two groups. Strategic decisions about management of the economy, foreign affairs, state security and major defence issues almost invariably involve the premier and Cabinet Secretary directly, together with relevant ministers and senior officials of the departments concerned. In most other policy areas Prime Ministerial involvement is highly selective and episodic, and Cabinet and its committees remain the crucial co-ordinating and decision mechanisms. Prime Ministerial interventions on non-strategic issues are more affected by personal style, but they are less likely to be decisive. The core executive's role in both policy areas is typically restricted to the policy formulation phase, with policy implementation resting primarily with Whitehall departments, quasi-governmental agencies, local authorities and other implementing agencies.

Bureaucratic Co-ordination Model

The core executive plays a secondary role in co-ordinating the state apparatus. Bureaucratic mechanisms in Whitehall provide the essential integration, especially the official committee system which parallels Cabinet committees, the key central departments (Treasury, Cabinet Office, Foreign Office), regular meetings of permanent secretaries, and multiple Whitehall information networks. This machinery delivers a different style of policy-making with governments from different parties, but there is never much political control over detailed policy-making. A 'presidential' role for the PM is fostered by Whitehall, since the machinery of bureaucratic co-ordination needs an ultimate tie-breaker to resolve inter-agency conflicts, and to provide a central reference point of policy definition to which civil servants in departments can appeal in by-passing troublesome Cabinet ministers. Bureaucratic co-ordination effectively excludes a role for collegial Cabinet decision-making. A variety of individual ministerial policy positions are tolerated on lesser issues, but major decisions require a standardised policy line across all departments which is more susceptible to civil service pressure.

* * * * * *

The debate about how the British core executive operates is one of the great 'chestnuts of the constitution' (Heclo and Wildavsky, 1974, p. 341). It has been dominated by a controversy about the relative power of the PM and the Cabinet, with strong prescriptive overtones but little contemporary evidence or systematic research. The fascination with this sterile dispute reflects the impact of the UK's unbalanced constitution in concentrating power in the hands of the executive, and the impact of strong partisanship in British politics in eliminating much rationalistic policy debate in Parliament or local communities. Consequently the crucial locus of argument and discussion about policy-making has been displaced into the core executive itself, far more so than in other liberal democracies. The Cabinet government model crystallises the traditional arguments about the special virtues of collegial decision-making, especially when set against the various mistakes which have been attributed to a dominant PM acting alone, or with only a small clique of advisors – such as Chamberlain's 1938 Munich agreement, Eden's conspiracy with the French and Israelis to invade Egypt in 1956, and Heath's decision to call a general election in February 1974. By contrast, the Prime Ministerial government thesis reflects a kind of institutional 'realism', popular with some pluralist authors in the 1950s and 1960s. It claims that Cabinet government does not fit the contemporary operations of the party system nor the decisive core executive style now required in external relations. Neither side of the debate has been significantly developed in the last twenty years. Most of the 'evidence' involved still consists of supposedly authoritative 'insider' judgments, scattered anecdotes from ministerial memoirs, or disorganised raw data such as the diaries of former Cabinet members.

One of the oddities of British thinking about the core executive is its very strong ethno-centrism: little or no reference is made to studies of the core executive in other liberal democracies. The segmented decision model, by contrast, asserts the fundamental similarity between British practice and those of other advanced industrial societies, implying that the unique characteristics of the constitution, party system or political culture have little impact on core executive operations. The model is widely accepted as oral folk-lore among political scientists but has yet to be given systematic statement.

A more fundamental question is raised by the bureaucratic co-ordination model. Edelman (1967, ch. 4) argues that in liberal democracies there is a strong psychological need among citizens to believe that someone is controlling political events, and there is every incentive for political leaders not to disillusion them. Hence the appearance of Prime Ministerial or Cabinet control may be a fiction, convenient alike for government, opposition and the mass media, each trying in different ways to sell their representation of reality to citizens – who in turn need to believe that someone is 'running the shop'.

The bureaucratic co-ordination model argues by contrast that no core executive, however structured, is able to fully control a modern state apparatus. Although scarcely addressed in the British literature, this theme is explored in some classic American studies of Presidential leadership, especially Graham Allison's accounts of the Cuban missile crisis, *Essence of Decision* (1973).

Topic 9: The Civil Service

Key question: Does the civil service role in government decision-making produce a particular slant in the advice given or the policies adopted?

Public Administration View

Every liberal democratic state has to manage a basic tension between building up a complex and expert bureaucracy and retaining control of this machinery by lay politicians. The British solution has been to interpose a cadre of 'generalist' administrators between specialist or technical staffs and ministers. Their key functions are to co-ordinate policy across the public sector, supporting ministers by pre-processing information, clarifying policy options, and managing relations with Parliament and the public. For a long period, civil service impartiality and anonymity obviated the need for political appointments of ministerial aides. Although this position is changing slightly, a permanent, non-partisan service still provides a key element of continuity between different administrations. Service-wide recruitment and limited transfers of personnel between departments (especially between line departments and the Treasury or Cabinet Office) both foster an orientation towards the overall public interest. Lifetime career paths to senior positions create a high level of civil service professionalism and independence from outside influences.

Power Bloc Model

The senior civil service is a very cohesive occupational group, overwhelmingly recruited from a privileged social background, educated preponderantly at public schools or Oxbridge, and extensively socialised into its own distinctive values – conservative, secretive, and centralising. Policy-level staffs in Whitehall have well-developed information networks and experience in managing troublesome ministers. They constitute a power bloc pressing for minimal change in the status quo which dominates advice-giving to elected ministers, and can undermine policy proposals threatening to civil service interests. This orientation may provoke occasional skirmishes with radical right-wing ministers, but has a much more general and restrictive impact upon governments of the left committed to social change.

Bureaucratic Over-Supply Model

Government agencies supply a whole block of outputs to the legislature in exchange for a departmental budget. Unlike a firm operating in a market, agencies do not have to reveal the costs of a unit of output at different levels of production, and the legislature may find it hard to ascertain the benefits of agency activities. Consequently bureaucrats have a monopoly of information about the costs and benefits of what they are doing, a strategic advantage which they use to advance their interests. Businessmen are profit-maximisers, and will not produce additional output whose benefit to consumers (and hence market price) is less than the costs of production. But bureaucrats' welfare goes up steadily with a rising agency budget. Hence rational officials maximise their personal welfare by expanding their budgets wherever they can, even if the value of extra public services provided is less than the costs of producing them. The only limit on budget maximisation is that an agency's activities cannot actively reduce the social welfare. But government bureaux still typically over-supply up to twice the level of outputs that would be produced by private firms under market competition – which in welfare economic terms can be seen as the social optimum. The best strategies for countering these inherent bureaucratic traits are to foster competition between different agencies, to create competition between agencies and the private sector, or to privatise functions completely.

* * * * * *

The conventional wisdom about the civil service has always been the public administration view in the post-war period. But the approach has been under more sustained attack in the 1980s, as the traditional left/liberal criticisms in the power bloc model have been supplemented by New Right attacks on wasteful bureaucracy. From being rather a bland position, the public administration view has become more internally divided about whether there should be a broader range of backgrounds for senior staff, better training in policy analysis or management, recruitment of outsiders to top departmental positions, more open government, and similar issues. A key issue under the Conservative governments has been the claim that senior civil servants are too orientated towards political and Parliamentary advice, and insufficiently concerned with efficient management of those functions which the civil service runs directly. In many ways these are the same criticisms made by the Fulton Committee (1968), which the civil service seemed to have successfully emasculated, only to find similar points rephrased by the Expenditure Committee (1978), and now used by the Thatcher government in promoting its search for greater civil service efficiency. Greenwood and Wilson (1984) provide an intelligent discussion of these debates within the orthodoxy.

The power bloc model, by contrast, remains largely unchanged since Harold Laski's critiques of the 1930s. Often inaccurately associated only with a Marxist position – for example, in the work of Miliband (1969) – the approach has always been fashionable among liberal writers as well. Like the PM vs. Cabinet power debate, the debate between the public administration and power bloc views relies a lot on anecdotes, 'insider dopesterism', and assorted ex-ministers' diaries. Crowther-Hunt and Kellner, *The Civil Servants* (1980), provide a good guide to this impoverished controversy.

The myth that the civil service has no enemies on the right continues to be promulgated by recent British politics textbooks (for example, Dearlove and Saunders, 1984, pp. 119–25). Yet for at least a decade, the main challenge to the civil service has come from the growth of the New Right, and their intellectual influence within Thatcherite conservatism. The bureaucratic over-supply model, formulated by William Niskanen's *Bureaucracy: Servant or Master?*, heavily influenced ministers' behaviour in the 1979–82 period. Acting on the assumption that all bureaucracies are wasteful, and spurred on by the premier's hostile reaction to civil service industrial action in 1981, the Conservatives set about cutting Whitehall's manpower and 'deprivileging' the service in pay and pension terms (see Chapter 4). But the performance of Whitehall contingency planning during the Falklands war in 1982 noticeably tempered hostility to the civil service in Downing Street, prompting the departure of disgruntled advisers such as Sir John Hoskyns, whose ideas for continued radical change were put on a back-burner. Instead, very conventional 'efficiency drive' measures dominated the second Thatcher administration, notably the Financial Management Initiative (Cmnd 9058). The waning practical influence of the over-supply model, at least with respect to central departments, has been matched by increasing academic criticism that the approach is empirically implausible (Goodin, 1982), or logically flawed.

Topic 10: Quasi-Governmental Agencies

Key question: Why have an increasing number of executive functions at national or regional level been administered by quasi-governmental agencies?

Administrative Accountability Model

There are two basic types of quasi-governmental agencies. The most important are public corporations, commercial companies controlling their own operations, with government-nominated boards, long-term plans and financial targets set by a minister, and some Treasury funding for investment. By competing in the market, public corporations are kept efficient and responsive to consumer demands. Second, there are numerous one-off bodies set up to tackle particular problems where national policy guidance is needed but detailed ministerial/civil service control is inappropriate. Virtually all have government-appointed executive committees, and receive partial or complete Exchequer funding. Key reasons for using the QGA form include divorcing government from policy implementation; creating a buffer body where state intervention is sensitive or controversial; disaggregating administration which would otherwise be too massive (such as the NHS); relieving the pressure of routine tasks on central departments; and integrating professions or interest groups into administration. Both types of QGA preserve clear lines of accountability running from public sector agencies to ministers, and via them to Parliament.

Privatisation Model

The growth of QGAs reflects two strong tendencies. First, especially under Labour governments, responsibility for producing purely private goods (such as coal, steel, or cars) has been inappropriately brought within the orbit of government decision-making. Public corporations are an unsuccessful attempt to manage the contradictions of running an efficient industry while cutting it off from market disciplines (such as the threat of a takeover or bankruptcy). Second, British politics has been characterised by an unfounded aversion towards using private sector firms or voluntary associations to achieve 'public goods' objectives. Instead, government intervention in any aspect of social development has implied new public sector organisations, even when funding private sector bodies could achieve similar results at less cost. Both tendencies can be effectively countered by privatisation. Public corporations and self-financing QGAs should be converted into normal companies and sold off on the Stock Market. Exchequer-financed QGAs, together with many areas of public service provision in central departments and local authorities, should be split up into those functions where only public agency implementation is appropriate, and those where private sector participation can be increased. Open tendering by firms or voluntary associations to carry out public policy functions should lead to extensive contracting-out, or the replacement of full public provision by subsidised private services. Hence the numbers of QGAs can and should be radically reduced.

Inter-Governmental Model

Modern policy-makers typically handle exceptionally complex problems with an overload of information but with underdeveloped technologies for reaching decisions. The growth of QGAs helps to create more interactive and expert decision-making in the following ways.

(1) The first is by 'factorising' administration so that different aspects of a single problem are handled by specialised agencies. Decisions then emerge from networks of inter-agency contacts and a structure of power/dependency relations, rather than being centrally made by unattainable synoptic policy planning. For example, Britain's civilian nuclear-energy programme involves separate agencies for developing nuclear-reactor technology, building nuclear power stations, ordering and operating power plants, supplying nuclear fuels, inspecting the safety of reactor designs, and controlling radiation emissions. Major developments in the use of nuclear technology additionally require massive public inquiries where non-governmental objections are weighed and the risks, co-ordination and feasibility of public policy goals are assessed.

(2) Second, is the creation of single-issue agencies particularly suitable for professional administration in very technical areas (for example, nuclear-reactors' safety) or where state agencies make sensitive personal decisions (as in health care). In neither case are normal line bureaucracies that are controlled by representative politics an appropriate form for safeguarding the public interest. Instead, general controls are operated by Parliament and Whitehall, while the internalised norms of professional staffs bear the chief load of protecting citizens from exploitation. Even if policy should deviate from a public-interest path, the damage will be limited, since no professional elite controls more than a single issue area.

Dual State Thesis

The problems of policy-making stem not from the intrinsic difficulties of decision-making but from contradictory pressures on the state apparatus under capitalism – on the one hand to foster business profitability (the accumulation function), and on the other to promote social stability (the legitimation function). State institutions must be arranged to insulate functions crucial for business from any effective popular control. Accordingly 'dual state' structures are set up in which strategic production or accumulation functions are concentrated in national or regional QGAs highly resistant to democratic capture (such as public corporations). QGAs can be used to create corporatist structures incorporating business (and sometimes trade-union) interest into policy-making. QGAs are also more dynamic and can undertake planning better than conventional line bureaucracies. Meanwhile, issues of lesser salience for capital interests, such as consumption policy, are concentrated in local government and politically visible areas of central government, where they absorb popular participation on secondary concerns.

* * * * * *

The state apparatus outside Whitehall and local government remains chronically under-researched. Until the late 1970s single-issue QGAs were hardly even counted with any reliability, let alone studied systematically. Public administration writers discussed the public corporation form devised by Morrison in the 1930s. But the actual diverse operations of state industries were rarely studied, and the debate about their effectiveness was conducted primarily in organisational terms, rather than by reference to economic criteria. In the 1970s an American literature on 'government by other means' was imported to update this picture and broaden debate, as in A. Barker (ed.), *Quangos in Britain* (1982). Greenwood and Wilson (1984) provide a useful summary of the mainstream approach.

The intellectual change promoted by the New Right has been nowhere more rapid, or more influential on Conservative government thinking in the 1980s, than on QGAs. The privatisation model criticises public corporations on grounds of economic theory rather than organisational design. It also dispenses with the timid reformism of earlier periods in favour of a root-and-branch change in the organisations' operating environments. On single issue QGAs the privatisation model is equally explicit, favouring simplified but more accountable central controls via competitive tendering, and strong corporate management structures. Both strategies are designed to foster more extensive privatisation and both are key features of Conservative policies in practice. Hiving-off and contracting-out have developed rather slowly even in the second Thatcher administration, but they at least have been more in evidence than the promised 'quangicide' of the 1979 manifesto. Much of the literature on privatisation remains very polemical (for or against), but some more academic treatments of its costs, benefits and mechanisms have begun to appear (Le Grand and Robinson, 1984; Ascher, 1986).

The intergovernmental model has not been by-passed like conventional public administration. Its approach stresses that the different elements in the 'toolkit of government', can achieve the same effects so there is nothing sacrosanct about existing institutional arrangements (Hood, 1983). The model reflects a neo-pluralist belief that the scale of government intervention cannot be significantly reduced in advanced industrial states. Consequently current changes (such as contracting-out) represent a search for new functional arrangements for achieving much the same goals as before. In the current climate QGAs and interactive decision-making have become more, not less, important. For example, most asset sales of public corporations involve the government only partially relinquishing its nominal ownership while increasing its regulatory apparatus, and bringing in private shareholders and finance institutions as additional elements in policy-making. In addition, selling off state companies continues a long post-war trend for policy-making powers to decentralise out of Whitehall and into the hands of specialised policy elites.

The dual state thesis offers the first plausible radical account of the reasons for the growth of QGAs (and also of the displacement of policy responsibilities away from local government). Cawson and Saunders (1983) link QGAs closely with the corporatist model of interest groups. An interesting literature has been spawned on why different kinds of functions are allocated to various tiers and sectors of the state apparatus (see Sharpe, 1984; Dunleavy, 1984). As yet only monograph studies of QGA's operations have been produced, however.

Topic 11: Local Government and Urban Politics

Key question: What do local political institutions contribute to the system of representative government in Britain?

Local Democracy Model

Decentralising policy responsibilities to local government pluralises the channels of representation open to citizens, and allows local knowledge and preferences to shape services within a framework of national legislation. The local election of councillors plays an important role in voting behaviour, party organisation and political recruitment. In a unitary state a government with a Parliamentary majority can impose new central controls over localities or reorganise local authorities completely. But such measures take time and effort to achieve, and councils still have their own sources of funding and control so much policy implementation that their scope for significant discretionary behaviour cannot be entirely eliminated. Control of local governments offers opposition parties a toehold from which to develop alternative policy strategies, and to influence the implementation of a wide range of central government's domestic policies. Mid-term swings against the national government commonly transfer control of many councils to the opposition parties, creating a significant countervailing power to the dominance of the Westminster executive.

Insulated Local Elite View

Local elections confer little citizen control over local policy-making, for four reasons. First, people vote for or against the Westminster government at local elections, on national lines not local issues. Council leaders know that their own policy performance has little influence on their re-election chances and so have little incentive to anticipate citizens' reactions to their decisions. Even in the new three-party system in local politics, distinctively local factors have only a minor impact on council election results. Second, the financial accountability of local governments is reduced, because much of the burden of increased local taxes is born by businesses (who do not command any votes directly) and lower-income voters who get local property tax rebates (and hence may be insensitive to the effect of tax rate changes). Third, the extent of party competition in council elections is less than at national elections, because many local authorities are normally controlled by one party. Fourth, local government administration is highly professionalised. Planners, teachers, engineers, social workers, etc. have a great deal of discretion in their decision-making, develop policy ideas in professional forums and journals, and orientate themselves to national-level ideas rather than local needs or preferences.

New Right View

Decentralisation of public spending functions to local governments can increase citizen control over public spending if municipalities are highly fragmented, as in the USA. Here people can 'vote with their feet' by moving out of localities which are inefficiently run or unresponsive to their preferences. But Britain's local government system has been reorganised to create a few very large municipalities in each area, especially at the top tier of local government. Citizens cannot easily 'exit' to escape from unpopular council policies, which devalues the usefulness of decentralisation. In addition, the organisation of local politics on national party lines has produced an inappropriate hyper-politicisation of straightforward local issues, and attempts by the Labour left to use municipalities as psuedo-Parliaments. The insulation of local elites from effective electoral control (see above), and the irresponsible financing of many local private goods from the national exchequer, have both created serious waste. Local government professionals have been able to over-supply services for many years, boosting their budgets out of all proportion to the growth in genuine needs.

Re-establishing citizen control over local councils requires basic changes in the existing system, such as: (i) the withdrawal of central government funding, except for minimal fiscal equalisation to protect the poorest localities; (ii) the imposition of a restrictive 'fiscal constitution' on councils to force them to improve efficiency and weigh priorities, to be achieved via tax-capping and expenditure limitations; (iii) the reform of local taxation to remove councils' powers to expropriate taxes from businesses and to broaden the base of residential taxpayers; (iv) improving citizens' ability to move between different public service areas; (v) subjecting municipal services to rigorous contracting-out requirements; (vi) privatising commercially viable services or transferring them to quasi-governmental agencies run on business lines.

Local State Model

'Local democracy' is a legacy from the ninteenth century, when municipalities were an important means for co-ordinating the interests of a territorially divided capitalist class. In the era of monopoly capital this function has disappeared, and the labour movement has used its electoral dominance in key areas to pioneer socialist strategies. Since it would be ideologically dangerous for the central state to abolish localities altogether, they have chiefly been by-passed by centralising key functions in a dual state structure (see above). Local politics and policies reflect the varying balance of class forces across different areas of the country. Ultimately the local state is tied into fostering capitalist state objectives. But the potential exists for the left to capture municipalities as 'stepping stones' to national power, using 'red islands' to demonstrate the viability of socialist strategies, or disruptive tactics to expose the oppressive character of central state controls.

* * * * * *

Urban politics has been one of the most intellectually vigorous areas of British political science for ten years, reflecting the stimulus produced by European neo-Marxist perspectives in the late 1970s, and American public choice theory in the early 1980s. Research access to local councils has been facilitated by more open government, and increasing controversy about local government in national politics has boosted academic interest in the field. Hence the conventional wisdom of the 1960s and 1970s is less discussed now than in any other area of British political science. The local democracy model still dominates traditional textbooks (such as Elcock, 1982) and academic debates close to practitioners' concerns (Jones and Stewart, 1983), but has received no systematic research restatement since Newton's *Second City Politics* (1976).

All three more contemporary models recognise serious deficiencies in local government accountability. The insulated local elite view has become more widely accepted in practitioner debates as well as academic discussion, paradoxically at about the time that increased Alliance has begun to have a lagged effect in reducing local one-party control and increasing the number of 'hung' councils where policy cannot be settled at meetings of the majority party group. Dunleavy, *Urban Political Analysis* (1980, especially ch. 5), provides a thoroughgoing statement, linked to a radical left perspective on urban policy-making.

The New Right critique also stresses the inadequacy of voting accountability in local politics. But its sources are quite different in American public choice theory. Tiebout's (1956) work provides a seminal statement of how citizens move between fragmented local governments in search of the right mix of services supplied and taxes levied for their individual preferences. The weakness of 'Tiebout forces' in Britain, which has the fewest and the most populated base local government units of any Western democracy, then provides a rationale for the New Right's emphasis instead upon privatisation and tax-capping. In addition, English neo-conservatives have strong centralist tendencies; Scruton (1981) suggests that municipal government should properly provide no more than a distracting spectacle to allow the emotional affirmation of localist sentiments. These diverse intellectual origins have combined in a rather explosive cocktail mix, with direct implications for the practical debates within the Conservative governments since 1979 (see Chapter 5). The New Right impetus has by no means run its course. For example, proposals to require competitive tendering across all areas of the local government service, or to fund schools run by boards of governors directly from Whitehall without the intermediation of local education authorities, may still affect practical politics significantly (Henney, 1984).

Practical politics have also had an immense impact on Marxist and radical views of urban politics. Writing in the mid-1970s about Labour-controlled Lambeth, Cockburn (1974) had no doubts that local authorities were but the fingertip of a unified capitalist state. This type of analysis has been one enduring strand of orthodox Marxist view. But under the Conservative governments from 1979, control of local government units was an important aspect of Labour resistance first to spending cuts and then to the abolition of metropolitan governments. The recruitment of a 'new urban left' into the Labour party in the mid-1970s meant that the strategies pursued by some Labour councils were more radical and (arguably) more successful than at any time since the inter-war experience of Poplarism. Boddy and Fudge's edited collection, *Local Socialism* (1984), provides a good picture of new urban left thinking.

16

Theories of the State in British Politics

PATRICK DUNLEAVY

The increasing ideological polarisation of British politics charted in Chapter 1 has been matched in academic work by a more vigorous debate between major theoretical positions. Three broad streams of thought project distinctive pictures of the British state and political system. The 'mainstream' liberal view is still a pluralist one, emphasising the dispersion of power in British society and the orientation of public policy towards improving overall social welfare. Ultimate control over government rests with citizens, and an open, many-sided process of political representation approximates as closely as is feasible to a working liberal democracy. Within this broad perspective, a neo-pluralist view emphasises the complexity of government in advanced industrial societies, the inevitability of an extended state role, and the need to supplement traditional representative politics with new kinds of control mechanisms in the public interest.

An increasingly influential New Right critique argues that British pluralism has been unstable. Many aspects of the political process have contributed to an excessive growth of government which threatens economic performance and the fabric of democratic freedoms themselves. In this perspective, the cessation of welfare state growth since the late 1970s is a belated sign of the health of British democracy, an indication that citizens can appreciate the defects in the political process which have underlain economic malaise and take steps to correct them.

By contrast, the radical and Marxist left see the current phase of welfare retrenchment as highlighting in a stark fashion the fundamentally class-based politics and oppressive inegalitarianism of British society. For a long period, democratic (as opposed to revolutionary) socialists had no developed theory of the state, and the only distinctive Left perspectives were orthodox Marxist explanations tightly linking defects in the political system to the essential nature of the capitalist mode of production. In the mid-1970s Marxist thought was extensively modernised under the influence of French, West German and American writers, creating a series of neo-Marxist positions which explained political and ideological developments more plausibly than the old orthodoxies. In the 1980s the development of all strands of Marxist political analysis has weakened. Instead, democratic socialists have begun to outline a radical theory of the state which combines elements of neo-Marxist thought with some critical liberal approaches.

For each of these perspectives, I briefly review the ways in which they approach the topics discussed in the previous chapter. Table 16.1 demonstrates that the vast majority of topic views do form building blocks in broader theories of the state, a testimony to the importance of these perspectives in structuring the analysis of contemporary British politics. I then examine in more detail how each overall approach explains the post-war growth of state intervention and the partial reversal of this long-term trend since the late 1970s.

Pluralism I: The Political Process

There has always been something of a divergence in pluralist views about the importance of constitutional factors in British politics (Table 16.1). Some conservative commentators define the constitution broadly to include such factors as the leadership selection process inside political parties, and hence maximise its significance in line with the constitutional/legal view (Bromhead, 1974). More sociologically inclined political scientists have adopted the behavioural view minimising the constitution's significance compared with political culture. But all pluralists agree on the institutional view of the law/politics relation, stressing the essential neutrality of the legal process as between social interests.

Pluralist views of input politics used to stress the importance of elections, voters' long-term identifications with the major parties, and the responsible party model of how political elites compete. In the 1970s a greater emphasis on issue voting was grafted onto the older view, and the growth of third-party voting produced a reappraisal by some pluralist writers of the defects of the two-party system on lines suggested by the adversary politics model. This readiness to criticise inadequacies in current institutions has produced a considerable reformulation of the rather complacent and optimistic pluralism of the 1960s. Pluralist writers are now less prone to accept lapses from the 'pure' representation of citizen views in trade-offs with other values, such as the 'stability' of the political system. Similarly, the group politics model still dominates pluralist accounts of the interest group process, but more recognition is now paid to the inherent difficulties of organising some groups and to the biases created by corporatist arrangements. However, most pluralist writers give a standard defence of existing mass-media arrangements on lines suggested by the public opinion control model.

Pluralist views of government institutions emphasise the even-handedness and basic 'public interest' orientation of administrative agencies. The civil service is seen as effectively subordinated to ministerial control while yet contributing impartial advice and continuity in policy-making, as the public administration model maintains. The administrative accountability of quasi-governmental agencies is also high, and their development betokens no essential change in the traditional ways of operating representative political control. Finally local democracy continues to play a valuable representative role, despite the remodelling of central–local relations to reflect new national government priorities in the 1980s. The major areas of debate within the pluralist perspective continue to focus on how political control is

TABLE 16.1 *Three overall views of British politics*

Topic	Pluralist/neo-pluralist approach	New Right view	Radical or Marxist approach
1. The Constitution	Institutional/legal view; or Behavioural view	—	Ideological front model
2. Law and politics	Institutional/pluralist view	Neo-conservative view	Radical view
3. Voting	Party identification view; or Issue voting model	Issue voting model	*Radical view
4. Party competition	Responsible party model; or Adversary politics model	Adversary politics model	*Radical view
5. The interest group process	Group politics model *Corporatism	Log-rolling model	Corporatism
6. The mass media	Public opinion control model	Market model	Dominant values model
7. Parliament	Westminster model; or Transformative model	—	—

8. The core executive	Cabinet government; or Prime Ministerial government; or Segmented decision models	—	Bureaucratic co-ordination model
9. The civil service	Public administration view	Bureaucratic over-supply model	Power bloc view
10. Quasi-government agencies	Administrative accountability model; *Inter-governmental model	—	*Dual state thesis
11. Local government and urban politics	Local democracy model	New Right view	Local state model *Insulated local elite view

Neo-pluralist views are marked by an * in the second column, and radical (i.e. non-Marxist) views are similarly designated in the fourth column.

asserted in the Westminster model. Authors who see the Commons as a partisan debating arena usually insist on the realities of collegial Cabinet control inside the core executive itself. By contrast, commentators who perceive a withering-away of Cabinet government and its replacement by a more 'presidential' style of monocratic or personal-clique politics, are more anxious to detect signs that the Commons is adopting a 'transformative' role to offset this change.

Pluralism II: Trends in State Intervention

Any overall approach to British politics has to be able to answer some basic questions about why the scope of government activities has changed over time. There is general agreement that the post-1960 period has seen a major expansion of public spending and greater government regulation, associated with the mixed economy and the welfare state. But there is no consensus on whether this process has simply stabilised or been put into reverse since the late 1970s, nor about what the future of 'big government' is likely to be.

Pluralists argue that the advent of an extended state in the post-war period reflected three basic forces. First, the political system responded after a long lag to the advent of adult suffrage, making full social and economic 'citizenship' a reality for all sections of the population in the aftermath of the Second World War. A consensus in favour of better welfare provision established a new political agenda, and re-drew the boundaries of 'public' and 'private' concerns (Rose, 1982). Second, the experience of government planning in wartime eroded hostility to state intervention in economic organisation and promoted the acceptance of a mixed economy. Similarly, experience of high wartime tax levels lessened citizens' resistance to greater public spending (Peacock and Wiseman, 1987). Keynesian demand-management policies produced benefits in terms of economic growth to finance welfare spending and ease income redistribution, and further legitimised an active governmental role. Finally, once a system of social and economic intervention was set up, trends in the social environment basically determined the path to higher spending. For example, demographic growth boosted education, housing and health care spending; changes in transport patterns dictated a motorway building programme; and the escalating costs of high-technology research upped the governmental support needed for growth industries or defence spending. Hence state intervention can be seen as fundamentally driven by public demands, environmental changes, and its own success.

State spending growth in Britain reached a plateau in the late 1970s for several reasons. Demographic trends turned down, so school populations dropped. Some public policies achieved their effect; further slum-clearance programmes for instance became unnecessary, and demand for new council housing decreased. Environmental changes made other policies less successful; for example, the worsening situation of inner-city areas made new towns policies unacceptable. Public opinion shifted modestly in favour of maintaining existing taxation rather than introducing more progressive measures in welfare payments. Economic conditions became more adverse, reducing the fiscal surplus available for new policy initiatives and causing a reappraisal of some past commitments. Public demand for some government outputs (such as health care, 'law and order' and defence spending) continued to grow, while other policy commitments expanded (such as welfare payments to the unemployed). Consequently a painful process of redirecting state spending via cutbacks in some areas and expansion in others has been undertaken. But the changes have been incremental in the main, and the level of consensus among citizens and political elites in support of the basic elements of a welfare state has hardly been reduced.

The debate about recent trends in state intervention is still a lively one, chiefly because of the considerable difficulty in devising any single index of 'intervention'. Public debate about the size of government is phrased chiefly in terms of dubious statistics, such as the proportion of GNP spent by government (Heald, 1983, ch .2). Not surprisingly, in view of their optimistic view of state growth, pluralists have been sceptical about claims of decisive cutbacks in government functions. Governments have a large 'toolkit' for intervening at their disposal, so public policy objectives can be pursued by means which do not necessarily involve a great deal of public expenditure (Hood, 1983).

Neo-Pluralism

Since the late 1960s a more extensive remodelling of pluralist political analysis has been undertaken, chiefly by key American writers such as John Galbraith (1969), Charles Lindblom (1977), Oliver Williamson (1973), Albert Hirschmann (1970) and Robert Dahl (1984). Their work has effectively constituted a novel liberal position which can conveniently be labelled neo-pluralism. Unlike previous pluralist approaches, neo-pluralism is primarily output-orientated, focusing on the public policies produced by state institutions rather than on the diverse channels by which citizens can influence government. Neo-pluralists believe that the growth of the state has reduced the scope of decision-making that can be successfully handled by traditional representative politics. The fundamental problems of government in advanced industrial societies are attributable chiefly to the inherent difficulties of making rational decisions. The traditional pluralist pre-occupations – with balancing majority and minority preferences, preventing the accretion of power in a monolithic state, and controlling the abuse of power for sectional interests – are no longer enough to ensure policy-making in the public interest.

Elections and party competition are important in laying down broad guidelines for public policy and in mapping the limits of manoeuvre within which governments must operate. But they do not give the mass of ordinary citizens any detailed control over public policy, nor could they do so given the nature of large, industrialised societies. Their most important role is in securing overall consent for government-formulated strategies, and maintaining a source of new ideas and solutions at the forefront of citizens' attention. However, neo-pluralists argue that liberal corporatist arrangements between government and major interest groups can play a more extended role, because they greatly enlarge the toolkit for tackling major problems that is available to governments; and because they can mobilise large groups of people behind public interest outcomes which governments acting alone could not achieve except at great cost, such as restraining the wage inflation within limits consistent with economic growth.

Neo-pluralist writers pay little attention to the detailed

institutional questions which preoccupy conventional pluralists. They are quite pessimistic about the scope of issues which can be successfully processed or resolved by representative institutions, such as Cabinet or Parliament. In the modern extended state, overt 'political' control must inevitably concentrate on fewer issues where all the options have anyway been extensively pre-digested by administrative agencies. The growth of quasi-government agencies is a fundamental trend, as their increased role under New Right governments notionally opposed to such institutions graphically demonstrates. The inter-governmental model explains this shift in state administration as an attempt to create interactive networks between fragmented agencies on the one hand, and to allow professional expertise to be more directly applied to problems on the other (Dunleavy, 1982).

Neo-pluralists generally endorse the conventional pluralist account of the development of the mixed economy and a welfare state. But they are even more sceptical about claims that state intervention is now declining. In an advanced industrial society it simply is not feasible to suppose that government can withdraw from a very extended role in fostering economic growth or guiding social development. None the less, in particular periods where the technology available to public-policy-makers is felt to be inadequate, futile attempts will be made to return to a mythical economic 'golden age'. *Laissez-faire* ideologies and a phobia about government spending are powerfully entrenched in Britain, so it is not surprising that in the worsening economic climate since the mid-1970s there has been large minority support for a policy of trying to dispense with societal corporatism and cut back decisively on government functions.

In practice these efforts have been almost completely counter-productive. Reducing state spending and raising real interest rates artificially deflates an already depressed economy, generating mass unemployment which constitutes a new drain on depleted government revenues. Since other countries continue to pursue state intervention, attempts to withdraw government support for new technology or cut back on training simply worsen Britain's competitive position in modern industrial markets. Various New Right policy prescriptions can have one-off effects in improving efficiency or eliminating small amounts of waste. For example, any large hierarchical organisation (whether a government agency or a private corporation) must continuously adjust the balance between contracting out for services or providing them in-house (Williamson, 1973). But this is a technical question which has nothing to do with advancing 'freedom' or supposedly immutable public/private sector differentials in effectiveness. As a continuing programme for government, monetarism and drastic public spending cutbacks have already been abandoned by erstwhile proponents. However, new kinds of economic intervention and social planning probably need to be invented before the public can be fully re-educated about alternative strategies, just as the 'Keynesian revolution' provided the basis for the post-war consensus on welfare state policies.

The New Right I: The Political Process

One of the most important changes in modern political science has been the evolution of a coherent New Right view of the political process, based largely on American public choice theory – which uses standard techniques from economics to analyse political processes (Mueller, 1979). Although some of this literature (such as Downs's work) operates within a pluralist approach, many key public choice writers (such as Buchanan, Tullock, Niskanen and Olson) have always been or have become identified with New Right views. In Britain, public choice approaches were for a long time promoted by bodies such as the Institute for Economic Affairs and utilised chiefly by conservative economists to criticise the welfare state. The British New Right in practical politics draws on a wide range of different influences (Bosanquet, 1983). However, in terms of academic political analysis, public choice approaches have remained intellectually dominant.

The New Right characteristically do not discuss the existing British constitution in any great detail, preferring to focus on normative questions about the ideal form of constitution for a liberal democracy. An important link to more-practical debates has been Robert Nozick's argument in *Anarchy, State and Utopia* (1971), that a just constitution for a society should simply lay down a set of general rights and then allow social arrangements to be constructed by individuals freely exercising these rights. British far-right authors such as Scruton (1980), otherwise very distant from the individualism and the liberal–rationalist commitments of public choice theory, have been quick to cite such ideas as justifying their neo-conservative view of the law/politics relationship.

The New Right analyse input politics much more critically than pluralist commentators (Table 16.1). The issue voting model of electoral behaviour is closely linked to the adversary politics account of party competition, in an argument that citizen control over government has heretofore been inadequate. The electoral 'market' bundles up too many issues for voters to decide at one go; it encourages politicians to forsake sound economic management for electoral priorities (Brittan, 1975); and under the two-party system it forces voters to choose between over-polarised approaches. Similarly the interest group process is dominated by log-rolling, in which sectional interests struggle to commit taxpayers' money to serving their own private purposes, and in which winning coalitions of groups can use the political process to exploit those left in the minority – usually the better-off. The New Right also use the market model to criticise the 'artificial' media controls currently operating over broadcast news.

The American origins of public choice theory have limited its application to some key British institutions. There is a great deal of US literature on how legislatures operate, designed to explain Congressional behaviour, very little of which can be successfully applied to the Commons. Similarly, there has as yet been no real attempt to translate public choice work on the theory of committees to help explain British cabinet government. So the New Right grip on the core executive and Parliament is still weak, and they are much more comfortable examining the operations of administrative agencies. The budget-maximising model of bureaucracy has been cited to explain Whitehall's budgeting round, with ministers and departments struggling for bigger shares of the public expenditure pie even under the supposedly different values of the Thatcher governments. Quasi-governmental agencies and local authorities are all fitted under the same account. But no explanation is provided of how the budget demands of numerous competing agencies are fitted together or of how relations between different tiers of government affect budget-maximisation. However, a more developed New Right literature on the limitations of local democracy in Britain has evolved, partly to justify the centralising tendencies of the Thatcher administrations.

The New Right II: Trends in State Intervention

The growth of the state has proved to be a fundamental pathology in pluralist ways of running liberal democratic arrangements. Politicians misrepresented what government can achieve, bidding up voters' demands to unrealistic levels and then bridging the gap between government revenues and public expectations by deficit financing. The increase in government debt triggers excessive money supply growth and hence fuels inflation, in addition to loading a debt mountain onto future generations of taxpayers. Log-rolling similarly boosts state spending as private goods are funded illegitimately out of general taxation. Vested interests become veto groups, slowing up responses to socio-economic change and reducing national economic competitiveness (Olson, 1982). Government agencies maximise their budgets and over-supply their output of services. The absence of bureaucratic competition allows public service professionals and unions to divert resources into feather-bedding their positions. Welfare state policies primarily benefit middle-income voters instead of achieving genuine social redistribution (Le Grand, 1982). Rising tax burdens reduce incentives for economic enterprise, and state borrowing and capital projects both squeeze out more productive private investments.

State spending follows a 'ratchet' pattern, with each burst of growth constituting a new benchmark from which future expansion can take place, and with veto groups preventing any attempts to cut back established policies. The normal budgetary process concentrates attention solely on new spending proposals, and the 'base budget' for ongoing programmes is left almost completely unanalysed. Since government is actually incompetent to guide an advanced industrial society or economy in any effective way, each wave of state intervention creates as many problems as it 'solves'. By introducing new distortions into economic markets, and by changing citizens' behaviour in countervailing ways, state intervention becomes a vicious circle, at once self-defeating and self-sustaining.

The New Right are far more comfortable denouncing the past evils of state growth than they are in analysing current trends under the Reagan or Thatcher administrations. One strand of New Right analysis is fatalistic about the prospects for altering the 'leviathan state', arguing that the trends established over previous decades cannot be easily reversed. Fatalists point to the failure to make dramatic cutbacks in public spending since 1979, which they attribute to 'weakness of will' on the part of the Thatcher administrations, which 'never really tried' to roll back the state because of political pragmatism and faint-hearted supporters. This approach is entirely consistent with the New Right's claim that state growth is an inherent by-product of pluralism. But it severely undermines the value of trying to implement New Right policy prescriptions, since more tough-minded leaders, prepared to weather greater unpopularity and more extreme transition costs than Thatcher or Reagan, are hard to envisage.

By contrast, the majority viewpoint within the British New Right is a 'heroic' interpretation which attributes much greater significance to the Thatcher or Reagan administrations. Since the mid-1970s there has been a fundamental shift in British public opinion and political life. Citizens are profoundly disillusioned with the 'collectivist ethic' of previous decades, and recognise that limits must be set to state spending if economic vitality is to be regained. The two Conservative victories in 1979 and 1983 reflected this new appreciation, and the two Thatcher governments have been committed to producing a 'new realism' among voters, union members and taxpayers. New Right policies have helped to achieve substantial changes away from further growth, especially in enforcing rigid limits on public service manpower, setting an overall ceiling on spending, and cutting back the most scandalously wasteful policies. Limits on state borrowing have been adhered to, reducing money supply growth and hence inflation. The formal public sector has been rolled back substantially by asset sales, contracting-out, and other forms of privatisation – reducing the size of the state-dependent workforce, liberalising initiative, and demonstrating the practicality of the New Right's policy agenda. The climate of public debate and the range of future options for public sector activity have all been moved sharply and irreversibly to the right.

The 'heroic view' sees the incorporation of New Right ideas into public policy-making as a turning-point where a real shift away from state growth can take place, hence the value of continuing with their policy prescriptions. However, the fact that such a radical alteration of the trajectory of state intervention can be achieved simply by a change of heart among a few leading politicians, casts doubt on the status of the New Right's theories about state growth as a fundamental pathology of pluralism. If such different results can follow just from bringing different personnel into the leadership of one political party, surely there was nothing deep-rooted or inevitable about state growth under pluralism in the first place?

Marxist and Radical Views I: The Political Process

Marxist political analysis focuses chiefly on macro-level theory, rather than on applied topics (Table 16.1). Orthodox Marxist theory asserts that class conflict underlies the apparent diversity of issues and groupings involved in liberal democratic politics. And it argues that capitalist domination of the political process is assured, despite the apparent separation of political and judicial institutions from direct control by capital-owners. The state apparatus remains an instrument of the ruling class, because of personnel interconnections, direct pressure by business on policy-making, and business domination of the mass media and input politics. The ideological front account of the British constitution, the critique of 'Parliamentarianism'

(Miliband, 1973), etc. are straightforward examples of this line.

More detailed Marxist work often takes the form of point-by-point critiques of pluralism (Miliband, 1969, 1983). However, the criticisms developed are rarely specific to a Marxist position: for example, the bureaucratic co-ordination model of core executive operations does not address a single issue in Marxist theory (see Benn, 1980). Similarly, critiques of the civil service or the judiciary as conservative power-blocs, recruited from a narrow class base and committed to obstructing radical change, can be enforced by critical liberal writers without in any way demonstrating the overall plausibility of a Marxist analysis of British politics. The instrumental approach has dramatically weakened since the development of neo-pluralist and new-right accounts has widened the range of available liberal views. A less orthodox approach to detailed analysis is exemplified by the Communist journal *Marxism Today*, which combines a theoretical attachment to Marxism with a readiness to use any available tool of analysis to pragmatically explain current political developments. For example, pluralist accounts of the decline of class voting have been simply accepted at face value by writers such as Hobsbawm (1981) in advocating that the Labour party adopt a 'popular front' strategy of coalition with the Alliance to prevent another Conservative electoral victory.

Empirical work more closely linked to theory has been produced mainly by neo-Marxist writers, influenced by 1970s writings in France, West Germany and the USA. These accounts remain macro-level but they have more successfully focused on particular aspects of the political process, notably developing Left analyses of corporatism in economic management (Panitch, 1979). There are numerous quite powerfully argued interpretations of recent British history informed by these approaches (for example, Gamble, 1985). Similarly neo-Marxist work on the importance of ideology in maintaining social stability has greatly influenced left analyses of mass-media outputs (for example, Glasgow University Media Group, 1976). Finally the 'local state' literature testifies to a new-found Marxist enthusiasm for local government.

Dissatisfaction with the lack of detailed application in Marxist work has been one influence behind the development of a distinctive radical but non-Marxist theory of the state. A second influence has been an unwillingness to resist the New Right intellectual advance by simply standing pat on a pluralist defence of existing institutions. Democratic socialists have increasingly been forced to develop their own models of what is valuable in the pluralist political process and the liberal welfare state, without at the same time whitewashing the status quo. One area of advance has been in interpreting voting and party competition (Dunleavy and Husbands, 1985). Some detailed analysis of state institutions has been stimulated by the dual state thesis (Cawson and Saunders, 1988), and work on professionalism, urban politics, and public administration.

Marxist and Radical Views II: Trends in State Intervention

Orthodox Marxist accounts used to suggest that welfare policies were no more than a cosmetic veneer on the reality of a coercive state committed to maintaining the predominant role of private capital in social and political life. But in the 1970s neo-Marxist work argued that a distinctive phase of capitalist development was inaugurated by the post-war welfare state, characterised by a much closer integration of monopoly capital and state modernisation efforts. In advanced industrial societies the profitability of monopoly firms and the rate of overall economic growth both depend on state policy initiatives. For instance, car manufacturers and the oil industry both rely on government highway programmes to boost consumption. Similarly advanced technology industries depend on better-educated workforces and consumers. And a consumer-durable boom requires the prior provision of modern housing in place of slum accommodation. Hence many 'welfare' policies directly foster greater capital accumulation. Others (such as social insurance) increase the 'social wage' and were important in reducing direct trade-union pressure on employers in the post-war boom period. Defence spending and support for advanced technology are clear-cut exercises in pump-priming and corporate subsidisation. Because more people now pay income taxes and indirect taxes there is no evidence that welfare state policies achieve any significant redistribution of income or wealth between social classes. Instead they switch resources within classes, from those in employment to the elderly, out-of-work, disabled and the sick (Westergaard and Resler, 1975).

All Marxist writers agree that in the mid-1970s the onset of a worldwide recession decisively terminated the post-war era of economic growth. An acute fiscal crisis in the advanced industrial countries (O'Connor, 1973) was crystallised by incidents such as the 1976 sterling crisis when the IMF demanded that the British Labour cabinet introduce major public spending cuts, or the virtual bankruptcy of New York in 1977. Dominant class strategies now focus on 'recapitalising capital' by cutting back on those welfare state policies which have the most indirect connections with business profitability. Hence resources have been pulled out of public housing and social insurance in an (ineffective) attempt to pump-prime private spending via defence spending. Mass unemployment has been deliberately re-created to curb working-class militancy and to reinstate the ideological importance of profit-making which has been eroded by welfare state policies. The new dominant class strategy displaces some crisis symptoms, but is itself riddled with contradictions. For example, measures to reassert 'market discipline' over trade unions and socialist local governments have led to an extension of state regulation, and a profound recentralisation of power in Whitehall. Similarly, although swelling dole queues have weakened the trade unions and allowed state workers' incomes to be eroded, it has simultaneously deepened the fiscal crisis and worsened the legitimacy of the British state. Fiscal strains are disguised only temporarily by sales of public assets and North Sea oil funding, while declining legitimacy cannot be tackled simply by increasing police powers to prevent inner-city riots.

Having worked out a set of functional explanations to demonstrate that welfare state growth is functional for capital, neo-Marxist authors have confronted more acute problems in explaining why cutbacks have ocurred. The danger is that their explanations can seem almost tautologous. State growth was functional for capital, otherwise it could never have taken place in a capitalist society. But the same must be just as true of public spending retrenchment – suggesting that whatever happens is functional for capital, a line of analysis more comfortable for orthodox Marxists than for neo-Marxists. The slackening of the European intellectual impetus behind all forms of Marxist analysis in the 1980s has left these problems still largely unaddressed (Jessop, 1982).

Radical approaches have coped with these difficulties by examining other possible limitations on welfare state spending growth. Some writers have argued that private sector forms of provision have decisive social and psychological advantages compared with public provision (Saunders, 1984). Others have suggested that welfare state growth might be politically self-stabilising, creating public/private sector cleavages which fragment the working-class coalitions that initially promoted reformist state intervention (Dunleavy, 1986b).

Guide to Further Reading

Students seeking further reading for the material discussed in Chapters 1 to 14 should first read the relevant section of Chapter 15. Some of the chapters are reviews of the literature on their subjects and to this extent contain their own suggestions of further reading. This is particularly true of Chapters 1, 2 and 14. The references are to books and papers listed in the Bibliography.

Chapter 1 Ideology

There is no reliable general account of the role of ideology in contemporary British politics. The student can get the flavour of the ideological debate within the parties by reading their (usually unofficial) periodical journals. *The New Socialist* conveys the confused open-mindedness of the Labour Party. *Marxism Today* is a much sharper journal which has had an influence within Labour on the specific question of a socialist strategy. The authoritarian wing of the Conservative Party is elegantly represented by the *Salisbury Review*. The anti-Thatcher wing of the Conservative Party is well represented in the political features of the *Financial Times* and the *Guardian*. Alliance thinking can also be found in the *Guardian*, and at greater length in the *New Democrat*.

More sustained Conservative ideas will be found in Cosgrave (1978) which is Thatcherist at its more self-confident. Though dated, Joseph (1976) shows what the Thatcher regime was trying to do. Gilmour (1983) is the best of the 'wet' critiques of that position. Scruton (1980) is the best statement of the authoritarian Conservative position.

Labour's now dominant 'inside left' is best represented by Hodgson (1984). A stunning left critique of this position can be found in Miliband (1985). The argument about whether Labour can ever again hope to form a majority government unaided is summarised in Jacques and Mulhern (1981).

The Social Democrats are represented by Owen (1981) and by Williams (1981). The thinking of the Liberal Party is not well represented in books, but some of the best of it can be found in Bogdanor (1983).

Chapter 2 Voting and the Electorate

There are a number of accounts of the 1983 election each of which has distinctive strengths and weaknesses. Dunleavy and Husbands (1985), Heath *et al.* (1985), McAllister and Rose (1984) all need to be studied.

Crewe (1984) has also updated his important work on dealignment, incorporating findings from the 1983 election. Crewe and Harrop (1986) is an edited book about the media and the polls in the campaign. McAllister and Rose (1984) examines voting patterns in 1983 in the different nations involved. The book by Butler and Kavanagh (1984) is an account of the campaign rather than an analysis of the results, though there is a masterly appendix on constituency voting patterns. With Jowett, Butler (1985) has written a similar book on the 1984 European elections.

There are three other important recent books on the electorate, though none is focused specifically on the 1983 election. Himmelweit *et al.* (1985) uses a long-term survey to develop a model of the voter as a consumer. Robertson (1984) develops an interesting classification of the electorate, while Jowell and Airey (1985) is the second in a series of annual surveys of public opinion.

Chapter 3 The Party System

Despite the recent changes in British party politics there have not been any sustained attempts to update the best books of the 1960s. Beer (1965) has been reissued, and a continuation volume produced (Beer, 1982). A review of the literature can be found in Finer (1980), and much background information can be found in Ball (1981). Finer (1975), Gamble and Walkland (1984), and Sharpe and Newton (1984) debate the importance of party competition in changing public policy. The last is much the most sophisticated of these. Moran (1985) is a reliable account of the current state of each of the parties.

Chapter 4 Inside Whitehall

For the central institutions of Cabinet and Prime Minister, the best book remains Mackintosh (1977). Of the personal memoirs of Cabinet Ministers the fullest are those by Richard Crossman (1975) and Barbara Castle (1980). Of the many books about the Thatcher government, that by Bruce-Gardyne (1984) has most coherence.

On the Civil Service, the Report of the Fulton Committee (Cmnd 3638, 1968) remains a landmark. Very good books by those committed to the Committee's programme and outlook include those by Crowther-Hunt and Kellner (1980) and Garrett (1980). A more sceptical attitude is to be found in Fry (1981, 1984a, 1985). This scepticism is present, too, in analyses of the Thatcher government's efforts at re-shaping the Civil Service and about its 'efficiency' programme. Two official publications give details of the Conservative Government's drive for 'efficiency': *Financial Management in Government Departments*, Cmnd 9058 (1983) and *Progress in Financial Management in Government Departments*, Cmnd 9297 (1984).

Of the academic journals which deal with developments in central government, the most helpful are *Public Administration*, *The Political Quarterly*, *Parliamentary Affairs*, and *Public Law*.

Chapter 5 Beyond Whitehall

The study of government beyond Whitehall remains depressingly sketchy, outside the sphere of urban and local politics. Elcock (1982) provides an institutional guide to local government, Gyford (1983) reviews the pluralist literature, while Dunleavy (1980) reviews the previous literature and raises some key questions about the interconnections of local and national policy-making. Rhodes (1983) and Newton and Karran (1985) both give clear accounts of the evolution of central–local relations from the 1970s up to the mid-1980s. The evolution of the new urban left in local government is well covered by Boddy and Fudge (1984) and Gyford (1985), while Parkinson (1985) provides an interesting analysis of Liverpool's troubles. More theoretical issues about why different tiers of government have separate responsibilities are raised by Sharpe (1984) and Dunleavy (1984). Bulpitt (1983) offers a stimulating interpretation of 'territorial management' in the UK as a whole.

Barker (1982) is a useful collection of articles on quasi-governmental agencies (which, however, are rather misleadingly described as 'governmental bodies' throughout). Ham (1985) gives a rather anodyne introduction to decision-making in the National Health Service. Some theoretical issues about the 'regional state' are raised by Saunders (1985). Steel and Heald (1984) discusses the problems of privatising public corporations. Unfortunately there are no good general studies of relations between public corporations and Whitehall; most of the literature consists of polemics for or against nationalisation, or of single-industry case-studies.

Chapter 6 The State and Civil Liberties

Because this chapter covers a wide range of themes, the suggestions for further reading are grouped thematically.

An excellent introduction to the problems of the British constitution is Marshall (1984). One politics textbook which pays particular attention to the themes raised in this chapter is Beloff and Peele (1985). Among the constitutional law books most useful to political scientists is Brazier and Street (1984).

On civil liberties there is a useful collection of essays edited by Wallington (1984).

On the police, there have been some good recent studies. Reiner (1985) is a balanced and comprehensive account. This may be supplemented by two more specialised studies: Jefferson and Grimshaw (1984) examines police accountability; Fine and Millar (1985) is of importance both for its subject-matter and the insight it provides into attitudes towards the police on the left of British politics.

There is little that is good on the security services, but mention should be made of Andrew (1985). Also of interest are the books by West, including his study of MI5 (1983). The topic of open government has, however, produced a burgeoning literature. Official secrecy is examined from a legal perspective in Williams (1965). Aubrey (1981) is of interest because the author was a defendant in the so-called 'ABC trial'. Wraith (1977) is a brief appraisal of the situation at that point. Useful material can be found in Michael (1984) and Boyle (1979).

Chapter 7 Economic Policy

Blackaby (1979a) is a collection of articles on the definition and measurement of Britain's industrial decline. Brittan (1977) reprints some of the authors' commentaries in the *Financial Times* together with some longer pieces on aspects of economic policy in the 1970s written from a monetarist point of view. Gould *et al.* (1981) provide a critical view of monetarism and advocates major changes in Britain's foreign economic policy. Hood and Wright (1981) is a collection of articles on the crisis of public expenditure in the 1970s and the new mechanisms of control.

A racy description of the making of economic policy when Labour was in power, can be found in Kegan and Pennant-Rae (1979). Young and Sloman (1984) gives an account of this process as it is seen from inside the Treasury. An academic account can be found in Mosley (1984a).

Chapter 8 Social Policy

For a basic textbook, there is Hill (1983). A book which looks at the record of the welfare state: George and Wilding (1984). Mack and Lansley (1985) follow Townsend's (1979) approach, reporting a major survey of household incomes, and is valuable in discussing the political context of dealing with poverty under the Conservatives. Two books make the connection between social need and the way it is tackled inside government: Donnison (1982) deals with the author's time as Chairman of the Supplementary Benefits Commission; Field (1982) is about the Child Poverty Action Group, a prominent pressure group. Finally, for an up-to-date analysis that touches on both theory and trends in public opinion, try Taylor-Gooby (1985).

Chapter 9 Foreign and Defence Policy

Capitanchik (1977 and 1983 (with Eichenberg)) provide an analysis of public attitudes, domestic and allied, towards defence issues. The 1977 book is a good, though inevitably uneven, collection of essays. The 1983 book is co-authored, in an excellent series of paperbacks produced by Chatham House. The article in *The Economist* (1982) is a very well-informed account of the differences between the FCO and the Prime Minister's Office in recent times.

The Falklands Campaign: The Lessons, Cmnd 8758 (1982) is the official view of the lessons to be learned. *The Handling of Press and Public Information During the Falklands Conflict* (1982) is an absorbing report on a relatively minor, but controversial, aspect of the campaign. House of Commons Defence Committee, Third Report 1983–4, HC 584, October 1984, *Ministry of Defence Reorganisation*, and articles in the

Journal of the Royal United Services Institute for March 1985 (vol. 130, no. 1) cover Mr Heseltine's managerial revolution.

Sir Geoffrey Howe's critique of the *Strategic Defence Initiative* is reprinted in the following number of the same journal, and also in 'Arms Control and Disarmament Newsletter No 23' (Jan–Mar 1985), ACDRU, Foreign and Commonwealth Office. David Watt, formerly the Director of Chatham House, writes an occasional column in *The Times* that is worth looking at. And for further reading: Burrows and Edwards (1982); Flynn (1982); Hagan (1982); Smith and Clarke (1985); Taylor T. (1978).

Chapter 10 Options for Northern Ireland

A good general introduction to Northern Ireland, written for outsiders, is Paul Arthur (1983). The historical background to the conflict is succinctly explained in Stewart (1977), and the 'conditional loyalty' of Ulster Protestants through the years is illuminated in Miller (1978). The record of the Ulster Unionist government from 1920 to 1972 is analysed from an academic point of view in Rose (1971) and from a nationalist point of view in Farrell (1976). The social and economic background to the conflict is explained in Darby (ed) (1983), and current proposals for resolving it in Boyle and Hadden (1985). Non-partisan commentary on all aspects of Northern Ireland affairs is provided in *Fortnight: An Independent Review for Northern Ireland* (Fortnight Publications, 7 Lower Crescent, Belfast 7).

Chapter 11 The Role of Law

Griffith (1977) remains the best introduction to the subject. While by no means totally convincing, it hits more targets than it misses. McAuslan and McEldoney (1985) is a collection of essays in the same tradition as Griffith's work, but from a more explicitly theoretical perspective. Another collection, Jowell and Oliver (1985) covers a wider range of subjects, but while always stimulating, never demonstrates the radical cutting edge of Griffiths's work.

Chapter 12 Industrial Relations

There is no good book-length treatment of the politics of industrial relations carrying the story into the second half of the 1980s. The best available introduction – doubly valuable because it is available in an inexpensive paperback – is Crouch (1982). This sketches in the whole of post-war history, links industrial relations to wider debates about the nature of British politics, and includes material on the early years of the first Thatcher administration. Bain (1983) is a collection of pieces describing succinctly the various recent developments in the structure and practice of industrial relations. The flavour of the period is captured by the Royal Commission Report on Industrial Relations (1968). Brittan's paper (1975) is the most stimulating statement of the thesis that unions are overwhelmingly powerful; Marsh and Locksley (1983) present the opposing view in an equally stimulating way. Coates (1980) is provocative and illuminating on the Labour government's industrial relations policies, 1974–9. Bell's collection (1985) summarises the record of the Conservatives in the 1980s. Taylor's (1984) book on the Yorkshire miners is much more useful than its narrow geographical range would suggest, because it examines the key group in the miners' disputes of the 1970s and 1980s.

Chapter 13 Blacks and Women in British Politics

During the past thirty years a number of important works about the position of black people in Britain have appeared. However, consideration of their role in politics is, with slight exceptions, relatively new. On the other hand, there are a number of useful books about race as an issue in British politics. Parts of Foot (1965) remain worth reading. Freeman (1979) is an important source, although its analysis does stop in the mid-1970s. Castles and Kosack (1973) remains the standard exposition of the thesis that the function of black people in Britain is akin to that of migrant labour on the continental model. More recently, Layton-Henry (1984) offers a thorough analysis of the race issue in British politics. Miles and Phizacklea (1984) covers similar ground, though less satisfactorily, since its tone is more strident and less scholarly.

On women in politics, the academic literature is a little more abundant. Two good overviews of various aspects of the subject are Randall (1982) and Stacey and Price (1981). Vallence (1979) remains the standard work on the subject. Since the early 1970s there has been an accumulating literature about women's political behaviour, both in this country and beyond. Lovenduski and Hills (1983) gives material for numerous different countries. Over the past few years, Jorgen Rasmussen (1981; 1983a; 1983b; 1984) has published in various journals a number of specialist articles about women in British politics since their attainment of the franchise.

Chapter 14 Mass Media

A comprehensive and up-to-date introduction is Tunstall (1983), which covers a wide range of media topics and not just its role in politics. Two early books which deal particularly with the media in politics are Blumler and McQuail (1968) and Seymour-Ure (1974). More recent is an account of the media's part in the 1979 election campaign, contained in Worcester and Harrop (eds) (1982).

Curran and Seaton (1985) contains a critique of the pluralist theory of the media, and is particularly good on the concentration of ownership and control. This critique is supplemented by the Glasgow University Media Group's books (1976, 1982).

A critique of the work of the Glasgow Media group is in Harrison (1985), which also contains the text of ITV reports on industrial disputes for the first four months of 1975. This book is at its best when considering the problem of TV news in general, rather than in its full blooded and emotionally charged critique of the Glasgow work. A defence of the pluralist position and of the high quality of the British media is presented in Hetherington (1985).

Tunstall (1983) contains a long and useful annotated bibliography.

Chapters 15 and 16 Topics in British Politics and Theories of the State in British Politics

There are now a number of useful sources which compare and contrast different approaches to the theory of the state. In general these sources are a much more useful introduction than monothematic books written from one perspective. Dunleavy (1981b) provides a short account of liberal and Marxist approaches. Self (1985) and Cox, Furlong and Page (1985) provide accessible coverage of pluralist, elite theory, corporatist and Marxist approaches, linked to public policy themes and issues. More purely theoretical and more difficult coverage is given by Alford and Friedland (1985). Dunleavy and O'Leary (1987) give a full account of all these approaches alongside neo-pluralist and new-right views.

Going beyond these sources usually entails raising your theoretical sights a bit. For pluralism, Dahl's (1956, 1982) treatments are still worth tackling. Nordlinger (1983) provides a good modern statement of elite theory, but is quite hard going. Jessop (1982) is a comprehensive survey of modern Marxism, but very difficult for beginners; try Dunleavy (1985) for a simplified but still useful guide.

Bibliography

(Official publications are normally listed under the name of their sponsoring department, such as 'Advisory Conciliation and Arbitration Service'. Reports of Select Committees are listed as 'House of Commons, Select Committee on . . .' However, when a Report is more likely to be known by its title (such as *The Falklands Campaign*) it is listed by title rather than by author.)

Advisory Conciliation and Arbitration Service (1985) *Annual Report 1984*, London, HMSO.

Alderson, J. (1985) 'Law and Order' in D. Bell (ed.), *The Conservative Government 1979–84: An Interim Report*, Beckenham, Croom Helm.

Alford, R. and Friedland, R. (1985) *Powers of Theory: Capitalism, the State and Democracy*, Cambridge, Cambridge University Press.

Allison, G. (1973) *The Essence of Decision: Explaining the Cuban Missile Crisis*, Boston, Little Brown & Co.

Amery, L. (1947) *Thoughts on the Constitution*, Oxford, Oxford University Press.

Amery, L. S. (1953) *Thoughts on the Constitution*, London, Oxford University Press.

Andrew, C. (1985) *Secret Service: The Making of the British Intelligence Community*, London, Heinemann.

Anwar, M. (1984) *Ethnic Minorities and the 1983 General Election*, London, Commission for Racial Equality.

Arthur, Paul (1980) *Ten Years on in Northern Ireland*, London, 1983.

Arthur, Paul (1983) *The Government and Politics of Northern Ireland*, London, Longman.

Ascher, K. (1986) *The Politics of Privatisation: Contracting Out in the NHS and Local Authorities*, London, Macmillan.

Aubrey, C. (1981) *Who's Watching You?: Britain's Security Services and the Official Secrets Act*, Harmondsworth, Penguin.

Bain, G. (ed.) (1983) *Industrial Relations in Britain*, Oxford, Basil Blackwell.

Baldwin, N. D. (1985) 'Behaviour Changes: A New Professionalism and a More Independent House', in Norton, P. (ed.), *Parliament in the 1980s*, Oxford, Blackwell, pp. 96–113.

Ball, A. R. (1981) *British Political Parties: The Emergence of a Modern Party System*, London, Macmillan.

Banton, M. (1983) 'The Influence of Colonial Status upon Black–White Relations in England, 1948–58', *Sociology*, vol. 17, no. 4, pp. 546–59.

Barker, A. (ed.) (1982) *Quangos in Britain*, Oxford, Oxford University Press

Barnett, J. (1982) *Inside the Treasury*, London, Andre Deutsch.

BBC (1985) *Annual Report and Handbook*, London, BBC.

Beer, S. H. (1965) *Modern British Politics: A Study of Parties and Pressure Groups*, London, Faber & Faber.

Beer, S. H. (1982) *Britain Against Itself: The Political Contradictions of Collectivism*, London, Faber & Faber.

Bell, D. (ed.) (1985) *The Conservative Government 1980–4*, London, Croom Helm.

Beloff, M. and Peele, G. (1985) *The Government of the United Kingdom: Political Authority in a Changing Society*, 2nd edn, London, Weidenfeld.

Benn, Tony (1980) 'The Case for a Constitutional Premiership', *Parliamentary Affairs*, vol. xxxiii, no. 1 (Winter).

Benn, Tony (1983) 'Spirit of Labour Reborn', *Guardian*, 20 June.

Blackaby, F. (ed.) (1979a) *British Economic Policy 1960–74*, Cambridge, NIESR/Cambridge University Press.

Blackaby, F. (ed.) (1979b) *De-industrialisation*, London, Heinemann.

Blumler, J. G. (1974) 'Mass Media, Votes and Reactions in the February Election', in H. R. Penniman (ed.), *Britain at the Polls: The Parliamentary Elections of 1974*, Washington, DC, A.E.I.

Blumler, J. and McQuail, D. (1968) *Television in Politics*, London, Faber & Faber.

Boddy, M. and Fudge, C. (eds) (1984) *Local Socialism*, London, Macmillan.

Bogdanor, V. (ed.) (1983) *Liberal Party Politics*, Oxford, Oxford University Press.

Bosanquet, N. (1983) *After The New Right*, London, Heinemann.

Boyce, G., Curran, J. and Wingate, P. (eds) (1978) *Newspaper History*, London, Constable.

Boyle, Andrew (1979) *Climate of Treason: Five Who Spied for Russia*, London, Hutchinson.

Boyle, K., Hadden, T. and Hillyard, P. (1980) *Ten Years On in Northern Ireland*, London, Cobden Trust.

Boyle, K. and Hadden, T. (1985) *Ireland: A Positive Proposal*, Harmondsworth, Penguin.

Bright, D., Embridge, D. and Rees, B. (1983) 'Industrial Relations of Recession', *Industrial Relations Journal*, no. 3.
Bristow, S. L. (1980) 'Women Councillors – An Explanation of the Under-representation of Women in Local Government', *Local Government Studies*, vol. 6, no. 3, pp. 73–90.
Brittan, S. (1975) 'The Economic Contradictions of Democracy', *British Journal of Political Science*, no. 5.
Brittan, S. (1977) *The Economic Consequences of Democracy*, London, Temple Smith.
Brittan, S. (1983) *The Role and Limits of Government*, London, Temple Smith.
Bromhead, P. A. (1974) *Britain's Developing Constitution*, London, Allen & Unwin.
Brown, Joan C. (1981) *Poverty and the Development of Anti-Poverty Policy in the United Kingdom*, London, Heinemann.
Brown, W. and Sison, K. (1984) 'Current Trends and Future Possibilities', in M. Poole *et al.*, *Industrial Relations In the Future: Trends And Possibilities In Britain Over the Next Decade*, London, Routledge & Kegan Paul.
Bruce-Gardyne, J. (1984) *Mrs Thatcher's First Administration: The Prophets Confounded*, London, Macmillan.
Bulpitt, J. (1983) *Territory and Power in the United Kingdom*, Manchester, Manchester University Press.
Burrows, B. and Edwards, G. (1982) *The Defence of Western Europe*, London, Butterworth.
Butler, D. and Jowett, P. (1985) *Party Strategies in Britain: A Study of the 1984 European Elections*, London, Macmillan.
Butler, D. and Kavanagh, D. (1984) *The British General Election of 1983*, London, Macmillan.
Butler, D. and Stokes, D. (1974) *Political Change in Britain*, London, Macmillan.
Capitanchik, D. (1977) 'Public Opinion and Popular Attitudes Towards Defence', in J. Bayliss (ed.), *British Defence Policy in a Changing World*, London, Croom Helm.
Capitanchik, D. and Eichenberg, R. C. (1983) *Defence and Public Opinion*, London, Royal Institute of International Affairs.
Castle, B. (1980) *The Castle Diaries 1974–6*, London, Weidenfeld & Nicolson.
Castle, B. (1984) *The Castle Diaries 1964–70*, London, Weidenfeld & Nicolson.
Castles, S. and Kosack, G. (1973) *Immigrant Workers and Class Structure in Western Europe*, Oxford, Oxford University Press.
Cawson, A. and Saunders, P. (1983) 'Corporatism, Competitive

Politics and the Class Struggle', in R. King (ed.), *Capital and Politics*, London, Routledge & Kegan Paul.
Civil Service Department (1981) *Non-Departmental Public Bodies: A Guide for Departments*, London, HMSO.
Clegg, T., Crouch, R., Dunleavy, P. and Harding, A. (1985) *The Future of London Government*, London, LSE Greater London Group.
Coates, D. (1980) *Labour in Power?*, London, Longman.
Cockburn, C. (1974) *The Local State*, London, Pluto Press.
Cockerell, M., Hennessey, P. and Walker, D. (1984) *Sources Close to the Prime Minister: Inside the Hidden World of the News Manipulators*, London, Macmillan.
Coopers & Lybrand (1984) *Streamlining the Cities: An Analysis of the Costs Involved in the Government's Proposal for Reorganising Local Government in the Six Metropolitan Counties*, London, Coopers & Lybrand Associates.
Cosgrove, P. (1978) *Margaret Thatcher*, London, Heinemann.
Cox, A., Furlong, P. and Page, E. (1985) *Power in Capitalist Society*, Brighton, Wheatsheaf.
CRC (Community Relations Commission) (1975), *Participation of Ethnic Minorities in the General Election, October 1974*, London, Community Relations Commission.
Crewe, I. (1983) 'How Labour Was Trounced All Round', *Guardian*, 14 June, p. 20.
Crewe, I. (1984) 'The Electorate: Partisan Dealignment Ten Years On', in H. Berrington (ed.), *Change in British Politics*, London, Cass, pp. 183–215.
Crewe, I. (1985) 'Great Britain', in Crewe, A. (ed.), *Electoral Change in Western Democracies: Patterns and Sources of Electoral Volatility*, Beckenham, Kent, Croom Helm, pp. 100–50.
Crewe, I. (1986) 'How to Win a Landslide Without Really Trying: Why the Conservatives Won in 1983', in H. Penniman and A. Ranney (eds), *Britain at the Polls, 1983*, Washington, DC, A.E.I.
Crewe, I. and Harrop, M. (1986) *Political Communications: The General Election Campaign of 1983*, Cambridge, Cambridge University Press.
Crossman, R. H. S. (1975–7) *Diaries of a Cabinet Minister*, 3 vols, London, Hamish Hamilton.
Crowther-Hunt, Lord and Kellner, P. (1980) *The Civil Servants: An Inquiry into Britain's Ruling Class*, London, Macdonald.
Crouch, C. (1982) *The Politics of Industrial Relations*, London, Fontana.
Curran, J. and Seaton, J. (1985) *Power Without Responsibility: The Press and Broadcasting in Britain*, London, Methuen.
Dahl, R. A. (1956) *Preface to Democratic Theory*, Chicago, University of Chicago Press.

Dahl, R. A. (1982) *Dilemmas of Pluralist Democracy*, New Haven, Yale University Press.

Dahl, R. A. (1984) *A Preface to Economic Democracy*, Oxford, Polity Press.

Darby, J. (ed.) (1983) *Northern Ireland: The Background to the Conflict*, Belfast, Appletree Press.

Dearlove, J. and Saunders, P. (1984) *Introduction to British Government*, Oxford, Polity Press.

Department of Health and Social Security (1985) *The Reform of Social Security*, Cmnd 9517, London, HMSO.

de Smith, S. A. (1975) *Constitutional and Administrative Law*, Harmondsworth, Penguin.

Dicey, A. V. (1959) *An Introduction to the Study of The Law of the Constitution*, 10th edn, London, Macmillan.

Dilnot, A. W., Kay, J. A. and Morris, C. N. (1984) *The Reform of Social Security*, Oxford, Clarendon Press.

Dix, S. and Perry, S. (1984a) 'Current Trends and Future Possibilities', in M. Poole *et al.* (eds), *Industrial Relations in the Future: Trends and Possibilities In Britain Over the Next Decade*, London, Routledge & Kegan Paul.

Dix, S. and Perry, S. (1984b) 'Women's Employment Changes in the 1970s', *Department of Employment Gazette*, no. 4.

Donnison, David (1982) *The Politics of Poverty*, Oxford, Martin Robertson.

Downs, A. (1957) *An Economic Theory of Democracy*, New York, Harper & Row.

Downs, S. J. (1985) 'Structural Changes: Select Committees: Experiment and Establishment', in Norton, P. (ed.), *Parliament in the 1980s*, Oxford, Blackwell, pp. 48–68.

Drucker, H. M. (1978) *Doctrine and Ethos in the Labour Party*, London, Allen & Unwin.

Drucker, H. M. *et al.* (eds) (1983) *Developments in British Politics*, London, Macmillan.

Dunleavy, P. (1979) 'The Urban Basis of Political Alignment: Social Class, Domestic Property Ownership and State Intervention in Consumption Processes', *British Journal of Political Science*, vol. 9, no. 4, pp. 409–43.

Dunleavy, P. (1980) *Urban Political Analysis*, London, Macmillan.

Dunleavy, P. (1981a) *The Politics of Mass Housing, 1945–75*, Oxford, Clarendon Press.

Dunleavy, P. (1981b) 'Alternative Theories of Liberal Democratic Politics: The Pluralist–Marxist Debate in the 1980s', in D. Potter (ed.), *Society and the Social Sciences*, London, Routledge & Kegan Paul.

Dunleavy, P. (1982) 'Is There a Radical Approach to Public Administration?', *Public Administration*, 60, pp. 25–33.

Dunleavy, P. (1983) 'Voting and the Electorate', in H. M. Drucker *et al.*, *Developments in British Politics*, London, Macmillan, pp. 30–58.

Dunleavy, P. (1984) 'The Limits to Local Government', in Boddy and Fudge.

Dunleavy, P. (1985a) 'Fleet Street: Its Bite on the Ballot', *New Socialist*, September.

Dunleavy, P. (1985b) 'Political Theory', in J. Short and Z. Bryzinski (eds), *Developing Contemporary Marxism*, London, Macmillan.

Dunleavy, P. (1986a) 'Explaining the Privatisation Boom', *Public Administration*, vol. 61.

Dunleavy, P. (1986b) 'The Growth of Sectoral Cleavages and the Stabilisation of State Expenditure', *Society and Space*, 4, pp. 129–44.

Dunleavy, P. and Husbands, C. T. (1985) *British Democracy at the Crossroads: Voting and Party Competition in the 1980s*, London, Allen & Unwin.

Dunleavy, P. and Rhodes, R. (1983) 'Beyond Whitehall', in H. M. Drucker *et al.* (eds), *Developments in British Politics*, London, Macmillan, pp. 106–33.

Dunleavy, P. and O'Leary, D. B. (1987) *Theories of the State*, London, Macmillan.

Dunn, S. and Gennard, J. (1984) *The Closed Shop in British Industry*, London, Macmillan.

The Economist (1982) 'Britain's Foreign Office', 27 November, pp. 25–9.

Edelman, M. (1967) *The Symbolic Uses of Politics*, London, University of Illinois Press.

Elcock, H. (1982) *Local Government: Politicians, Professionals and the Public in Local Authorities*, London, Methuen.

Elliott, M. J. (1983) *The Role of Law in Central/Local Relations*, London, SSRC.

Ermisch, John (1983) *The Political Economy of Demographic Change*, London, Heinemann.

Expenditure Committee (1977) *The Civil Service: 11th Report and Volumes of Evidence 1976–77*, HC 535, London, HMSO.

Falklands Campaign: The Lessons, The (1982) Cmnd 8758.

Farrell, M. (1976) *Northern Ireland: The Orange State*, London, Pluto.

Field, Frank (1982) *Poverty and Politics*, London, Heinemann.

Fine, B. and Millar, R. (1985) *Policing the Miners' Strike*, London, Lawrence & Wishart.

Finer, S. E. (1973) 'The Political Power of Organised Labour', *Government and Opposition*, no. 9.

Finer, S. E. (1975) *Adversarial Politics and Electoral Reform*, London, Anthony Wigram.
Finer, S. E. (1980) *The Changing Party System: 1945–1979*, Washington DC, American Enterprise Institute.
Fitzgerald, M. (1983) 'Are Blacks an Electoral Liability?', *New Society*, 8 December, pp. 394–5.
Fitzgerald, M. (1984) *Political Parties and Black People: Participation, Representation and Exploitation*, London, Runnymede Trust.
Flynn, G. (ed.) (1982) *The Internal Fabric of Western Europe*, London, Butterworth.
Foot, P. (1965) *Immigration and Race in British Politics*, Harmondsworth, Penguin.
Forrester, A., Lansley, S. and Pauley, R. (1985) *Beyond Our Ken*, London, Fourth Estate.
Franklin, M. (1985) *The Decline of Class Voting in Britain*, Oxford, Oxford University Press.
Franklin, M. and Page, E. (1984) 'A Critique of the Consumption Cleavage Approach in British Voting Studies', *Political Studies*, vol. 32, pp. 521–36.
Freeman, Gary (1979) *Immigrant Labour and Racial Conflict in Industrial Societies: The French and British Experience, 1945–75*, Princeton, Princeton University Press.
Fry, G. K. (1981) *The Administrative 'Revolution' in Whitehall*, London, Croom Helm.
Fry, G. K. (1984a) 'The Development of the Thatcher Government's "Grand Strategy" for the Civil Service: A Public Policy Perspective', *Public Administration*, vol. 62.
Fry, G. K. (1984b) 'The Attack on the Civil Service and the Response of the Insiders', *Parliamentary Affairs*, vol. 37.
Fry, G. K. (1985) *The Changing Civil Service*, London, Allen & Unwin.
Fulton, (1968) *Report of the Committee on the Civil Service* (The Fulton Report), Cmnd 3628, London, HMSO.
Galbraith, J. K. (1969) *The New Industrial State*, Harmondsworth, Penguin.
Gamble, A. M. (1983) 'Economic Policy', in H. M. Drucker *et al.* (eds), *Developments in British Politics*, London, Macmillan, pp. 134–55.
Gamble, A. M. (1985) *Britain in Decline*, 2nd edn, London, Macmillan.
Gamble, A. M. and Walkland, S. A. (1984) *The British Party System and Economic Policy 1945–83*, Oxford, Clarendon Press.
Garrett, J. (1980) *Managing the Civil Service*, London, Heinemann.
Gennard, J. (1984) 'The Implications of the Messenger Newspaper Group Dispute', *Industrial Relations Journal*, no. 3.

George, V. and Wilding, P. (1984) *The Impact of Social Policy*, London, Routledge & Kegan Paul.

Gilmour, Sir I. (1978) *Inside Right: A Study of Conservatism*, London, Quartet Books.

Gilmour, Sir I. (1983) *Britain Can Work*, Oxford, Martin Robertson.

Gimson, A. (1985) 'The Guardians of Thatcherism', *The Spectator*, 8 June.

Glasgow University Media Group (1976) *Bad News*, London, Routledge & Kegan Paul.

Glasgow University Media Group (Members of) (1982) *Really Bad News*, London, Writers and Readers.

Goodin, R. E. (1982) 'Rational Politicians and Rational Bureaucrats in Washington and Whitehall', *Public Administration*, vol. 60, pp. 23–41.

Gould, B. *et al.* (1981) *Monetarism or Prosperity*, London, Macmillan.

GPI (Gallup Political Index) February 1967 and February 1975.

Gray, A. and Jenkins, W. (1982) 'Policy Analysis in British Central Government: The Experience of PAR', *Public Administration*, vol. 60.

Greenleaf, W. H. (1983) *The British Political Tradition*, vol. II, London, Methuen.

Greenwood, J. R. and Wilson, D. J. (1984) *Public Administration in Britain*, London, Allen & Unwin.

Greenwood, R. (1982) 'Pressure from Whitehall', in R. Rose and E. Page (eds), *Fiscal Stress in the Cities*, London, Cambridge University Press.

Griffith J. A. G. (1974) *Parliamentary Scrutiny of Government Bills*, London, Allen & Unwin.

Griffith, J. A. G. (1977) *The Politics of the Judiciary*, London, Fontana.

Gunter, B., Svennevig, M. and Weber, M. (1984) *Television Coverage of the 1983 Election*, London, BBC/IBA.

Gyford, J. (1983) *Local Politics in Britain*, London, Croom Helm.

Gyford, J. (1985) *The Politics of Local Socialism*, London, Allen & Unwin.

Hagan, L. S. (ed.), (1982) *The Crisis of Western Security*, London, Croom Helm.

Hailsham, Lord (1978) *The Dilemma of Democracy*, London, Collins.

Hain, P. (1984) *The Democratic Alternative*, Harmondsworth, Penguin.

Ham, C. (1985) *Health Policy in Britain*, 2nd edition, London, Macmillan.

Harrison, Martin (1984) 'Whose Bias? Strikes, TV News and Media Studies', Keele Research Papers, No. 19, Department of Politics, University of Keele.

Harrison, Martin (1985) *TV News: Whose Bias?*, Hermitage, Berks, Policy Journals.

Harrop, M. (1982) 'Labour-Voting Conservatives: Policy Differences Between the Labour Party and Labour Voters', in R. Worcester and M. Harrop (eds), *Political Communications: The General Election Campaign of 1979*, London, Allen & Unwin, pp. 152–63.

Harrop, M. (1986) 'Press Coverage of Post-War British Elections', in I. Crewe and M. Harrop (eds), *Political Communications: The General Election Campaign of 1983*, Cambridge, Cambridge University Press.

Hattersley, R. *The Role of Markets in a Socialist Society*, London, Routledge & Kegan Paul.

von Hayek, F. A. (1944) *The Road to Serfdom*, London, Routledge & Kegan Paul.

Heald, D. (1983) *Public Expenditure*, Oxford, Martin Robertson.

Healey, Denis (1981) *Managing the Economy* (City Association Lecture for 1981) London, Association of Chartered Accountants and City of London Polytechnic.

Heath, A., Jowell, R. and Curtice, J. (1985) *How Britain Votes*, Oxford, Pergamon.

Heclo, H. and Wildavsky, A. (1974) *The Private Government of Public Money*, London, Macmillan.

Hechter, M. (1975) *Internal Colonialism: The Celtic Fringe in British National Development, 1536–1966*, Berkeley, University of California Press.

Henney, A. (1984) *Inside Local Government: A Case for Radical Reform*, London, Sinclair Browne.

Hetherington, Alastair (1985) *News, Newspapers and Television*, London, Macmillan.

Hill, Michael (1983) *Understanding Social Policy*, Oxford, Basil Blackwell.

Hills, J. (1981) 'Britain', in J. Lovenduski and J. Hills (eds), *The Politics of the Second Electorate: Women and Public Participation*, London, Routledge & Kegan Paul, pp. 8–32.

Himmelweit, H., Humphries P. and Jaeger, M. (1985) *How Voters Decide*, Milton Keynes, Open University Press.

Hindess, B. (1983) *Parliamentary Democracy and Socialist Politics*, London, Routledge & Kegan Paul.

Hirsch, Fred (1977) *Social Limits to Growth*, London, Routledge & Kegan Paul.

Hirschmann, Albert O. (1970) *Exit, Voice, and Loyalty: Responses to Decline in Firms, Organisations and States*, Cambridge, Mass., Harvard University Press.

HM Treasury (1985) *The Government's Expenditure Plans 1985–86 to 1987–88*, Cmnd 9428.

Hobsbawm, E. J. (1981) 'The Forward March of Labour Halted?', in

M. Jacques and F. Mulhern (eds), *The Forward March of Labour Halted?*, London, Verso.

Hodgson, G. (1984) *The Democratic Economy*, Harmondsworth, Penguin.

Hood, C. (1983) *The Tools of Government*, London, Macmillan.

Hood, C. and Wright, D. (1981) *Big Government in Hard Times*, Oxford, Martin Robertson.

Hoskyns, Sir J. (1983) 'Whitehall and Westminster: An Outsider's View', *Parliamentary Affairs*, vol. 36.

Hoskyns, Sir J. (1984) 'Conservatism is Not Enough', *Political Quarterly*, vol. 55.

House of Commons (1981) *Monetary Policy: Third Report of the Treasury and Civil Service Committee for 1980–81*, London, HMSO.

House of Commons (1982) *The Handling of Press and Public Information During the Falklands Conflict.*

House of Commons (1984) *Third Report of the Defence Committee*, HC 584.

Hunt, Sir John (1982) 'Access to Previous Government's Papers', *Public Law*, pp. 514–18.

Husbands, C. T. (1979) 'The "Threat" Hypothesis and Racist Voting in England and the United States', in R. Miles and A. Philzacklea (eds), *Racism and Political Action in Britain*, London, Routledge & Kegan Paul, pp. 147–83.

Husbands, C. T. (1983a) 'Race and Immigration', in J. A. G. Griffith (ed), *Socialism in a Cold Climate*, London, Allen & Unwin, pp. 161–83.

Husbands, C. T. (1983b) *Racial Exclusionism and the City: The Urban Support of the National Front*, London, Allen & Unwin.

Husbands, C. T. (1985) 'Government Popularity and the Unemployment Issue, 1966–83', *Sociology*, vol. 19, no. 1, pp. 1–18.

Hutchinson, (1985) 'The Death of Administrative Law', *Modern Law Review*.

Jacques, M. and Mulhern, F. (eds) (1981) *The Forward March of Labour Halted?*, London, Verso.

James, Michael (1982) *The Politics of Secrecy*, Harmondsworth, Penguin.

Jefferson, T. and Grimshaw, R. (1984) *Controlling the Constitution: Police Accountability in England and Wales*, London, Muller/The Cobden Trust.

Jenkins, S. (1985) 'The "Star Chamber", PESC and the Cabinet', *Political Quarterly*, vol. 56.

Jenkins, S. and Sloman, A. (1985) *With Respect Ambassador: An Inquiry into the Foreign Office*, London, BBC Publications.

Jennings, Sir I. (1933) *The Law and the Constitution*, London, University of London Press.
Jessop, R. (1982) *The Capitalist State*, Oxford, Martin Robertson.
Johnson, L. (1985) *A Season of Inquiry: The Senate Intelligence Investigation*, Louisville, University of Kentucky Press.
Johnson, R. (1985) *The Geography of English Politics: The 1983 General Election*, Beckenham, Kent, Croom Helm.
Johnson, T. J. (1977) *Professions and Power*, London, Macmillan.
Jones, G. W. (1983) 'Prime Ministers' Departments Really Create Problems: A Rejoinder to Patrick Weller', *Public Administration*, vol. 61.
Jones, G. W. and Stewart, J. (1983) *The Case for Local Government*, London, Allen & Unwin.
Joseph, Sir K. (1975) *Reversing the Trend*, London, Centre for Policy Studies.
Joseph, Sir K. (1976) *Stranded in the Middle Ground*, London, Centre for Policy Studies.
Joseph, Sir K. and Sumption, J. (1979) *Equality*, Chatham, John Murray.
Jowell, J. and Oliver, D. (eds) (1985) *The Changing Constitution*, Oxford, Oxford University Press.
Jowell, R. and Airey, C. (1985) *British Social Attitudes: The 1985 Report*, Aldershot, Gower.
Keays, S. (1985) *A Question of Judgement*, London, Quintessential Press.
Keegan, W. (1984) *Mrs Thatcher's Economic Experiment*, Harmondsworth, Penguin.
Keegan, W. and Pennant-Rae, R. (1979) *Who Runs the Economy?*, London, Temple Smith.
Kellner, P. (1984) 'Are Markets Compatible with Socialism?', in B. Pimlott (ed.), *Fabian Essays in Socialist Thought*, London, Heinemann.
Kettle, M. (1985) 'The National Reporting Centre and the 1984 Miners' Strike', in Fine and Millar (1985) pp. 65–78.
King, A. (ed.) (1976) *Why Is Britain Becoming Harder To Govern?*, London, BBC.
King, D. S. (1984) 'The New Right and the Public Sector in Britain and the USA', ECPR paper, pp. 1–25.
The Law of Public Order (1984), Cmnd 9510.
Layton-Henry, Z. (1983) 'Immigration and Race Relations: Political Aspects – No 9.', *New Community*, vol. 9, nos 1–2, pp. 109–16.
Layton-Henry, Z. (1984) *The Politics of Race in Britain*, London, Allen & Unwin.
Le Grand, Julian (1982) *The Strategy of Equality*, London, George Allen & Unwin.

Le Grand, J. and Robinson, R. (1984) *Privatisation and the Welfare State*, London, Allen & Unwin.

Leigh, David (1980) *The Frontiers of Secrecy*, London, Junction Books.

Le Lohé, M. J. (1983) 'Voter Discrimination Against Asian and Black Candidates in the 1983 General Election', *New Community*, No. 9, nos 1–2, pp. 101–8.

Lievesley, D. and Waterton D. (1985) 'Measuring Individual Attitude Change', in R. Jowell and S. Witherspoon (eds), *British Social Attitudes: The 1985 Report*, Aldershot, Hants, Gower.

Lindblom, C. (1977) *Politics and Markets*, New York, Basic Books.

Lloyd, C. (1985) 'A National Riot Police: Britain's Third Force?', in Fine and Millar (1985).

Lovenduski, J. and Hills, J. (eds) (1983) *The Politics of the Second Electorate: Women and Public Participation*, London, Routledge & Kegan Paul.

Luard, E. (1979) *Socialism Without the State*, London, Macmillan.

McAllister, I. (1986) 'Social Context, Turnout and the Vote: The Australian and British Comparison', *Political Geography Quarterly*, vol. 4.

McAllister I. and Rose, R. (1984) *The Nationwide Competition for Votes: The 1983 British General Election*, London, Pinter.

McAuslan, P. and McEldoney, J. (eds) (1985) *Law, Legitimacy, and the Constitution*, London, Sweet & Maxwell.

McGahey, M. (1985) 'Introduction', in Fine and Millar (1985).

McKenzie, R. T. (1965) *British Political Parties: The Distribution of Power Within the Conservative and Labour Parties*, London, Heinemann.

Mackintosh, J. P. (1977) *The British Cabinet*, 3rd edn, London, Stevens.

McQuail, Denis (1977) 'The Influence and Effects of Mass Media', in J. Curran, M. Gurevitch and Janet Wollacott (eds), *Mass Communication and Society*, London, Arnold, pp. 70–94.

Mack, Joanna and Lansley, Stewart (1985) *Poor Britain*, London, George Allen & Unwin.

Maitland, F. W. (1888) *The Constitutional History of England*, London, Cambridge University Press.

Marquand, D. (1981) 'Review of *The Socialist Agenda*', *London Review of Books*, February.

Marsh, D. and Locksley, G. (1983) 'Labour: The Dominant Force in British Politics?', in D. Marsh (ed.), *Pressure Politics*, London, Junction Books.

Marsh, D, and King, J. (1985) 'The Trade Unions Under Thatcher', paper to the Annual Conference of the Political Studies Association.

Marshall, G. (1984) *Constitutional Conventions: The Rules and Forms of Political Accountability*, Oxford, The Clarendon Press.

Mason, C. (1984) 'YTS and Local Education Authorities – A Context', *Local Government Studies*, vol. 10, no. 1, pp. 63–73.

Metcalf, D. and Nickell, S. (1985) 'Jobs and Pay', *Midland Bank Review*, Spring issue.

Michael, James (1982) *The Politics of Secrecy*, Harmondsworth, Penguin.

Middlemas, K. (1979) *Politics in Industrial Society*, London, Deutsch.

Middlemas, K. and Barnes, J. (1969) *Baldwin: A Biography*, London, Weidenfeld & Nicolson.

Miles, R. and Phizacklea, A. (1984) *White Man's Country: Racism in British Politics*, London, Pluto.

Miliband, R. (1969) *The State in Capitalist Society*, London, Weidenfeld & Nicolson.

Miliband, R. (1973) *Parliamentary Socialism*, London, Merlin.

Miliband, R. (1983) *Capitalist Democracy in Britain*, Oxford, Oxford University Press.

Miliband, R. (1985) 'The New Revisionism', *New Left Review*, 150, March/Apr.

Millar, R. (1985) in Fine and Millar (1985) pp. 23–33.

Miller, D. W. (1978) *Queen's Rebels*, London, Gill & Macmillan.

Miller, W. (1980) 'What Was the Profit in Following the Crowd? The Effectiveness of Party Strategies on Immigration and Devolution', *British Journal of Political Science*, vol. 10, no. 1, pp. 15–38.

Miller, W. (1983) 'Testing the Power of Media Consensus', Strathclyde Papers on Government and Politics No. 17.

Miller, W, (1984) 'There Was No Alternative: The British General Election of 1983', *Parliamentary Affairs*, vol. 37, no. 4, pp. 364–84.

Miller, W. and Taylor, C. (1985) 'Structured Protest: A Comparison of Third Force Challenges in Britain, Germany and France during the Eighties', paper presented to the International Political Science Association Congress, Paris.

Minford, Patrick (1984) 'State Expenditure: A Study in Waste', *Economic Affairs*, vol. 4, no. 3.

Mole, S. (1983) 'Community Politics', in V. Bogdanor (ed.), *Liberal Party Politics*, Oxford, The Clarendon Press.

Moon, J. and Richardson, J. J. (1984) 'Policy-Making With a Difference?: The Technical and Vocational Education Initiative', *Public Administration*, vol. 62, pp. 32–3.

Moran, M. (1985) *Politics and Society in Britain*, London, Macmillan.

Mosley, Paul (1984a) *The Making of Economic Policy*, Brighton, Harvester Press.

Mosley, Paul (1984b) ' "Popularity Functions" and the Role of the Media: A Pilot Study of the Popular Press', *British Journal of Political Science*, vol. 14, no. 1, pp. 117–28

Mueller, D. (1979) *Public Choice*, Cambridge, Cambridge University Press.

Murray, R. (1984) 'New Directions in Municipal Socialism' in Pimlott, B. (ed.), *New Directions in Socialist Thought*, London, Heinemann.

Naylor, P. (1984) *A Man and an Institution: Sir Maurice Hankey, The Cabinet Secretariat and the Custody of Cabinet Secrecy*, Cambridge, Cambridge University Press.

New Ireland Forum (1984) *The Cost of Violence Arising from the Northern Ireland Crisis Since 1969*, Dublin, Stationery Office.

Newton, K. (1976) *Second City Politics*, London, Oxford University Press.

Newton, K. and Karran, T. J. (1985) *The Politics of Local Expenditure*, London, Macmillan.

Niskanen, W. (1973) *Bureaucracy: Servant or Master?*, London, Institute of Economic Affairs.

Nordlinger, E. (1983) *On the Autonomy of the State in Democratic Society*.

Norton, P. (1982) *The Constitution in Flux*, Oxford, Martin Robertson.

Norton, P. (ed.) (1985) *Parliament in the 1980s*, Oxford, Blackwell.

Norton-Taylor, R. (1985) *The Ponting Affair*, London, Cecil Woolf.

Nove, A. (1983) *The Economics of Feasible Socialism*, London, George Allen & Unwin.

Nozick, R. (1971) *Anarchy, State and Utopia*, New York, Basic Books.

O'Connor, J. (1973) *The Fiscal Crisis of the State*, London Macmillan.

O'Leary, D. B. (1985) 'Is There a Radical Public Administration?', *Public Administration*, vol. 63, pp. 345–52.

Olson, M. (1968) *The Logic of Collective Action*, Cambridge, Mass., Harvard University Press.

Olson, M. (1982) *The Rise and Decline of Nations*, New Haven, Yale University Press.

Orridge, A. (1981) 'Uneven Development and Nationalism', *Political Studies*, vol. 29, pp. 1–15; 181–90.

Owen, D. (1981) *Face the Future*, Oxford, Oxford University Press.

Panitch, L. V. (1979) *Social Democracy and Industrial Militancy*, Cambridge, Cambridge University Press.

Parkin, F. (1968) *Middle-Class Radicalism*, Manchester, Manchester University Press.

Peacock, A. T. and Wiseman, J. (1967) *The Growth of Public Expenditure in the United Kingdom*, London, Allen & Unwin.

Peele, G. (1978) 'Britain's Developing Constitution', in J. Ramsden and C. Cook (eds), *Trends in British Politics Since 1945*, London, Macmillan.

Peele, G. (1983) 'Government at the Centre', in H. M. Drucker *et al.* (eds), *Developments in British Politics*, London, Macmillan, pp. 83–105.

Pirie, M. (1985) *Privatisation*, London, Adam Smith Institute.

Plant, R. and Hoover, K. (forthcoming) *The Rise of Conservative Capitalism*, London, Methuen.

Pliatsky Report (1980) *Report on Non-Departmental Public Bodies*, Cmnd 7797, London, HMSO.

Pliatsky, Sir Leo (1982) *Getting and Spending*, London, Hamish Hamilton.

Polsby, N. (1980) 'The News Media as an Alternative to Party in the Presidential Selection Process', in R. A. Goodwin, *Political Parties in the Eighties*, Washington, American Enterprise Institute.

Ponting, C. (1985) *The Right to Know: The Inside Story of the Belgrano Affair*, London, Sphere Books.

Purcell, J. and Sisson, K. (1983) 'Strategies and Practices in the Management of Industrial Relations', in G. Bain (ed.), *Industrial Relations In Britain*, Oxford, Blackwell.

Pyper, Robert (1985) 'Sarah Tisdall, Ian Willmore and the Civil Servant's "Right to Leak" ', *Political Quarterly*, vol. 56.

Randall, V. (1982) *Women and Politics*, London, Macmillan.

Ranson, S. (1982) 'Central–Local Planning in Education', in C. R. Hinnings, S. Leach, S. Ranson and C. K. Sketcher, *Policy Planning Systems in Central–Local Relations*, Final Report to the SSRC Panel on Central–Local Government Relations, Appendix C.

Rasmussen, J. S. (1981) 'Female Political Career Patterns and Leadership Disabilities in Britain: The Crucial Role of Gatekeepers in Regulating Entry to the Political Elite', *Polity*, vol. 13, no. 4, pp. 600–20.

Rasmussen, J. S. (1983a) 'The Political Integration of British Women: The Response of a Traditional System to a Newly Emergent Group', *Social Science History*, vol. 7, no. 1, pp. 61–95.

Rasmussen, J. S. (1983b) 'Women's Role in Contemporary British Politics: Impediments to Parliamentary Candidature', *Parliamentary Affairs*, vol. 36, no. 3, pp. 300–15.

Rasmussen, J. S. (1984) 'Women in Labour: The Flapper Vote and the Party System Transformation in Britain', *Electoral Studies*, vol. 3, no. 1, pp. 47–63.

Reform of the Official Secrets Act (1978) Cmnd 7285.

Reiner, Robert (1985) *The Politics of the Police*, Brighton, Wheatsheaf.

Report of the Departmental Committee on s2 of the Official Secrets Act 1911 (1972) (Franks), Cmnd 5104.

Rhodes, R. A. W. (1983) 'Continuity and Change in British Central/Local Relations', *British Journal of Political Science*.

Rhodes, R. A. W. (1985a) ' "A Squalid and Politically Corrupt Process"? Intergovernmental Relations in the Post-War Period', *Local Government Studies*, vol. 11, no. 6.

Rhodes, R. A. W. (1985b) 'Power-Dependence, Policy Communities and Intergovernmental Networks', *Public Administration Bulletin*, no. 50, April.

Rhodes, R. A. W. (1986) *The National World of Local Government*, London, Allen & Unwin.

Richardson, J. J. and Jordan, G. (1979) *Governing Under Pressure*, Oxford, Martin Robertson.

Riddell, P. (1983) *The Thatcher Government*, Oxford, Martin Robertson.

Robertson, D. (1984) *Class and the British Electorate*, Oxford, Blackwell.

Rose, R. (1971) *Governing Without Consensus*, London, Faber & Faber.

Rose, R. (1980) *Do Parties Make a Difference?*, London, Macmillan.

Rose, R. (1982) *Understanding the United Kingdom*, London, Longman.

Rosenberg, M. (1968) *The Logic of Survey Analysis*, New York, Basic Books.

Royal Commission on the Press Industrial Relations in the National Newspaper Industry (1976) *Report*, Cmnd 6680, London, HMSO.

Royal Commission on the Press (1977) *Final Report*, Cmnd 6810, London, HMSO.

Royal Commission on Trade Unions and Employers' Associations (1968) *Report*, Cmnd 3623, London, HMSO.

Rubery, J. *et al.* (1983) 'Industrial Relations Issues in the 1980s: An Economic Analysis', in M. Poole *et al.*, *Industrial Relations in the Future: Trends and Possibilities In Britain Over the Next Decade*, London, Routledge & Kegan Paul.

Särlvik, B. and Crewe, I. (1983) *Decade of Dealignment: The Conservative Victory of 1979 and Electoral Trends in the 1970s*, Cambridge, Cambridge University Press.

Saunders, P. (1984) 'Beyond Housing Class', *International Journal of Urban and Regional Research*, VIII.

Saunders, P. (1985) 'The Forgotten Dimension of Central–Local Relations: Theorising the "Regional State" ', *Environment and Planning C: Government and Policy*, 3, 149–62.

Scarborough, E. (1984) *Political Ideology and Voting: An Exploratory Study*, Oxford, Oxford University Press.

Schmitter, P. C. and Lehmbruch, G. (1979) *Trends Towards Corporatist Intermediation*, London, Sage.

Schoen, D. S. (1977) *Enoch Powell and the Powellites*, London, Macmillan.

Schumpeter, P. (1944) *Capitalism, Socialism, and Democracy*, London, George Allen & Unwin.

Scruton, R. (1980) *The Meaning of Conservatism*, Harmondsworth, Penguin.
Self, P. (1985) *Political Theories of Modern Government*, London, Allen & Unwin.
Seymour-Ure, Colin (1974) *The Political Impact of the Mass Media*, London, Constable.
Sharpe, L. J. (1984) 'Functional Allocation in the Welfare State', *Local Government Studies*, Jan/Feb., pp. 27–45.
Sharpe, L. J. and Newton, K. (1984) *Does Politics Matter?: The Determinants of Public Policy*, Oxford, Clarendon Press.
Sisson, K. and Brown, W. (1983) 'Industrial Relations in the Private Sector: Donovan Revisited', in G. Bain (ed.), *Industrial Relations in Britain*, Oxford, Blackwell.
Smith, David *et al.* (1983) *Police and People in London*, London Policy Studies Institute.
Smith, S. and Clarke, M. (eds) (1985) *Foreign Policy Implementation*, London, George Allen & Unwin.
Spencer, Sarah (1985), 'The Eclipse of the Police Authority', in Fine and Millar (1985) pp. 34–54.
Stacey, M. and Price, M. (1981) *Women, Power and Politics*, London.
Stanworth, M. (ed.) (1984) *Women and the Public Sphere: A Critique of Sociology and Politics*, London, Hutchinson.
Steel, D. and Heald, D. (1984) *Privatising Public Enterprise*, London, RIPA.
Stevas, N. St John (1974) 'Introduction to Walter Bagehot', *The English Constitution* in *The collected Works of Walter Bagehot*, vol. 5, London, The Economist.
Stewart, A. T. Q. (1977) *The Narrow Ground: Aspects of Ulster 1609–1969*, London, Faber & Faber.
Stewart, J. D. (1983) *Local Government: The Constitution of Choice*, London, Allen & Unwin.
Street, H. and Brazier, R. (1984) *De Smith's Constitutional and Administrative Law*, Harmondsworth, Penguin. Fourth edition.
Studlar, D. T. (1977) 'Social Context and Attitudes Toward Coloured Immigrants', *British Journal of Sociology*, vol. 28, no. 2, pp. 168–84.
Studlar, D. T. (1983) 'The Ethnic Vote, 1983: Problems of Analysis and Interpretation', *New Community*, vol. 9, nos 1–2, pp. 92–100.
Taylor, A. (1984) *The Politics of the Yorkshire Miners*, London, Croom Helm.
Taylor, A. (1985) 'The Politics of Coal: Some Aspects of the Miners' Strike', *Politics*, no. 1.
Taylor, S. (1978) 'Parkin's Theory of Working Class Conservatism:

Two Hypotheses Investigated', *Sociological Review*, vol. 26, no. 4, pp. 827–42.

Taylor, T. (ed.) (1978) *Approaches and Theory in International Relations*, London, Longman.

Taylor-Gooby, Peter (1985) *Public Opinion, Ideology and State Welfare*, London, Routledge & Kegan Paul.

Thatcher, M. (1977) *Let Our Children Grow Tall*, London, Centre for Policy Studies.

Tiebout, C. (1956) 'A Pure Theory of Local Expenditure', *Journal of Political Economy*, vol. 64, pp. 416–24.

Todd, J. and Butcher, B. (1982) *Electoral Registration in 1981*, London, Office of Population Censuses and Surveys.

Townsend, Peter (1979) *Poverty in the United Kingdom*, Harmondsworth, Penguin.

Townsend, Peter and Davidson, Nick (eds) (1982) *Inequality and Health* (the Black Report), Harmondsworth, Penguin.

Tredinnick, D. (1985) 'The Police and Ethnic Minorities in Britain Since 1945 with Special Reference to the Metropolitan Police District', unpublished M.Litt thesis, University of Oxford.

Tunstall, J. (1983) *The Media in Britain*, London, Constable.

Vallance, E. (1979) *Women in the House: A Study of Women Members of Parliament*, London, Athlone Press.

Vallance, E. (1981) 'Women Candidates and Electoral Preference', *Politics*, vol. 1, no. 2, pp. 27–31.

Vallance, E. (1984) 'Women Candidates in the 1983 General Election', *Parliamentary Affairs*, vol. 37, no. 3, pp. 301–9.

Wallington, P. (1984) *Civil Liberties 1984*, Oxford, Martin Robertson/ The Cobden Trust.

Walsh, K., Dunne, R., Stolen, R. and Stewart, J. D. (n.d.) *Falling School Rolls and the Management of the Teaching Profession*, Windsor, Berks, NFER-Nelson.

Ward, T. (1983) 'Cash Planning', *Public Administration*, vol. 61.

Ward, T. (1986) 'Memorandum by Mr Terry Ward, Specialist Adviser to the Committee', in Treasury and Civil Service Committee, Third Report, *The Government's Expenditure Plans 1984–5 to 1986–7*, HC 285, Session 1983–4, London, HMSO, Appendix 1.

Wass, Sir Douglas (1983) 'The Public Service in Modern Society', *Public Administration*, vol. 61.

Wass, Sir Douglas (1984) *Government and the Governed*, London, Routledge & Kegan Paul.

Watt, David (1985a) 'Why CND Could Mushroom Again', *The Times*, 12 April.

Watt, David (1985b) 'Mrs Thatcher's Nationalisation', *The Times*, 14 June.

Webb, A. and Wistow, G. (1983) 'Public Expenditure and Policy Implementation: The Case of Community Care', *Public Administration*, vol. 60, pp. 21–44.

Webster, B. (n.d., but 1984 or 1985) *Bearing the Burden: Women's Work and Local Government*, London, Local Government Campaign Unit.

Weller, P. (1983) 'Do Prime Ministers' Departments Really Create Problems?', *Public Administration*, vol. 61.

West, Nigel (1983) *MI5: British Security Operations 1909–1945*, London, Triad/Panther.

Westergaard, J. and Resler, H. (1975) *Class in a Capitalist Society*, Harmondsworth, Penguin.

Wilkinson, B. (1983) *The Shopfloor Politics of New Technology*, London, Heinemann.

Williams, D. (1965) *Not in the Public Interest*, London, Hutchinson.

Williams, S. (1981) *Politics is for People*, Harmondsworth, Penguin.

Williamson, O. (1973) *Markets and Hierarchies*, New York, Free Press.

Wilsher, P., McIntyre, D. and Jones, M. (1985) *Strike: Thatcher, Scargill and the Miners*, London, Coronet.

Wilson, Des (1984) *The Secrets File: The Case for Freedom of Information in Britain Today*, London, Hutchinson.

Wilson, J. H. (1976) *The Governance of Britain*, London, Weidenfield & Nicolson.

Worcester, R. M. and Harrop, M. (eds) (1982) *Political Communications: the General Election Campaign of 1979*, London, Allen & Unwin.

Wraith, R. (1977) *Open Government: The British Interpretation*, London, RIPA.

Young, H. and Sloman, A. (1984) *But, Chancellor: An Inquiry into the Treasury*, London, BBC.

Index